12
Topics in Medicinal Chemistry

Aims and Scope

Drug research requires interdisciplinary team-work at the interface between chemistry, biology and medicine. Therefore, the new topic-related series *Topics in Medicinal Chemistry* will cover all relevant aspects of drug research, e.g. pathobiochemistry of diseases, identification and validation of (emerging) drug targets, structural biology, drugability of targets, drug design approaches, chemogenomics, synthetic chemistry including combinatorial methods, bioorganic chemistry, natural compounds, high-throughput screening, pharmacological in vitro and in vivo investigations, drug-receptor interactions on the molecular level, structure-activity relationships, drug absorption, distribution, metabolism, elimination, toxicology and pharmacogenomics.

In general, special volumes are edited by well known guest editors.

In references *Topics in Medicinal Chemistry* is abbreviated *Top Med Chem* and is cited as a journal.

More information about this series at
http://www.springer.com/series/7355

Peter H. Seeberger • Christoph Rademacher
Editors

Carbohydrates as Drugs

With contributions by

J. Bouckaert · A. Braem · D.R. Bundle · W.-S. Chen ·
P.P. Deshpande · B.A. Ellsworth · M. Frank · S.G. Gouin ·
G. Horne · H. Leffler · M. Leibeling · A. Mackinnon ·
U.J. Nilsson · N. Panjwani · G. Roos · H. Schambye ·
T. Sethi · A. Titz · W.N. Washburn · D.B. Werz

Springer

Editors
Peter H. Seeberger
Christoph Rademacher
Max Planck Institute of Colloids
and Interfaces
Department of Biomolecular Systems
Potsdam
Germany

ISSN 1862-2461 ISSN 1862-247X (electronic)
ISBN 978-3-319-08674-3 ISBN 978-3-319-08675-0 (eBook)
DOI 10.1007/978-3-319-08675-0
Springer Cham Heidelberg New York Dordrecht London

Library of Congress Control Number: 2014954644

Printed on acid-free paper

Springer is part of Springer Science+Business Media (www.springer.com)

Preface

Carbohydrates are the most abundant class of biomolecules and nature has made use of these structurally complex molecules for many applications. Every living cell is covered by oligosaccharides contributing to their stability, promoting cell–cell communication, or control cell migration and tissue homing. The diversity of this biopolymer exceeds that of any other organic structure. Consequently, the number of glycosylation-associated genes is high, with approximately 2% of the human genome encoding for these proteins. With novel tools for the synthesis and analysis of glycans becoming available to the broader community, the role of glycosylation-associated genes in health and disease is beginning to unravel. These advances have spurred increasing interest in carbohydrates and their derivatives as active ingredients in pharmaceuticals such as inhibitors of carbohydrate-binding or -processing proteins, or novel adjuvants and vaccines. However, in contrast to the potential of glycans for the treatment of human disease, the number of carbohydrate-based drugs is rather limited. The fact that each glycosidic linkage constitutes a new stereogenic center renders the synthesis of carbohydrates challenging. Carbohydrates are hydrophilic, metabolically unstable and sometimes promiscuous, as in many cases low affinity recognition is essential for some of their biological functions. Efficient synthetic access to carbohydrate-derivatives enables us to overcome these obstacles. Stability and hydrophilicity can be controlled and specificity can be established by evolution of the carbohydrate scaffold. Although a monosaccharide often has a rather low affinity to its target receptor, it provides high ligand efficiency, because of its low molecular weight. Moreover, the enthalpy-driven interaction is based on directed interactions, an exquisite starting point for small molecule evolution. Otherwise inaccessible binding sites can addressed that do not accommodate flat and aromatic molecules.

This volume is comprised of a collection of chapters addressing the problems and the potentials of carbohydrates as drugs from different perspectives. The opening chapter describes the target-oriented synthesis of carbohydrate-derivatives, in particular highlighting the synthesis of C-glycosides to overcome metabolic instability of carbohydrates. Moreover, the authors explore domino carbopalladation to transform carbohydrates into drug-like scaffolds. The improved synthetic accessibility of carbohydrates and their mimetics is then further explained in a chapter on iminosugars. Two marketed drugs based on this class of molecules are available and more are currently in clinical trials. Graeme Horne also raises an important question regarding the use of carbohydrates in drug design, asking why the glycospace has not been explored in medicinal chemistry. Martin Frank provides insights into the potential and limitations of computational approaches such as molecular docking to understand carbohydrate–receptor interaction. The team from Bristol–Myers Squibb brings together key aspects raised in the previous chapters. The C-glycoside Dapagliflozin for the treatment of diabetes type II is discussed. This arylglycoside is highly selective for the sodium-dependent glucose transporter 2 and blocks renal glucose reabsorption. Combined with its favorable stability and ADME profile, it was recently approved by FDA and EMA. Away from transporters towards human lectins as target for carbohydrate-based drugs, galectin-3 is then the subject of the following chapter. Galectin-3 plays an important role in metastasis, chronic inflammation, and fibrosis. Hence, the rational design of nanomolar inhibitors based on a galactose scaffold is discussed. The next two chapters turn the focus towards bacterial adhesins. First, FimH antagonists based on mannose are described, followed by work from Alexander Titz summarizing the work on the inhibition of biofilm formation of *Pseudomonas aeruginosa*. In the latter case, the lectins LecA and LecB are inhibited using galactose as a starting scaffold aiming for the treatment of chronic infections. Finally, the contribution from David Bundle reviews the structure-based, rational design of a carbohydrate-based vaccine against *Candida albicans*. The application of defined carbohydrate substructures of polysaccharides as vaccines is a rapidly growing market and harbors great potential for carbohydrates in the pharmaceutical industry.

All chapters emphasize that carbohydrates are the central scaffold that can be elaborated into a pharmaceutically active molecule. The potential of glycosylation-associated genes as targets for therapy has just begun to be fully elucidated and we are looking forward to see how this rapidly evolving field progresses. We thank all the authors for their contributions to this development and for providing their insight into the advancement of carbohydrates as drugs in this volume.

Potsdam, Germany
May 2014

Christoph Rademacher
Peter H. Seeberger

Contents

Top Med Chem (2014) 12: 1–22
DOI: 10.1007/7355_2014_43

Published online: 20 July 2014

Carbohydrate-Based Synthetic Chemistry in the Context of Drug Design

Markus Leibeling and Daniel B. Werz

Abstract Various carbohydrate building blocks serve as versatile and valuable precursors for modern target-orientated synthesis. The large number of different carbohydrates, their structural diversity with respect to functional groups and stereogenic centers allows the facile synthesis of a broad variety of possible target molecules. In a personal selection the authors demonstrate how to make use of carbohydrates as a starting point for Pd-catalyzed coupling reactions to obtain *C*-glycosides, how to link carbohydrate and cyclopropane chemistry, and how efficient carbopalladation cascades fuse carbohydrates with aromatic units.

Keywords Anthracyclines, Carbohydrate mimics, Carbohydrates, Catalysis, *C*-Glycosides, Chromans, Isochromans, Spiroketals

Contents

M. Leibeling and D.B. Werz (✉)
Technische Universität Braunschweig, Braunschweig, Germany

Abbreviations

Ac	Acetyl
anhyd	Anhydrous
Ar	Aryl
Bn	Benzyl
Bu	Butyl
Bz	Benzoyl
cat	Catalyst
concd	Concentrated
Cp	Cyclopentadienyl
d	Day(s)
DIBALH	Diisobutylaluminum hydride
DMAP	4-(dimethylamino)pyridine
DMDO	Dimethyldioxirane
DMF	Dimethylformamide
DMSO	Dimethyl sulfoxide
ee	Enantiomeric excess
equiv.	Equivalent(s)
Et	Ethyl
i-Pr	Isopropyl
KHMDS	Potassium hexamethyldisilazide, potassium bis(trimethylsilyl)amide
LHMDS	Lithium hexamethyldisilazide, lithium bis(trimethylsilyl)amide
Me	Methyl
min	Minute(s)
mol	Mole(s)
NBS	*N*-bromosuccinimide
NCS	*N*-chlorosuccinimide
Nu	Nucleophile
Ph	Phenyl
PPTS	Pyridinium *p*-toluenesulfonate
Pr	Propyl
py	Pyridine
rt	Room temperature
s	Second(s)
TBAF	Tetrabutylammonium fluoride
TBDMS	*tert*-butyldimethylsilyl
t-Bu	*tert*-butyl
THF	Tetrahydrofuran
TIPS	Triisopropylsilyl
TMS	Trimethylsilyl
Tol	4-Methylphenyl
Ts	Tosyl, 4-toluenesulfonyl

1 Introduction

Carbohydrates belong to the three most abundant classes of organic biomolecules. They can be found as monosaccharides, oligosaccharides, and polysaccharides. In addition, often combinations with other types of biomolecules are observed leading to glycoconjugates, glycolipids, glycopeptides, and glycosylated natural products. As a result from photosynthesis carbohydrates constitute the largest part of bio-derived compounds and comprise various functions in living organisms. Polysaccharides primarily serve as material for energy storage (such as starch or glycogen) as well as structural components providing stability to cell walls (such as cellulose in plants or chitin in arthropods and fungi) [1].

The concept of carbohydrate mimics includes organic compounds with carbohydrate-related structures to imitate their mode of action. Naturally occurring carbohydrates are often restricted for the use as potential drug, due to several drawbacks which can partially be suppressed by utilization of respective mimics. The high polarity of carbohydrates often represses the resorption of pharmaceuticals when they are substituted by sugars. Furthermore, the lability of the *O*-glycosidic bond hampers an appropriate transport of the molecule to the site of pharmacological action. This bond is often cleaved, resulting in a release of the agent too early and hence impaired pharmacodynamics [2]. Therefore it is of essential importance to find more stable and less hydrophilic analogs than common carbohydrates.

2 Carbohydrate Mimics

Carbohydrate mimics is a huge area. One class ranges from subtle modifications of sugars to carbasugars (with missing endocyclic oxygens), iminosugars (in which the endocyclic oxygen is replaced by nitrogen, see [3] of this chapter) and thiasugars (with an endocyclic sulfur). Another class focuses on linkage modifications. *C*-Glycosides commonly replace the linking oxygen by a saturated carbon (in the simplest case a methylene unit), whereas *N*-glycosides replace it by an amino group and *S*-glycosides by sulfur. Besides such modifications the skeleton itself might be varied, e.g. via attachment of rigidifying elements or the annulation of further ring systems. In this short report, we focus on work that was mainly carried out in our own laboratories dealing with Pd-catalyzed reactions to yield *C*-glycosides and chroman-like structures. In addition, a paragraph of the union of carbohydrates with cyclopropanes is added because these three-membered rings add stiffness to the system without changing the size of the six-membered ring to a major extent. Since most of the synthetic studies have just been carried out, biological data are almost missing so far. Thus, the review focuses on the synthetic issues.

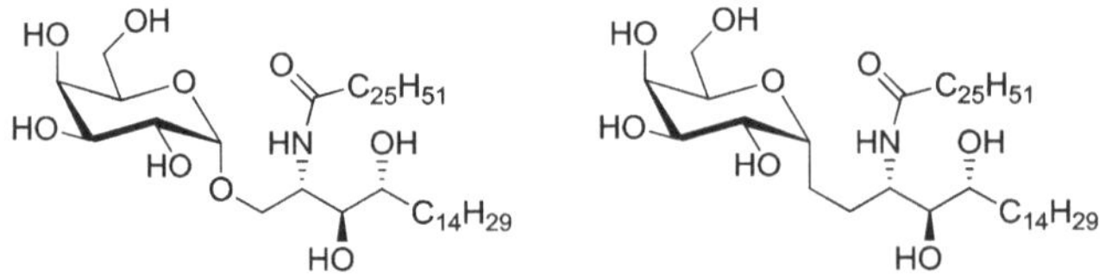

1a **1b**

Fig. 1 Agelashin KRN7000 (**1a**) and the *C*-glycosidic analog **1b**

2.1 C-Glycosides via Pd-Catalyzed Cross-Couplings

C-glycosides are an essential class of carbohydrate mimics. The linkage between both monosaccharide subunits is established by a methylene unit instead of an oxygen atom. Since such a linkage does not exist in Nature the stability towards enzymatic and chemical cleavage is dramatically increased. Due to that effect this class of carbohydrate mimics found broad application in medicinal chemistry and drug design. However, the synthesis is often a difficult and very time-consuming endeavor. Most of the methods developed to access this structural motif are confined to elaborate building blocks and provide only specific linkages [4–6].

A very prominent example for a beneficial therapeutical effect of *C*-glycosides is the Agelashin derivative KRN7000 **1** (Fig. 1) [7–9]. This α-galactosylceramide consists of a carbohydrate head group, a sphingosine derivative, and a fatty acid being linked via an amide bond. Preliminary investigations revealed that the *C*-glycosidic analog showed a 100-times higher activity against malaria in mice than the analogous *O*-glycoside. Kinetic studies illustrated that the *C*-glycoside experienced a fourfold slower off-rate compared to its *O*-glycoside analog.

Since most methods for the synthesis of *C*-glycosides are only useful for special linkages, a modular and robust synthetic route for the assembly of a variety of *C*-glycosidic linkages was highly desirable. For this purpose, palladium-catalyzed cross-coupling reactions were used to link the two monosaccharidic subunits.

The (1 → 6)-linkage was obtained by a Sonogashira–Hagihara-type reaction from a suitable substituted 1-iodoglycal **2a** and respective alkynyl glycosides **3** [10]. The iodoglucal is easily available from D-glucose [11], whereas the 6-alkynyl glycosides were synthesized under Ohira–Bestmann conditions from the corresponding aldehydes [12, 13]. Since iodogalactals are difficult to prepare respective triflates were chosen as coupling partners instead of the iodogalactals. Therefore, perbenzylated galactal was converted via an acid-mediated addition of water to the double bond to the corresponding hemiacetal [14]. Successive oxidation of the α:β-mixture furnished the lactone [15], which was enolized with KHDMS and $PhNTf_2$ to afford the desired triflate **2b** [16, 17]. Because of the low stability of such triflates **2b** they were directly reacted in a Cacchi-coupling with a variety of different alkynyl glycosides **3** to yield enynes **4** (Scheme 1).

A crucial step of this synthetic approach is the refunctionalization of the enyne system to the fully deprotected *C*-disaccharide. During this process, also the

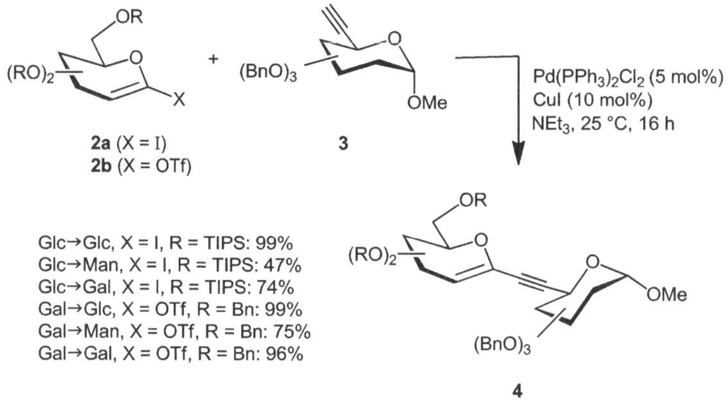

Scheme 1 Sonogashira–Hagihara and Cacchi-coupling for the synthesis of different carbohydrate-derived enynes

2-hydroxyl and the correct anomeric stereochemistry has to be built up. Promising results were obtained by selective reduction of the triple bond of **4** in presence of the electron-rich enol ether. Raney-nickel revealed to be the reagent of choice for this step to furnish **5**. A formal regio- and stereoselective addition of one equivalent of water to the enol ether moiety was necessary to build up the native hydroxyl pattern. Investigations with several borane complexes showed that hydroboration by utilization of $BH_3{\cdot}THF$ and hydrogen peroxide gave good results with respect to β-configured *gluco* products (Variant A) [18, 19]. As alternative an epoxidation-reduction sequence could be used to provide access to both, the α- and the β-configuration at the pseudoanomeric center depending on the hydride reagent utilized in the second step. DMDO-mediated epoxidation of the enol ethers gave high facial selectivities in favor of the *gluco* form (>20:1) [20]. The opening of the epoxides was either performed with DIBAL as hydride transfer reagent (Variant B) yielding exclusively β-products [21] or with superhydride ($LiBHEt_3$) which selectively forms an α-configured product (Variant C) [22, 23]. By a permutation of these methodologies diverse (1 → 6)-linked disaccharides could be obtained after global deprotection of the respective permanent protective groups (Scheme 2).

These results set the stage for the synthesis of further (1 → n)-linked *C*-glycosides (n = 2, 3, 4) [24]. In contrast to a Sonogashira–Hagihara reaction a Stille cross-coupling employing 1-stannylglycals **7** and exocyclic bromoolefins **8** proved to be the method of choice to link the two subunits. The resulting *pseudo-C*-disaccharides **9** were refunctionalized to *C*-glycosides with the native hydroxyl group pattern.

The synthesis of the 1-stannylglycals **7** was performed according to literature-known procedures [25]. The exocyclic bromoolefins **8** were prepared from respective ketones by a Wittig-type reaction using Ph_3PCHBr. An exception was made for the 2-substituted bromoolefins. Cyclopropanation of the corresponding 1,2-bromoglycals and subsequent opening of the three-membered ring furnished

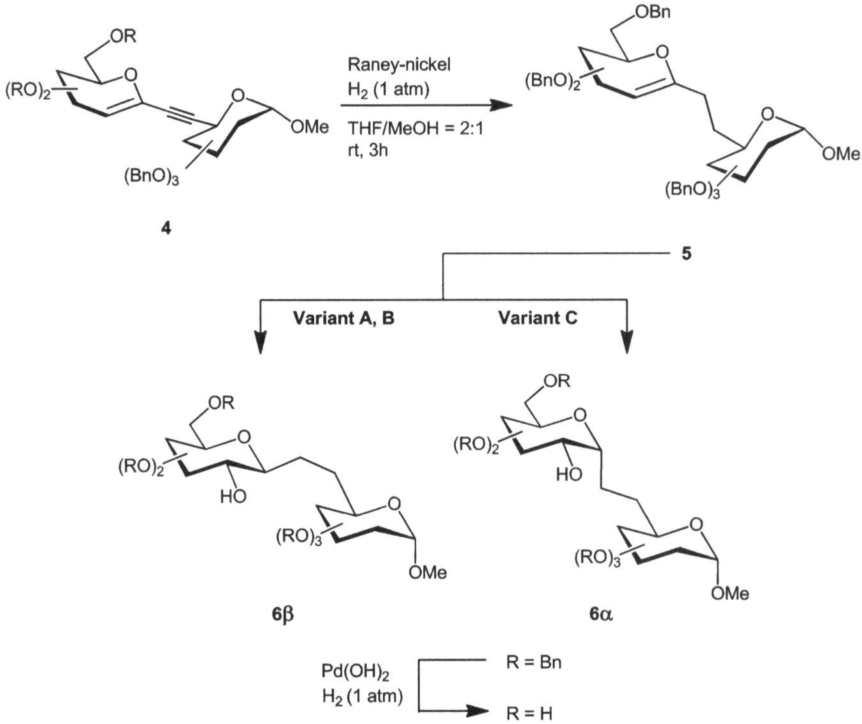

Scheme 2 Refunctionalization of the enyne systems in **4** via three different variants (*A*, *B*, *C*). Variant *A*: (1) BH_3·THF, THF, 0°C → 25°C, 12 h, (2) H_2O_2, NaOH, 2 h. Variant *B*: (1) DMDO, CH_2Cl_2, −78°C → 0°C, 30 min, (2) DIBAL, CH_2Cl_2, −78°C → 25°C, 1 h. Variant *C*: (1) DMDO, CH_2Cl_2, −78°C → 0°C, 30 min, (2) $LiBHEt_3$, THF, 0°C, 3 h

the desired products [26–30]. In all cases the cross-coupling reactions proceeded smoothly with yields up to 88%. The reductive-oxidative refunctionalization sequence which was successfully used for the (1 → 6)-linked *C*-glycosides only worked in rare cases. In contrast to (1 → 6)-linkages the systems setting the stage for (1 → n)-linkages (n = 2, 3, 4) consist of a diene subunit differing strongly from the enyne motif employed in the prior case. Another major obstacle is the construction of a further stereocenter that results from the hydrogenation of the exocyclic double bond whereas in the reduction of the exocyclic alkyne in (1 → 6)-*C*-disaccharides only an ethano unit was formed.

Therefore, an alternative strategy was designed relying on the oxidative functionalization of the enol ether moiety of **9** as the first step. DMDO as epoxidizing agent proved to be highly chemo- and diastereoselective. The use of only one equivalent of the reagent led to the exclusive epoxidation of the more electron-rich enol ether while the exocyclic double bond remained untouched. The opening of the epoxide was performed as previously described for the (1 → 6)-linked substrates. DIBAL led to the β-anomer whereas the strongly nucleophilic superhydride ($LiBHEt_3$) generated the α-product. The reduction of the exocyclic double bond

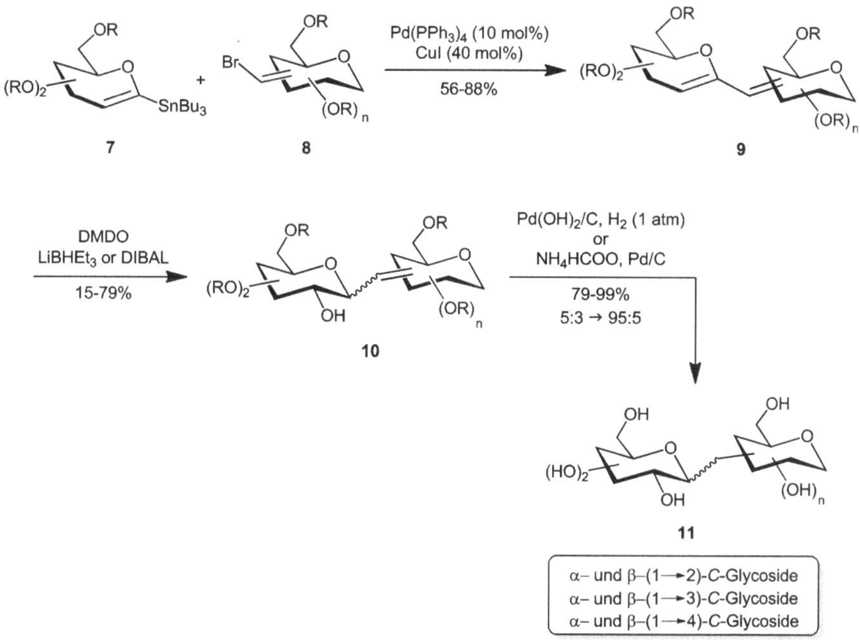

Scheme 3 Synthesis of (1 → 2)-, (1 → 3)- and (1 → 4)-linked *C*-glycosides using dienes as crucial intermediates

of **10** was performed by either hydrogenolysis using Pearlman's catalyst or diastereoselective transfer hydrogenation with ammonium formate [31] (Scheme 3).

2.2 *Carbohydrates with Reduced Conformational Flexibility of the 6-Hydroxyl Group*

The installation of cyclopropyl subunits provides an interesting approach for the synthesis of natural product mimics. Saturated C_2 moieties which are flexible units can be rigidified by the formal addition of only one carbon atom yielding a three-membered ring. Such a change adds only a minimum of steric impact to the system. Reiser demonstrated that the implementation of cyclopropane units has a stabilizing effect to the secondary structure of oligopeptides [32, 33].

Such an approach was also used to reduce the conformational flexibility of the 6-hydroxyl group in carbohydrates [34]. The small cyclopropyl ring has only a minimal effect on the structural integrity of the carbohydrate backbone and therefore represents an excellent strategy to lock the 6-hydroxyl moiety into a certain direction.

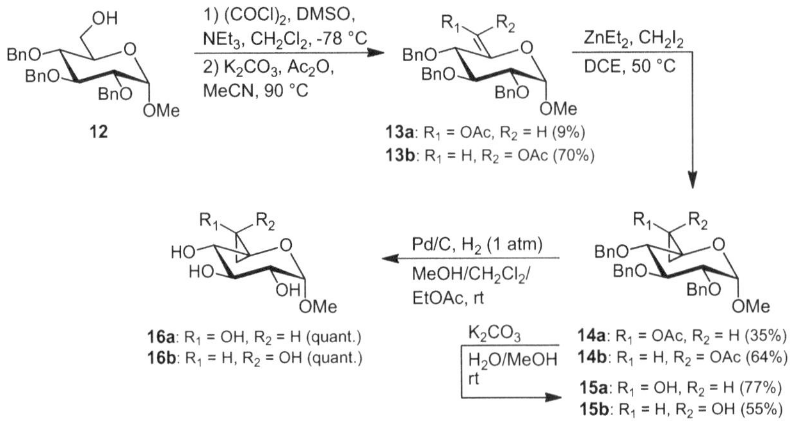

Scheme 4 Preparation of glucose derivative **16** with locked 6-hydroxyl group

The general synthetic route to access such systems consists of enolization of an aldehyde in 6-position of the monosaccharide derived from **12**. A Swern oxidation of the primary alcohol moiety and subsequent enolization of the aldehyde with potassium carbonate in acetic anhydride furnished the respective exocyclic double bond [35]. The corresponding enol acetates were obtained in a ratio of 8:1 in favor of the diastereoisomer **13b**. Subsequent cyclopropanation under Simmons–Smith–Furukawa conditions formed the three-membered ring [36–39]. Deacetylation of **14** furnished **15** either as versatile building block for further glycosylation steps or as starting material for unprotected methyl glucoside **16** (Scheme 4).

Similar procedures were applied for the synthesis of mannose units [40]. Respective glycosyl donors as well as glycosyl acceptors were prepared. With donor **17** several glycosylation studies were undertaken; good yields were observed and the cyclopropyl moiety remained untouched under the reaction conditions. The analysis of the diastereoselectivity revealed that in all cases the β-mannoside was the major product. This result is noteworthy and it seems that the three-membered ring shows a similar effect as a 4,6-benzylidene acetal. Although the selectivities were worse than in the pioneering work of Crich [41–43], which used 4,6-benzylidene acetal as prerequisite for a high β-selectivity, this procedure gave significant better results compared to the 2,3,4-tribenzylated-6-acetylated mannosyl trichloracetimidate. The spiroannulation may lead to a fixation of the chair-like conformation as a 4,6-benzylidene protecting group does.

In addition both, the α- and the β-mannoside, **19** were used for a further glycosylation step. Therefore the acetate protecting group was cleaved using potassium carbonate to yield acceptor **20**, which was converted with an appropriate glucoside donor [44, 45] and under the influence of trimethylsilyl triflate into the trisaccharide **21**. The global deprotection of the benzoyl and benzyl protecting groups furnished the naked carbohydrate **22** in moderate yield (Scheme 5).

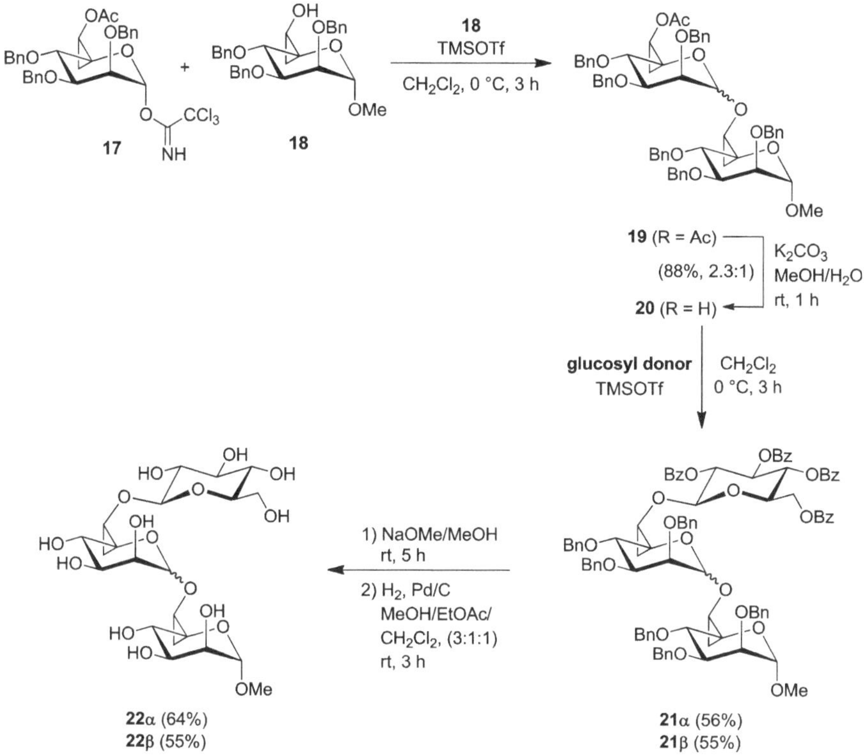

Scheme 5 Synthesis of cyclopropyl-modified trisaccharide **22**

2.3 *Synthesis of Carbohydrate-Derived* [n,5]-*Spiroketals*

Spiroketals are found in a plethora of natural products. Their semirigid characteristic moiety represents one main feature in many different toxins and pheromons [46–51]. Classical approaches consist of Brønsted or Lewis acid-mediated dehydrative cyclization of an appropriate ketodiol.

Recently, a novel gold rush in the chemistry of donor–acceptor cyclopropanes has started. Our group has developed several ring-enlargement reactions leading to various heterocycles based on the special features of these three-membered rings [52–56]. Thus, we used a similar protocol to furnish spiroketals starting from carbohydrates [57].

The synthetic approach consists of three major steps: As starting material one uses an exocyclic enol ether easily available either via Petasis olefination [58] using Cp_2TiMe_2 from the sugar lactone or via elimination of HI or HCl [59] from the 6-halogenated carbohydrates. In both cases a corresponding electron-rich double bond was formed. Such exocyclic glycals are especially suited to react with ethyl

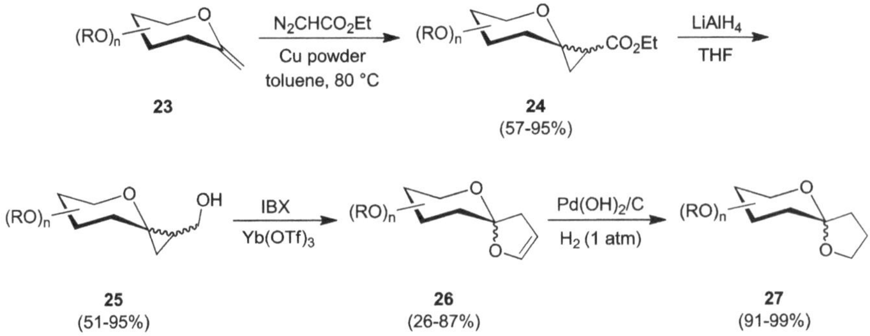

Scheme 6 Carbohydrate-derived spiroketals via donor–acceptor cyclopropanes

diazoacetate under rhodium or copper catalysis to afford the cyclopropane derivatives **24** [60]. The resulting cyclopropyl esters were reduced with $LiAlH_4$ to the respective alcohols **25** in very good yield. The oxidation to the aldehyde moiety is immediately followed by the rearrangement reaction to the desired spiroketal **26**. The choice of the oxidation agent proved to be crucial. In our hands the hypervalent iodine reagent IBX proved to be the best choice. This reagent might also act as Lewis acid coordinating with the aldehyde oxygen and thus weakening the bond in the three-membered ring. Sometimes better results were obtained by adding $Yb(OTf)_3$ as external Lewis acid. The sensitive enol ether moiety is easily removed by Pearlman's catalyst and an atmosphere of hydrogen (Scheme 6).

2.4 *Hybrids of Carbohydrates and Arenes*

The fusion of carbohydrate ring systems and an aromatic core is still a challenging task in organic synthesis. The annulation of a benzene moiety to the pyran system of the carbohydrate would lead to a variety of chromans and isochromans. These structural motifs are widespread elements of different natural products with a variety of biological and pharmaceutical activities [61, 62] (Fig. 2).

2.4.1 Synthesis of Chromans and Isochromans via Domino Carbopalladation Reactions

The key step of this highly effective approach is a palladium-catalyzed domino carbopalladation sequence starting from suitable substituted bromoglycals **33** [63–66]. 2-Bromoglycals **31** are accessed from commercially available glycals via a bromination-elimination procedure [67]. After the installation of suitable protecting groups at 4- and 6-hydroxyls attachment of various dialkyne chains **32**

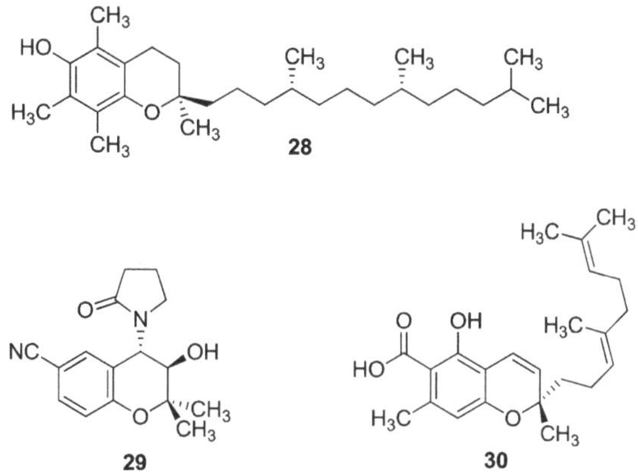

Fig. 2 Three examples of chromans revealing biological activity

was carried out yielding a broad variety of different domino precursors **33**. The synthesis of the corresponding propargylic halides **32** was achieved starting from symmetrical dialkynes, e.g. from heptadiyne or octadiyne. The coupling was conducted either by nucleophilic substitution of the free 3-hydroxyl group with the propargylic halide or via Ferrier reaction (in the case of isochroman precursors) [68]. Optimal conditions for the domino reaction comprise the use of a palladium(0) catalyst such as ($Pd(PPh_3)_4$) and an additional electron-rich ligand ($HP(t\text{Bu})_3{\cdot}BF_4$) under microwave irradiation. The reaction cascade is initiated by an oxidative addition of the Pd-catalyst into the C-Br bond and followed by two consecutive carbopalladation steps forming a triene system [69, 70]. The final step forms the benzene moiety and may be regarded as 6π-electrocyclization, followed by a β-hydride elimination. In most cases the domino reaction proceeds smoothly; different monosaccharide building blocks and various substituted diyne chains were employed. Alkyl-, aryl-, and silyl-substituted triple bonds were successfully tested and even terminal acetylene units led to the desired products **34**. A similar synthetic route yielded the respective isochromans **36**.

The studies summarized in Scheme 7 show limitations with respect to the substitution pattern of the chroman and isochroman framework. Therefore, a more flexible approach is desirable [71]. By omitting the tether between both triple bonds it was possible to perform the second carbopalladation with an external alkyne in an intermolecular fashion.

The attachment of the first alkyne to the sugar moiety was performed as already described above. Best results for the domino reaction were obtained using a large excess (10–20-fold) of the external alkyne in a highly concentrated reaction mixture. The variation of the carbohydrate part and the differentiation of the attached alkyne as well as the external alkyne provided a broad range of different products. To demonstrate the generality and applicability of this approach besides chromans isochromans were accessed in such a way. Glucose- and galactose-derived carbohydrates moieties as well as alkynes with silyl-, alkyl-, and aryl-substitution were utilized. Good yields up to 69% were obtained for symmetrical external alkynes

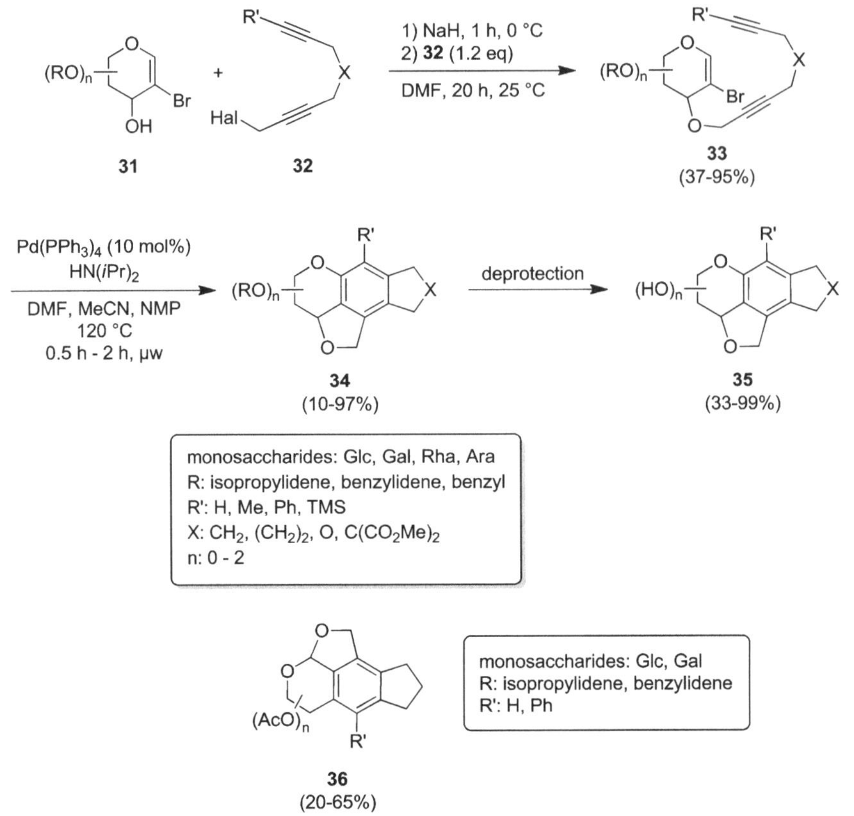

Scheme 7 Chroman and isochroman synthesis via a domino carbopalladation approach

such as bis(trimethylsilyl)acetylene or 3-hexyne. The yield dramatically dropped down when sterically encumbered alkynes such as tolane were employed as reaction partners (Scheme 8).

The extension to a tandem-domino process opened the possibility to form biphenyl structures with a chiral axis [72]. Therefore, the same domino sequence as described before took place at both sides of a linear C_2-symmetric molecule. The reacting centers are linked by a 1,3-butadiyne unit; thus, every separate process starts from its own vinyl bromide and dialkyne. The final cyclization step of the second domino reaction generates a chiral axis via the twofold benzene formation in the respective domino reaction.

The tandem domino approach to access carbohydrate-derived biphenyls **43e–f** was investigated with glucose and galactose units as well as varying tether lengths. Best yields (61–66%) and diastereoselectivities (only one diastereoisomer) were obtained with galactose units bearing benzylidene protecting groups. In the case of glucose units protected with isopropylidene groups only moderate yields and worse diastereoselectivities could be observed (Scheme 9).

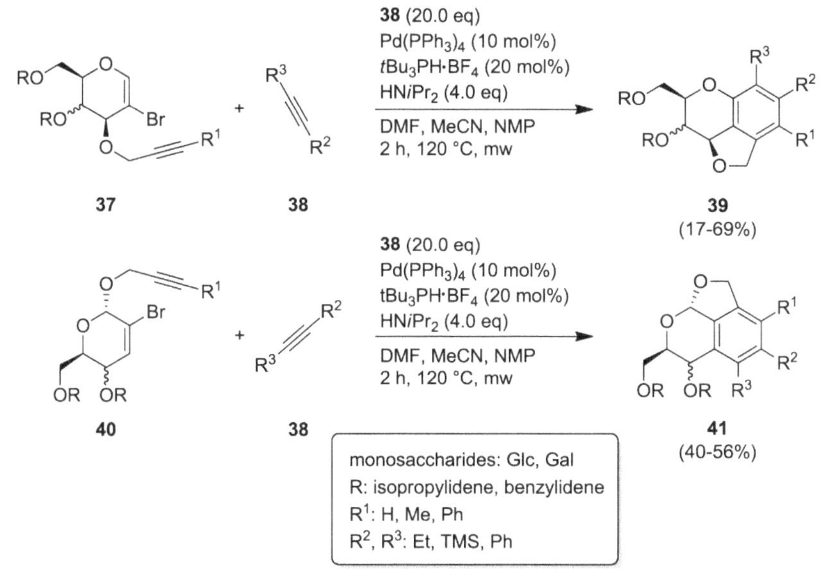

Scheme 8 Intermolecular domino carbopalladation approach using external alkynes

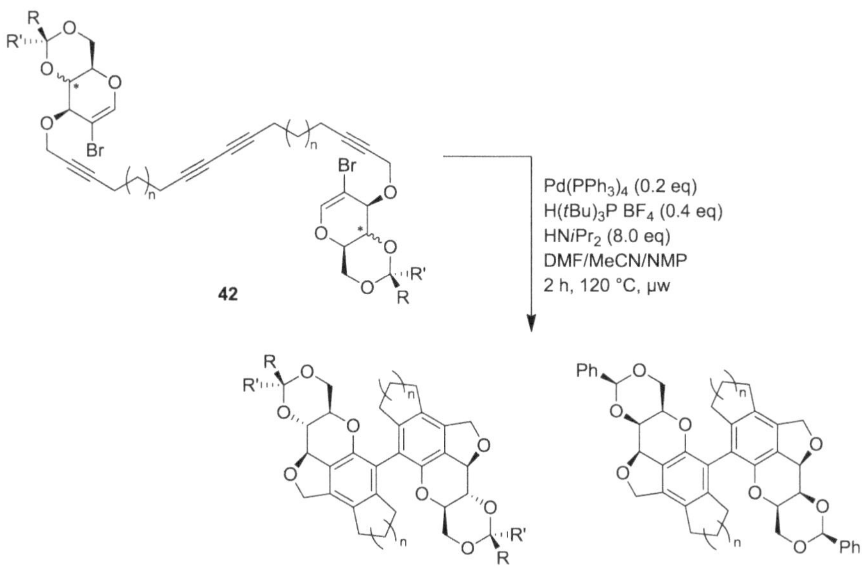

43a (Glc, R = R': CH_3, n: 1, 30%, *dr* = 1:1)
43b (Glc, R = R': CH_3, n: 2, 38%, *dr* = 4:1)
43c (Glc, R: H, R': Ph, n: 1, 44%, *dr* = 2:1)
43d (Glc, R: H, R': Ph, n: 2, 60%, *dr* = 10:1)

43e (Gal, n: 1, 66%) *(1 diastereoisomer)*
43f (Gal; n: 2: 61%) *(1 diastereoisomer)*

Scheme 9 Tandem domino process for the synthesis of carbohydrate-derived biphenyls

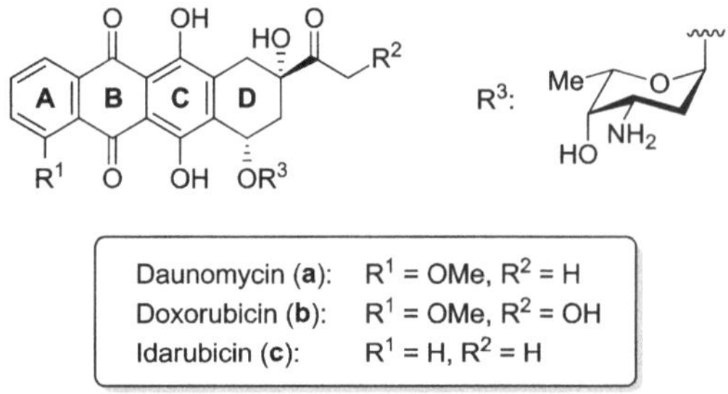

Fig. 3 Several naturally occurring anthracycline antibiotics

2.4.2 Synthesis of Anthracycline Derivatives

Anthracyclines belong to the natural product class of aromatic polyketides [73]. They were first isolated from the order of *Streptomycetales* by Brockmann in 1963, who described them as red to orange dyes [74]. Their structural features consist of a fourfold annulated ring system including two benzene moieties. The substitution of the D-ring bares several functional groups, i.e. two alcohol groups and a glycosylated 2,6-didesoxy sugar [75]. These carbohydrate functionalities are of highest importance for the biological and pharmaceutical activity of these natural products [76–78]. Their main application relies on the treatment of different types of cancer such as leukemia, lymphomas, breast, uterine, ovarian, and lung cancer (Fig. 3) [79].

Many groups faced the challenge of employing suitable pathways for the synthesis of anthracyclines and their derivatives. Thus, most of these approaches involve the application of a Diels–Alder reaction as the key step [80–83], while industrial approaches made use of recombinant microorganisms with mutated genes of the anthracycline metabolism [84].

The results of the investigations with respect to the synthesis of chromans and isochromans provided a promising fundament for the application to the synthesis of anthracycline mimics starting from carbohydrates. Therefore, the D-ring was exchanged by a 2-bromoglycal and instead of an aliphatic dialkyne chain an aromatic diyne was employed. Furthermore a silyl ether moiety should serve as tethering unit between both building blocks. The fourfold annulated ring system should be obtained in one single step, by applying the domino carbopalladation procedure as already used for the synthesis of chromans. This powerful transformation allows the formation of the B- and C-ring and consequently the annulation of all four cycles (Scheme 10) [85].

Whereas the carbopalladation cascade as the key reaction did not cause any difficulties several other transformations proved to be rather challenging. In a very first step compounds of type **47** had to be synthesized (not shown). For this purpose, two different silylalkynyl groups were installed to appropriate benzylic iodides [86].

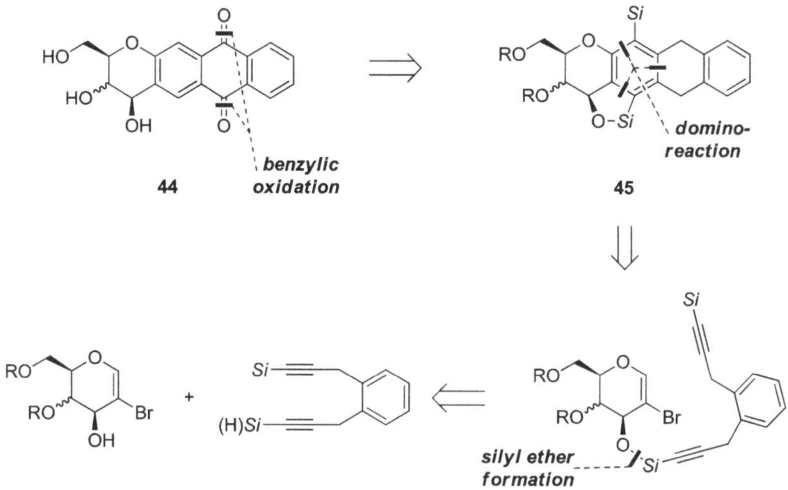

Scheme 10 Retrosynthetic analysis of anthracycline aglycone mimics. Si: any silyl group

In order to run an intramolecular and regiochemically defined carbopalladation process the 2-bromoglycal and the dialkyne have to be linked to each other. This transformation proved to be a highly demanding task; a huge number of reaction conditions were explored, but only one led to the desired product in sufficient yield. Therefore, the terminal silane **47** was converted into the respective silabromide with a solution of elementary bromine in tetrachloromethane [87, 88]. The so formed reactive species was treated with the corresponding alcohol and triethylamine as base to furnish the coupling product **46**. Utilization of the already known conditions for such a domino reaction provided the fourfold annulated ring system in only one step and very good yield. It is noteworthy that the best yields were obtained for the TMS-substituted substrates.

The silyl groups at the newly formed benzene ring were cleaved by the action of acetyl chloride in anhydrous methanol and tetrabutylammonium fluoride. A Tamao-Fleming [89–91] like oxidation to furnish the respective phenol derivatives did not take place probably due to the sp^2-hybridized carbon of the arene moiety. To install the anthraquinone moiety by oxidation of the benzylic positions it was necessary to reprotect the alcohol functionalities. After TBS-protection of the hydroxyl moieties an iron(III)-catalyzed benzylic oxidation proceeded smoothly with yields up to 70% [92]. Finally, hydrolysis with hydrochloric acid of the protection groups afforded the desired anthracycline derivatives based on a carbohydrate skeleton (Scheme 11) [85].

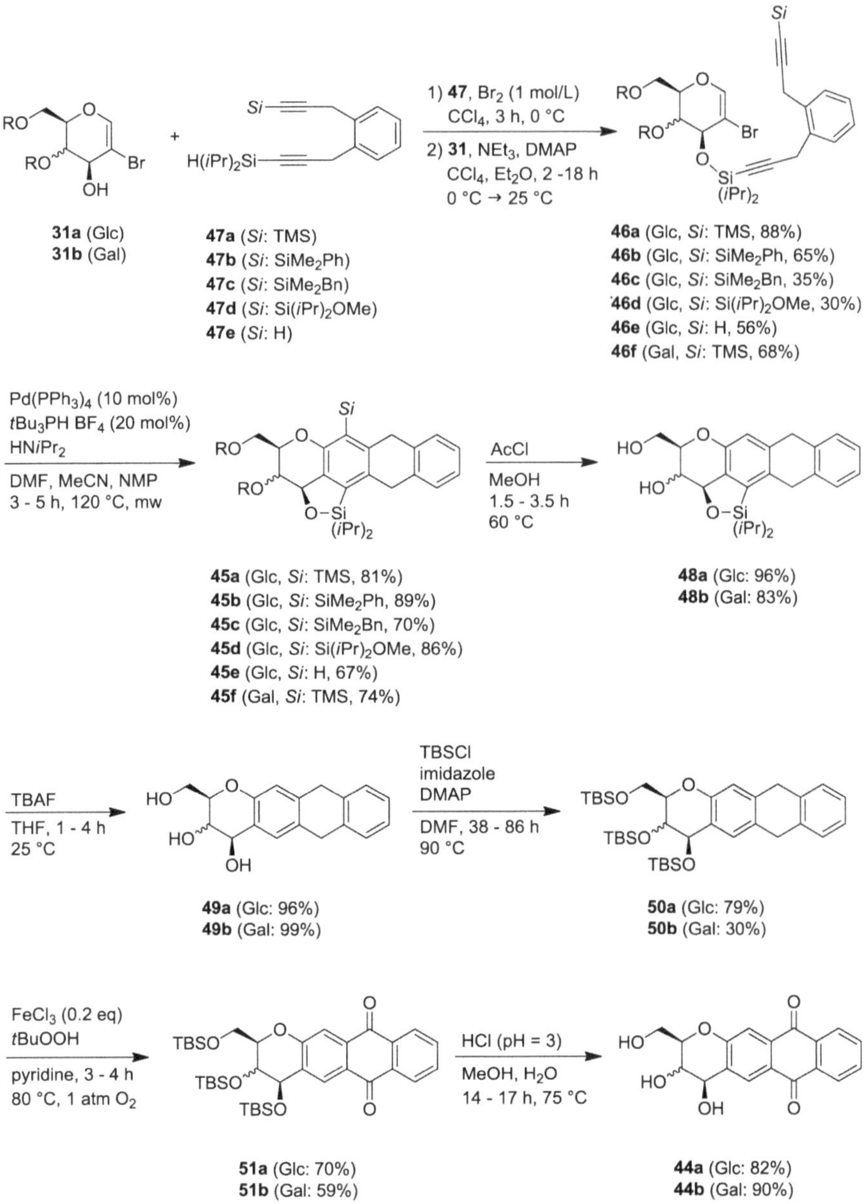

Scheme 11 Silyl ether coupling, domino carbopalladation reaction, and derivatization to anthracycline derivatives

3 Conclusion

The chemistry of carbohydrate mimics has seen much progress during the last decade. Key to success is always novel synthetic methodologies or at least the application of known methods developed in other fields to carbohydrate chemistry. Pd-catalyzed reactions have proven to be a reliable and robust method to access *C*-glycosides in a highly modular way. In addition, Pd-catalyzed domino sequences were shown to be remarkably efficient tools to fuse arene systems with carbohydrate scaffolds. In such a way, numerous chroman and isochroman systems were built up starting from sugars. Merging carbohydrates with three-membered ring chemistry pave the way for sugar-derived spiroketals and rigidified systems with locked 6-hydroxyl groups. Most of these systems are still awaiting biological investigation. Even if synthetic problems in the chemistry of carbohydrate mimics might be solved and will find medical applications, Nature will challenge future chemists as well since microorganisms will never rest to evade our tailor-made molecules to kill them.

References

1. Lindhorst TK (2007) Essentials of carbohydrate chemistry and biochemistry. Wiley, Weinheim
2. Koester DC, Holkenbrink A, Werz DB (2010) Recent advances in the synthesis of carbohydrate mimics. Synthesis 3217–3242
3. Horne G (2014) Iminosugars: therapeutic applications and synthetic considerations. Top Med Chem. doi:10.1007/7355_2014_50
4. Giese B, Witzel T (1986) Synthesis of "*C*-Disaccharides" by radical C–C bond formation. Angew Chem Int Ed Engl 25:450–451
5. Schmidt RR, Preuss R (1989) Synthesis of carbon bridged *C*-disaccharides. Tetrahedron Lett 30:3409–3412
6. Patro B, Schmidt RR (2000) (1-1)-Linked *C*-disaccharides – synthesis of bis(β-D-Galactopyranosyl)methane. J Carbohydr Chem 19:817–826
7. Yang G, Schmieg J, Tsuji M, Franck RW (2004) The *C*-glycoside analogue of the immunostimulant α-galactosylceramide (KRN7000): synthesis and striking enhancement of activity. Angew Chem Int Ed 43:3818–3822
8. Yang G (2010) Synthesis of C-glycosides via Ramberg-Bäcklund reaction: synthesis of C-glycosides KRN-7000. LAP LAMBERT Academic Publishing
9. Chaulagain MR, Postema MHD, Valeriote F, Pietraszkewicz H (2004) Synthesis and antitumor activity of β-*C*-glycoside analogs of the immunostimulant KRN7000. Tetrahedron Lett 45:7791–7794
10. Koester DC, Leibeling M, Neufeld R, Werz DB (2010) A Pd-catalyzed approach to (1–6)-linked *C*-glycosides. Org Lett 12:3934–3937
11. Potuzak JS, Tan DS (2004) Synthesis of *C*1-alkyl and *C*1-acylglycals from glycals using a *B*-alkyl Suzuki–Miyaura cross coupling approach. Tetrahedron Lett 45:1797–1801
12. Lehmann J, Thieme R (1986) Synthese von 6-*C*-Azi-6-desoxy-D-glucose und -D-galactose für die Photoaffinitätsmarkierung von kohlenhydratbindenden Proteinen. Liebigs Ann:525–532
13. Roth GJ, Liepold B, Müller SG, Bestmann HJ (2004) Further improvements of the synthesis of alkynes from aldehydes. Synthesis 59–62

14. Wild R, Schmidt RR (1995) Synthesis of sphingosines, 11. Convenient synthesis of phytosphingosine and sphinganine from D-galactal and D-arabitol. Liebigs Ann 755–764
15. Jensen HH, Bols M (2003) Steric effects are not the cause of the rate difference in hydrolysis of stereoisomeric glycosides. Org Lett 5:3419–3421
16. Fujiwara K, Tsunashima M, Awakura D, Murai A (1995) Stereoselective synthesis of Δ^5-oxonene and its novel ring contraction to Δ^4–oxocene. Tetrahedron Lett 36:8263–8266
17. Sasaki M, Ishikawa M, Fuwa H, Tachibana K (2002) A general strategy for the convergent synthesis of fused polycyclic ethers via B-alkyl Suzuki coupling: synthesis of the ABCD ring fragment of ciguatoxins. Tetrahedron 58:1889–1911
18. Hanessian S, Martin M, Desai RCJ (1986) Formation of *C*-glycosides by polarity inversion at the anomeric centre. J Chem Soc Chem Commun 926–927
19. Schmidt RR, Preuss R, Betz R (1987) C-1 lithiation of c-2 activated glucals. Tetrahedron Lett 28:6591–6594
20. Halcomb RL, Danishefsky SJJ (1989) On the direct epoxidation of glycals: application of a reiterative strategy for the synthesis of β-linked oligosaccharides. J Am Chem Soc 111: 6661–6666
21. Majumder U, Cox JM, Johnson HWB, Rainier JD (2006) Total synthesis of Gambierol: the generation of the A–C and F–H subunits by using a *C*-glycoside centered strategy. Chem Eur J 12:1736–1746
22. Inoue M, Yamashita S, Tatami A, Miyazaki K, Hirama MJ (2004) A new stereoselective synthesis of ciguatoxin right wing fragments. J Org Chem 69:2797–2804
23. Krishnamurthy S, Schubert RM, Brown HC (1973) Lithium triethylborohydride as a convenient reagent for the facile reduction of both hindered and bicyclic epoxides prone to electrophilically induced rearrangement. J Am Chem Soc 95:8486–8487
24. Koester DC, Kriemen E, Werz DB (2013) Flexible synthesis of 2-deoxy-C-glycosides and (1-2)-, (1-3)-, and (1-4)-linked *C*-glycosides. Angew Chem Int Ed 52:2985–2989
25. Jarowicki K, Kilner C, Kocienski P, Komsta Z, Milne J, Wojtasiewicz A, Coombs V (2008) A synthesis of 1-lithiated glycals and 1-tributylstannyl glycals from 1-phenylsulfinyl glycals via sulfoxide-lithium ligand exchange. Synthesis 2747–2763
26. Lu H, Silverman RB (2006) Fluorinated conformationally restricted γ-aminobutyric acid aminotransferase inhibitors. J Med Chem 49:7404–7412
27. Ramana CV, Murali R, Nagarjan M (1997) Synthesis and reactions of 1,2-cyclopropanated sugars. J Org Chem 62:7694–7703
28. Sato K, Sekiguchi T, Hozumi T, Yamazaki T, Akai S (2002) Improved synthetic method for preparing spiro α-chloroepoxides. Tetrahedron Lett 43:3087–3090
29. Seyferth D, Heeren JK, Grim SO (1961) The action of phenyllithium on bromomethyl- and iodomethyltriphenylphosphonium halides. J Org Chem 26:4783–4784
30. Hewitt RJ, Harvey JE (2010) Synthesis of Oxepines and 2-branched pyranosides from a D-glucal-derived gem-dibromo-1,2-cyclopropanated sugar. J Org Chem 75:955–958
31. Li X, Li L, Tang Y, Zhong L, Cun L, Zhu J, Liao J, Deng J (2010) Chemoselective conjugate reduction of α, β-unsaturated ketones catalyzed by rhodium amido complexes in aqueous media. J Org Chem 75:2981–2988
32. Koglin N, Zorn C, Beumer R, Cabrele C, Bubert C, Sewald N, Reiser O, Beck-Sickinger AG (2003) Analogues of Neuropeptide Y containing β-Aminocyclopropane carboxylic acids are the shortest linear peptides that are selective for the Y1 receptor. Angew Chem Int Ed 42: 202–205
33. De Pol S, Zorn C, Klein CD, Zerbe O, Reiser O (2004) Surprisingly stable helical conformations in α/β-peptides by incorporation of *cis*-β-aminocyclopropane carboxylic acids. Angew Chem Int Ed 43:511–514
34. Brand C, Granitzka M, Stalke D, Werz DB (2011) Reducing the conformational flexibility of carbohydrates: locking the 6-hydroxyl group by cyclopropanes. Chem Commun 47: 10782–10784

35. Takahashi H, Kittaka H, Ikegami S (2001) Novel synthesis of enantiomerically pure natural inositols and their diastereoisomers. J Org Chem 66:2705–2716
36. Simmons HE, Smith RD (1959) A new synthesis of cyclopropanes. J Am Chem Soc 81: 4256–4264
37. Furukawa J, Kawabata N, Nishimura J (1967) Synthesis of cyclopropanes by the reaction of olefins with dialkylzinc and methylene iodide. Tetrahedron 24:53–58
38. Furukawa J, Kawabata N, Nishimura J (1966) A novel route to cyclopropanes from olefins. Tetrahedron Lett 7:3353–3354
39. Song Z, Lu T, Hsung RP, Al-Rashid ZF, Ko C, Tang Y (2007) Stereoselective Simmons–Smith cyclopropanation of chiral enamides. Angew Chem Int Ed 46:4069–4072
40. Brand C, Kettelhoit K, Werz DB (2012) Glycosylations of cyclopropyl-modified carbohydrates: remarkable β-selectivity using a mannose building block. Org Lett 14:5126–5129
41. Crich D, Sun S (1996) Formation of β-mannopyranosides of primary alcohols using the sulfoxide method. J Org Chem 61:4506–4507
42. Crich D, Sun S (1997) Are glycosyl triflates intermediates in the sulfoxide glycosylation method? a chemical and 1H, 13C, and 19F NMR spectroscopic investigation. J Am Chem Soc 119:11217–11223
43. Crich D, Sun S (1998) Direct chemical synthesis of β-mannopyranosides and other glycosides via glycosyl triflates. Tetrahedron 54:8321–8348
44. Egusa K, Kusumoto S, Fukase K (2003) Solid-phase synthesis of a phytoalexin elicitor pentasaccharide using a 4-azido-3-chlorobenzyl group as the key for temporary protection and catch-and-release purification. Eur J Org Chem 3435–3445
45. Schmidt RR, Michel J, Moos M (1984) Glycosylimidate, 12 Direkte synthese von *O*-α- und *O*-β-Glycosyl-imidaten. Liebigs Ann Chem 1343–1357
46. Francke W, Kitching W (2001) Spiroacetals in insects. Curr Org Chem 5:233–251
47. Mead KT, Brewer BN (2003) Strategies in spiroketal synthesis revisited: recent applications and advances. Curr Org Chem 7:227–256
48. Aho JE, Pihko PM, Rissa TK (2005) Nonanomeric spiroketals in natural products: structures, sources, and synthetic strategies. Chem Rev 105:4406–4440
49. Brimble MA, Fares FA (1999) Synthesis of bis-spiroacetal ring systems. Tetrahedron 55: 7661–7706
50. Brasholz M, Sörgel S, Azap C, Reissig H-U (2007) Rubromycins: structurally intriguing, biologically valuable, synthetically challenging antitumour antibiotics. Eur J Org Chem 3801–3814
51. Rizzacasa MA, Pollex A (2009) The hetero-Diels–Alder approach to spiroketals. Org Biomol Chem 7:1053–1059
52. Schneider TF, Kaschel J, Dittrich B, Werz DB (2009) Anti-oligoanellated THF moieties: synthesis via Push-Pull-substituted cyclopropanes. Org Lett 11:2317–2320
53. Kaschel J, Schmidt CD, Mumby M, Kratzert D, Stalke D, Werz DB (2013) Donor-acceptor cyclopropanes with Lawesson's and Woollins' reagents: formation of bisthiophenes and unprecedented cage-like molecules. Chem Commun 49:4403–4405
54. Kaschel J, Schneider TF, Kratzert D, Stalke D, Werz DB (2012) Domino reactions of donor-acceptor-substituted cyclopropanes for the synthesis of 3,3′-linked oligopyrroles and pyrrolo [3,2-e]indoles. Angew Chem Int Ed 51:11153–11156
55. Kaschel J, Schneider TF, Kratzert D, Stalke D, Werz DB (2013) Symmetric and unsymmetric 3,3′-linked bispyrroles via ring-enlargement reactions of furan-derived donor-acceptor cyclopropanes. Org Biomol Chem 11:3494–3509
56. Kaschel J, Schneider TF, Schirmer P, Maaß C, Stalke D, Werz DB (2013) Rearrangements of Furan-, Thiophene- and *N*-Boc-pyrrole-derived donor-acceptor cyclopropanes: scope and limitations. Eur J Org Chem 4539–4551
57. Brand C, Rauch G, Zanoni M, Dittrich B, Werz DB (2009) Synthesis of [n,5]-spiroketals by ring enlargement of donor-acceptor-substituted cyclopropane derivatives. J Org Chem 74: 8779–8786

58. Petasis NA, Bzowej EI (1990) Titanium-mediated carbonyl olefinations. 1. Methylenations of carbonyl compounds with dimethyltitanocene. J Am Chem Soc 112:6392–6394
59. Martin OR, Xie F (1994) Synthesis and spontaneous dimerization of the tri-*O*-benzyl derivative of"2-keto-1-*C*-methylene-D-glucopyranose" (2,6-anhydro-4,5,7-tri-*O*-benzyl-1-deoxy-D-*arabino*-hept-1-en-3-ulose). Carbohydr Res 264:141–146
60. Bluechel C, Ramana CV, Vasella A (2003) Synthesis of monosaccharide-derived spirocyclic cyclopropylamines and their evaluation as glycosidase inhibitors. Helv Chim Acta 86: 2998–3036
61. Ellis GP, Lockhart IM (2007) The chemistry of heterocyclic compounds, chromenes, chromanones, and chromones. Wiley, New York
62. Shen HC (2009) Asymmetric synthesis of chiral chromans. Tetrahedron 65:3931–3952
63. Leibeling M, Koester DC, Pawliczek M, Schild SC, Werz DB (2010) Domino access to highly substituted chromans and isochromans from carbohydrates. Nature Chem Biol 6:199–201
64. Tietze LF, Brasche G, Gericke KM (2006) Domino reactions in organic synthesis. Wiley, Weinheim
65. Leibeling M, Koester DC, Pawliczek M, Kratzert D, Dittrich B, Werz DB (2010) Hybrids of sugars and aromatics: A Pd-catalyzed modular approach to chromans and isochromans. Bioorg Med Chem 18:3656–3667
66. Tietze LF (1996) Domino reactions in organic synthesis. Chem Rev 96:115–136
67. Yoshimoto K, Kawabata H, Nakamichi N, Hayashi M (2001) Tris(2,4,6-trimethoxyphenyl) phosphine (TTMPP): a novel catalyst for selective deacetylation. Chem Lett 30:934–935
68. Ferrier RJ, Overend WG, Ryan AE (1962) The reaction between 3,4,6-tri-*O*-acetyl-D-glucal and *p*-nitrophenol. J Chem Soc 3667–3670
69. Meyer FE, de Meijere A (1991) Palladium-catalyzed polycyclizations of enediynes: a convenient one-step synthesis of polyfunctional angularly bisanellated benzene derivatives. Synlett 777–778
70. Blond G, Bour C, Salem B, Suffert J (2008) A new Pd-catalyzed cascade reaction for the synthesis of strained aromatic polycycles. Org Lett 10:1075–1078
71. Leibeling M, Milde B, Kratzert D, Stalke D, Werz DB (2011) Intermolecular twofold carbopalladation/cyclization sequence to access chromans and isochromans from carbohydrates. Chem Eur J 17:9888–9892
72. Leibeling M, Werz DB (2012) Winding up alkynes: a Pd-catalyzed Tandem-Domino reaction to chiral biphenyls. Chem Eur J 18:6138–6141
73. Laatsch H, Fotso S (2008) Naturally occurring anthracyclines. Top Curr Chem 282:3–74
74. Brockmann H (1963) Anthracyclinone und anthracycline. Fortschr Chem Org Naturst 21:121–182
75. Grynkiewicz G, Wieslaw S (2008) Synthesis of sugar moieties. Top Curr Chem 282:249–284
76. Leng F, Savkur R, Fokt I, Przewloka T, Priebe W, Chaires JB (1996) Base specific and regioselective chemical cross-linking of daunorubicin to DNA. J Am Chem Soc 118: 4731–4838
77. Chaires JB, Satyanarayana S, Suh D, Fokt I, Przewloka T, Priebe W (1996) Parsing the free energy of anthracycline antibiotic binding to DNA. Biochemistry 35:2047–2053
78. Menna P, Salvatorelli E, Gianni L, Minotti G (2008) Anthracycline cardiotoxicity. Top Curr Chem 283:21–44
79. Cortés-Funes H, Coronado C (2007) Role of anthracyclines in the era of targeted therapy. Cardiovasc Toxicol 7:56–60
80. Vogel P (2008) Combinatorial synthesis of linearly condensed polycyclic compounds, including anthracyclinones, through Tandem Diels–Alder additions. Top Curr Chem 282:187–214
81. Gupta RC, Harland PA, Stoodley RJ (1983) A new strategy for the enantiocontrolled synthesis of anthracyclines resulting in a practical route to (+)-4-demethoxydaunomycinone. J Chem Soc Chem Commun 754–756
82. Tamariz J, Vogel P (1984) A doubly-convergent and regioselective synthesis of ($\pm$)-daunomycinone. Tetrahedron 40:4549–4560

83. Carrupt P-A, Vogel P (1979) A new, doubly convergent synthesis of anthracyclinones. Diels–Alder additions to 2,3,5,6-tetrakis(methylene)-7-oxanorbornane. Tetrahedron Lett 20: 4533–4536
84. Filippini S, Lomovskaya N, Fonstein L, Colombo AL, Hutchinson CR, Otten SL, Breme U (2001) Process for preparing doxorubicin. Patent No.: US 6,210,930 B1:1–22
85. Leibeling M, Werz DB (2013) Flexible synthesis of anthracycline aglycone mimics via domino carbopalladation reactions. Beilstein J Org Chem 9:2194–2201
86. Takahashi T, Li S, Huang W, Kong F, Nakajima K, Shen B, Ohe T, Kanno K-I (2006) Article homologation method for preparation of substituted pentacenes and naphthacenes. J Org Chem 71:7967–7977
87. Petit M, Chouraqui G, Aubert C, Malacria M (2003) New and efficient procedure for the preparation of unsymmetrical silaketals. Org Lett 5:2037–2040
88. Pichaandi KR, Mague JT, Fink MJ (2011) Synthesis of a *tert*-butyl substituted bis(silirane) and comparison with its methyl and phenyl analogs. J Organomet Chem 696:1957–1963
89. Tamao K, Akita M, Kumada M (1983) Silafunctional compounds in organic synthesis: XVIII. Oxidative cleavage of the silicon-carbon bond in alkenylfluorosilanes to carbonyl compounds: Synthetic and mechanistic aspects. J Organomet Chem 254:13–22
90. Tamao K, Ishida N, Kumada M (1983) (Diisopropoxymethylsilyl)methyl Grignard reagent: a new, practically useful nucleophilic hydroxymethylating agent. J Org Chem 48:2120–2122
91. Fleming I, Henning R, Plaut HJ (1984) The phenyldimethylsilyl group as a masked form of the hydroxy group. J Chem Soc Chem Commun 29–31
92. Nakanishi M, Bolm C (2007) Iron-catalyzed benzylic oxidation with aqueous *tert*-butyl hydroperoxide. Adv Synth Catal 349:861–864

Top Med Chem (2014) 12: 23–52
DOI: 10.1007/7355_2014_50

Published online: 10 May 2014

Iminosugars: Therapeutic Applications and Synthetic Considerations

Graeme Horne

Abstract Iminosugars, carbohydrate mimetics in which the endocyclic oxygen of the parent carbohydrate is replaced with nitrogen, are the most important class of carbohydrate mimetic reported to date with two marketed drugs and several in clinical development. Since their first isolation in the 1960s iminosugars have captured the imagination of both synthetic and medicinal chemists alike with recent therapeutic developments highlighting the need for improved routes of synthesis. The resurgence in the therapeutic application of iminosugars has arisen as a consequence of our growing understanding on the role that glycobiology plays in disease and development. There are myriad possible individual targets encompassing a range of therapeutic areas, all of which can potentially be addressed by iminosugars. This chapter presents the historcial development of this compound class before discussing some of the issues that iminosugars present as synthetic targets. The therapeutic potential of this class of compound with specific reference to the development of modulators of glucocerebrosidase activity is discussed.

Keywords Drug discovery, Glycobiology, Glycospace, Iminosugar, Synthesis

Contents

G. Horne (✉)
Summit Plc, 85b Park Drive, Milton Park, Abingdon, Oxfordshire, UK
e-mail: graeme.horne@summitplc.com

1 Introduction

The importance of nucleic acids and proteins in providing biochemical targets in the search for new drugs is well established and has been successfully exploited over many years leading to many marketed drugs. However, these biopolymers do not account for the development, function and dysfunction of pathways, cells, tissues and biological systems alone. Carbohydrates comprise another major molecular class incorporating wide structural diversity that underlies a vast array of biological function [1–15]. The role that carbohydrates play in major disease such as cancer, diabetes and infection has been understood for many years [16–23]. More recently this understanding has extended to other diseases of high unmet medical need including neurodegeneration and rare diseases [24–28]. Approximately 80% of rare diseases have a genetic component ranging from small mutations in a single gene to the addition or loss of an entire chromosome or set of chromosomes. With over 2% of the genome believed to encode enzymes associated with glycan formation and degradation, it should be of no surprise that disorders of glycosylation or glycan degradation comprise a noteworthy proportion of rare diseases. With many of these disorders sharing pathways and exhibit phenotypes associated with more common ailments, developing a drug for a carbohydrate-related disorder may have the added benefit of potentially leading to a treatment that is suitable for more common indications. This highlights the growing need to develop agents targeting the synthesis, metabolism and degradation of carbohydrates and presents considerable opportunity to identify new drugs, drug classes and therapeutic strategies.

Owing to their relative metabolic instability and degradation in vivo, carbohydrates themselves present significant concerns for employment as therapeutic agents. In light of this, researchers have focused on the design and synthesis of mimetics that offer greater stability, affinity, selectivity and efficacy. Of the many scaffolds investigated iminosugars are the most attractive class of carbohydrate mimetic reported to date and are ideally positioned to take advantage of our increasing understanding of this area. Historically, the attention paid to iminosugars lay in their powerful inhibition and modulation of carbohydrate-processing enzymes, alterations which are implicated in a wide variety of diseases. However, the therapeutic application of this compound class extends far beyond this and as such recent work has focused on developing syntheses towards both the naturally occurring iminosugars and their synthetic analogues. The scope of this chapter is to provide an overview of the classification, occurrence and synthetic strategies employed towards the synthesis of both natural and unnatural iminosugar scaffolds set against the backdrop of therapeutic utility and drug discovery advances.

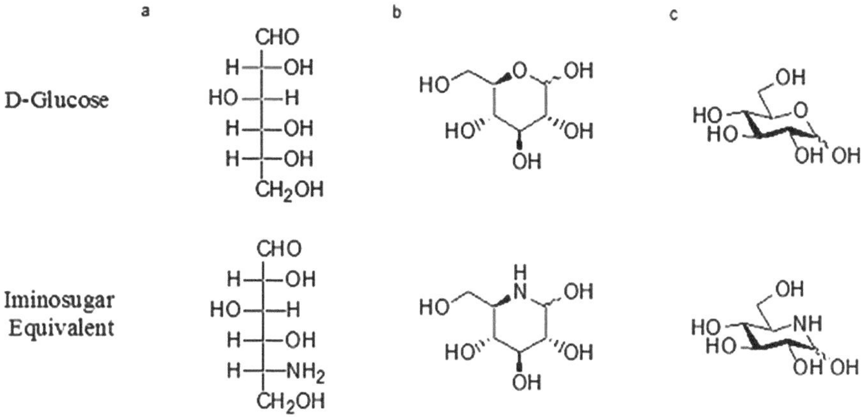

Fig. 1 Monosaccharide and iminosugar equivalence demonstrated through (**a**) Fisher (**b**) Haworth and (c) chair configurations

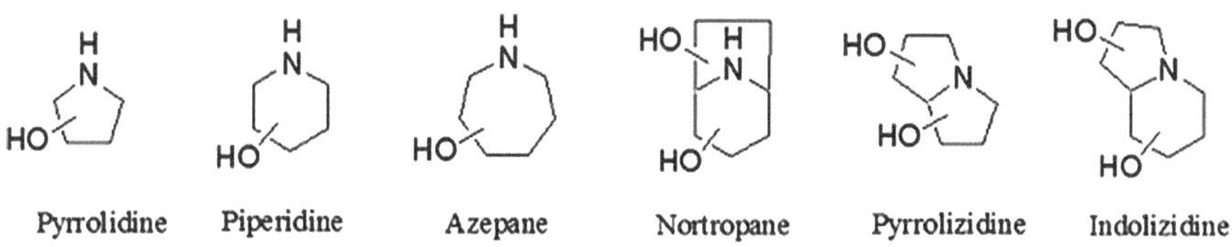

Fig. 2 Common iminosugar ring motifs

2 Iminosugars: Classification, Occurrence and Historical Context

Iminosugars, whether of natural or synthetic origin, are small organic compounds that mimic carbohydrates or their hydrolysis transition states but contain a nitrogen atom instead of oxygen in the ring system template (Fig. 1) [29–35]. Such substitution enables not only monocyclic but also bicyclic scaffolds and while the synthetic analogues may be represented by numerous ring forms, the naturally occurring iminosugars are classified into five structural classes: pyrrolidines, piperidines, indolizidines, pyrrolizidines and nortropanes (Fig. 2). Ring substitution is typically by hydroxyls, but other functional groups such as carboxylic acids and amides are found in nature, with more varied substitutions observed in synthetic analogues (Fig. 3).

The first iminosugar to be isolated was the piperidine 5-amino-5-deoxy-glucopyranose (1) from *Streptomyces* [36]. This was given the trivial name of nojirimycin following structural characterisation after isolation from *Streptomyces nojiriensis* [37–39]. The finding of two related compounds nojirimycin B (2) and galactostatin (3) followed soon after [40, 41]. The isolation and

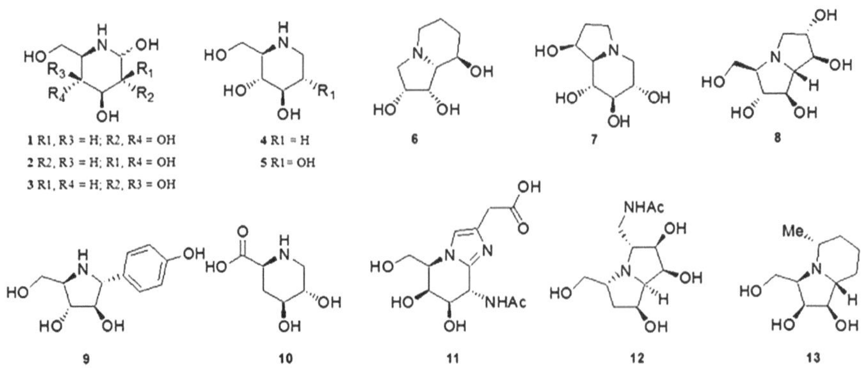

Fig. 3 Naturally occurring iminosugars

identifcation of iminosugars from plants proceeded these initial discoveries from micro-organisms, firstly with the monocyclic structures fagomine (4) and moranoline (5, deoxynojirimycin) followed by the isolation of bicyclic structures. The first members of this class were the indolizidines swainsonine (6) [42] and castanospermine (7) [43]. Pyrrolizidines including the alexines [44, 45] casuarine (8) [46], and the hyacinthacines [47] followed with the nortropane calystegines first reported in 1988 [48]. Although the structural diversity of the early compounds was limited, their biological activity was sufficient to elicit interest in their therapeutic potential (vide infra). To date over 100 iminosugar structures have been isolated from natural sources exhibiting wide structural variety and functionality (9–13) [49–53].

3 Glycospace and the Drug Target Landscape

Target druggability can be defined as the likelihood that a particular molecular target is amenable to functional modulation in vivo by a "drug-like" molecule. Assessment of target druggability has advanced significantly over recent years with numerous methods available to the medicinal chemist [54–58]. The description of what identifies a molecule as "drug like" has, to some extent, been defined by a set of guidelines known as the "rule-of-five" that define physical property requirements for maximal drug absorption [59, 60]. These guidelines, originally derived from analysis of the physical attributes of clinical candidates that had reached Phase 2 and beyond, suggest that poor absorption is more likely when cLog P is >5; molecular mass is >500 daltons; the number of hydrogen-bond donors (OH plus NH count) is >5; and the number of hydrogen-bond acceptors (O plus N atoms) is >10. This simplicity in definition has made the "rule-of-five" the leading measure of drug-likeness. Several studies have reported on the physicochemical properties of oral drugs demonstrating that in the majority they adhere to the criteria set suggesting that the range of acceptable oral properties is independent of synthetic complexity or targeted receptor [61–63]. However, absorption is only one activity that a drug must satisfy to mediate its effect. Once the drug has reached the desired target it is then required to interact with the said target to illicit the desired response. The majority of drugs will express their therapeutic activity through interaction with specific molecular targets. Consequently, for a drug to be biologically active it should share certain physicochemical attributes with the natural ligand(s). Although accordance with the "rule-of-five" for ligand and associated target is generally acknowledged as acceptance criteria for discovery programs, the absence of a "rule-of-five" compliant molecule (and/or target) does not necessarily make a target undruggable or compound unsuitable for drug development.

3.1 Glycospace: A Source of New Medicines

Drug discovery is dominated by a remarkably small number of molecular scaffolds, with many areas of chemical space remaining relatively unexplored [64–67]. Glycospace, the structural and functional diversity of carbohydrates and carbohydrate-containing molecules is one area that still remains to be fully exploited in the pursuit of new medicines [68]. On the one hand this is surprising when one considers the role that carbohydrates play in development and disease, and the structural diversity afforded by this class of molecule. However, when one considers the "rule-of-five" criteria, carbohydrate structures and associated targets may be viewed as unsuitable for drug development due to limited compliance with these guidelines. While functional knowledge and an understanding of the endogenous ligands of a protein target can enable the druggability of that target to be ascertained a more accurate assessment will come through consideration of the structure and properties of the target-binding site.

Exploiting glycospace offers substantial opportunity in identifying new molecular scaffolds for developing therapies relating to the processing, sensing and recognition of carbohydrates. At the simplest level glycospace can be considered as the constituent monosaccharides which make up the entire complement of carbohydrates of a cell, tissue or organism, whether free or present in more complex structures. For example, if we take the hexose sugars, there are eight stereochemistries that may be defined in one of two configurations (D-/L-enantiomers), leading to a total of 16 monosaccharide scaffolds. With each additional stereogenic carbon, the number of possible monosaccharides increases twofold such that the number of scaffolds under ten-carbons is >500 structures. However, carbohydrate recognition is not limited to the interactions of the hydroxyl group. Substitution and modification of the ring with amides, amines and acids, as well as sites of deoxygenation and branching, are all important in defining carbohydrate function and purpose in nature (Fig. 4).

For example, the sialic acids themselves, of which *N*-acetylneuraminic acid (18) is a member, comprise a naturally occurring family of over 50 structures

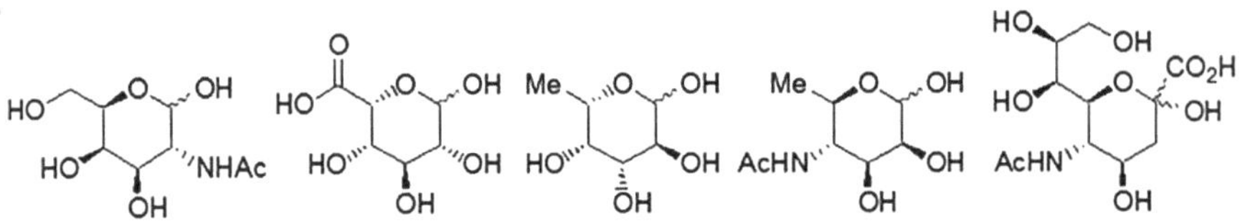

N-Acetyl-D-galactosamine (14) | L-Iduronic acid (15) | L-Fucose (16) | 4-Acetamido-4-deoxy-D-rhamnose (17) | *N*-Acetylneuraminic acid (18)

Fig. 4 Structural diversity of naturally occurring carbohydrates

[69]. Additional consideration of the inventory of carbohydrate structures present in prokaryotes (a higher content of > six-carbon structures, increased substitution and unusual deoxygenations) greatly increases the structural repertoire [70]. This leads to additional levels of structural complexity and an increased number of monosaccharides occupying glycospace.

3.2 Glycomimetics as Modulators of Carbohydrate Biology

By there very nature carbohydrate receptors, hydrolases and transferases mediate their specific activities through interactions with a structurally diverse collection of highly polar and stereochemically defined ligands. While carbohydrates are excellent recognition molecules and mediators of cellular activity and response, binding affinities for the mono-valent structures tend towards the milli- to micromolar range. Coupled with rapid renal clearance and the disadvantages identified above, rationally designed "drug-like" molecules are essential for successful prosecution against carbohydrate-related targets.

Conventional non-carbohydrate screening collections have failed to reflect the natural chemical complimentarity of glycospace and consequently have had limited success against carbohydrate-related drug-targets. The comparatively planar and stereochemically indistinct structures obtained from traditional drug scaffolds may not offer the target selectivity necessary for drug development. For example, the structural nuances required to differentiate between the enzymes α-*N*-acetylglucosaminidase [71], β-hexosaminidase [72], nucleocytoplasmic β-*N*-acetylglucosaminidase [73–75] and α-*N*-acetylgalactosaminidase [76] are more likely to be attainable in a stereochemically diverse and defined scaffold such as a carbohydrate or mimetic than with a planar and structurally in-distinct scaffold. These enzymes all hydrolyse *N*-acetylhexosamine residues from a range of substrates and all are implicated in disease with selectivity key to therapeutic development. A key challenge in attaining specificity and potency is in using an appropriate scaffold able to capture the structural and functional diversity of glycospace. As previously mentioned, carbohydrates are not the most suitable candidates for drug development owing to their relative metabolic instability and degradation in vivo. One approach for targetting glycospace is to use a carbohydrate mimetic that is able to attain higher affinity with the target than the natural ligand and attain an effective conformation through appropriate synthetic design. Strategies directed at enhancing oral bioavailability and minimising metabolism while maintaining potency are also more achievable with a mimetic than the natural carbohydrate ligand.

3.3 Glycomimetics: Stereochemical and Structural Considerations

To be functionally active a carbohydrate mimetic needs to replicate the key stereochemical and structural motifs of the carbohydrate being mimicked. Direct replication of the carbohydrate structure may give rise to shared functional attributes such as at the level of interaction with a biological target in vivo. However, to attain the desired biological effect the mimetic may also need to contain additional functional groups or motifs related to, but distinct from, those of the carbohydrate they mimic. These may confer certain attributes that, for example, enable the molecule to competitively inhibit the enzyme for which the carbohydrate mimicked is a substrate for in vivo or impart target selectivity.

Although functional and stereochemical equivalence between the natural ligand and mimetic may not always be required for activity, an understanding of how ligand stereochemistry impacts on interactions with the target-binding site is important in aiding inhibitor design. For example, the four enzymes referred to in the previous section (α-*N*-acetylglucosaminidase, β-hexosaminidase, nucleocytoplasmic β-*N*-acetylglucosaminidase and α-*N*-acetylgalactosaminidase) all share similarity in the monsaccharide unit that is hydrolysed. An understanding of the key stereochemical and functional interactions between the ligand and the active site of each enzyme enables selectivity to be attained whilst retaining potency. Structural comparison of the *N*-acetyl-binding pockets from *Clostridium perfringens* nucleocytoplasmic β-*N*-acetylglucosaminidase, a homolog of the human enzyme, and β-hexosaminidase reveals subtle differences. In the case of β-hexosaminidase this pocket is both narrower and shallower than for the nucleocytoplasmic β-*N*-acetylglucosaminidase and has been exploited in the design of selective inhibitors [77]. An understanidng of carbohydrate stereochemistry is essential when designing molecules targeted at carbohydrate processes. If we consider the pentose and hexose sugars, their stereochemistry can be represented through consideration of the contiguous chiral centres (Fig. 5). This analysis is non-limiting, intended to be illustrative only of a wider principle and can be equally applied to both lower sugars (e.g. tetroses) and higher sugars (e.g. heptoses), as well as to ketoses and the like.

A carbohydrate mimetic can be considered as being a structural mimic of a target monosaccharide, disaccharide or oligosaccharide when stereochemical and functional comparisons between the mimetic and the relative carbohydrate stereochemistry exhibited by the carbohydrate scaffold reveal shared stereochemical motifs. As monosaccharides can exist in both acyclic and several cyclic forms, the relative

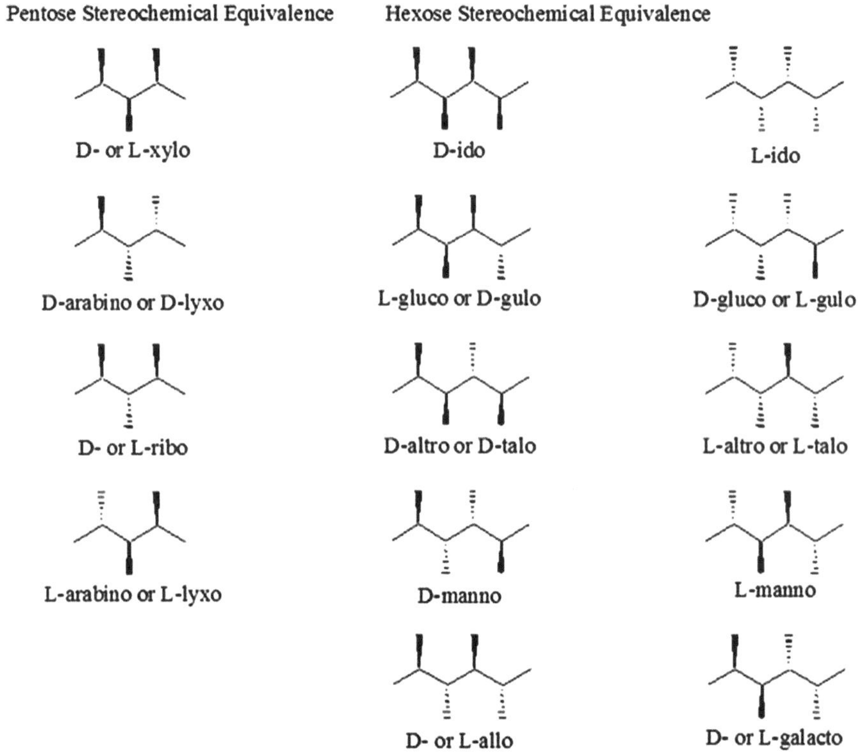

Fig. 5 Relative carbohydrate stereochemistry

stereochemical relationship between the mimetic and the parent monosaccharide is not necessarily fixed to one structural class or type or to the contiguous sequence depicted in Fig. 5. However, unlike their monosaccharide counterparts the drug candidate should not interconvert. This aspect is important when considering the design of the mimetic scaffold. Furthermore, stereochemical considerations are not only restricted to the design of the mimetic alone but are also important in the synthetic approach to be considered towards the target.

3.4 Iminosugars: The Glycomimetic of Choice

The biological activity and therapeutic utility of iminosugars is largely founded on them being able to replicate the transition state that is considered to be generated during the hydrolysis of carbohydrates and carbohydrate-containing structures (Fig. 6).

Glycosidases promote the hydrolysis of α- and β-glycosidic bonds via nucleophilic substitution reactions that result in either inversion or retention of the

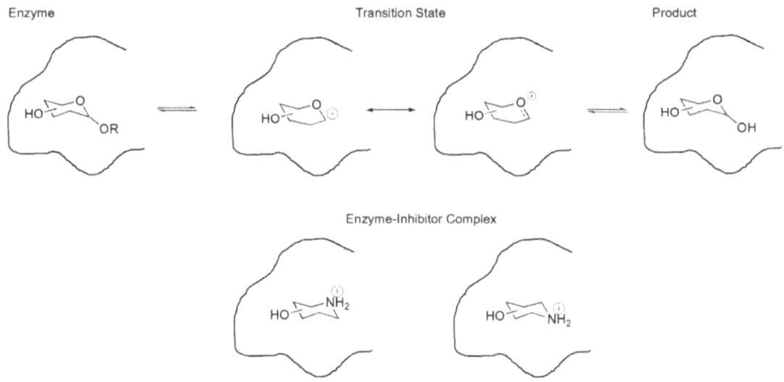

Fig. 6 Carbohydrate transition state and iminosugar mimicry

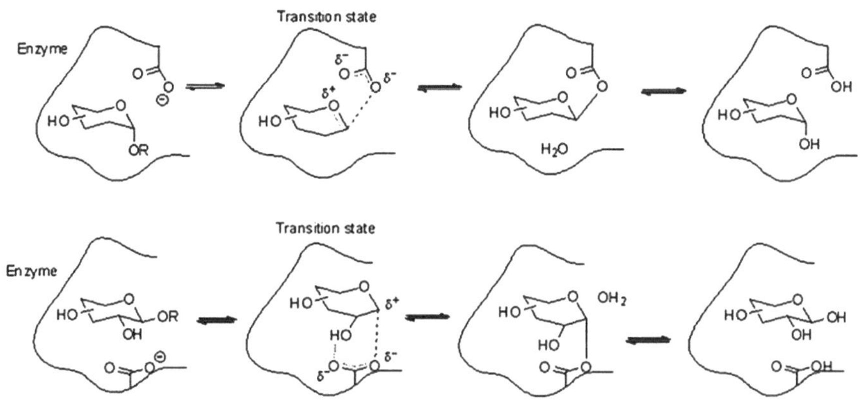

Fig. 7 General mechanism for retaining α-glycosidases (*top*) and β-glycosidases (*bottom*)

anomeric configuration (Fig. 7). Differences in enzymatic hydrolysis can be rationalised through consideration of the charge distribution effect. The hydrolysis mechanism of the retaining glycosidases [78] involves glycosylation of a nucleophilic catalytic residue of the active site, commonly a carboxylate, as the rate-determining step. Hydrolysis then results by displacement of the catalytic residue with water. The anomeric configuration is inverted twice in this mechanism, leading to net retention of configuration. In the case of the α-glycosidases, the nucleophilic residue is positioned in such a way that a build-up of positive charge at the endocyclic oxygen is favoured as the anomeric group is ejected, resulting in a transition state with oxocarbenium character. With β-glycosidases the nucleophilic residue is positioned in such a way that it interacts predominantly with the anomeric centre and the adjacent exocyclic hydroxyl. This configuration favours carbocation character at the anomeric centre of the transition state.

As iminosugars can be considered to be in the protonated state at physiological pH, the resulting ammonium ions mimic the oxocarbenium character of the hydrolysis transition states. Activity, potency and enzymatic preference of an iminosugar

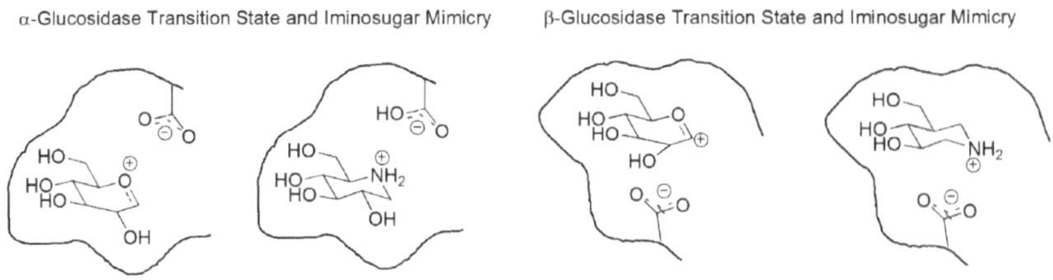

Fig. 8 α- and β-Glycosidase iminosugar specificity

inhibitor is not only dependent upon hydroxyl group configuration and exocyclic substituents but is also heavily influenced by charge distribution. For example, replacement of the endocyclic oxygen with nitrogen as occurring in deoxynojirimycin (5) will generally result in a specific inhibitor of α-glycosidase. Conversely, positioning the iminosugar endocyclic nitrogen at the anomeric centre is likely to result in β-glycosidase inhibition (Fig. 8).

Replication of charge distribution and hydroxyl group positioning is not restricted to the piperidine ring types as shown in Fig. 8. Pyrrolidines have often been found to be more potent inhibitors of glycosidases than their piperidine congeners. This has been rationalised through greater resemblance that the pyrrolidine envelope configuration has to the half-chair conformation of the glycosyl transition state. Bicyclic structures offer another degree of structural complexity that can also elicit significant potency and selectivity that may not be attainable with a monocyclic scaffold. Although an understanding of enzymatic mechanism is important in designing a specific and selective inhibitor, it should not necessarily be implied that inhibition can only be achieved with comparable ligand and inhibitor ring scaffolds. The structural variety and versatility present within the iminosugars greatly endears the class to the design of potent, specific and selective inhibitors of carbohydrate processing.

3.5 *Iminosugars: Therapeutic Perspective*

Iminosugars have a number of positive attributes that delineate them from traditional drug classes. Chemical and metabolic stability and high aqueous solubility are all favourable characteristics associated with this compound class. The problem created by lead molecule solubility has occupied a prominent place in drug discovery and development with many potent classical small molecules suffering from marginal water solubility. Enormous resources have been deployed to address this issue both in medicinal chemistry and in formulation [79]. No such problem impedes the iminosugar field where adequate solubility is a key benefit. In cases where polarity impedes absorption, administration as prodrugs is feasible. Compared to most small polar molecules, iminosugars are well absorbed having high bioavailability and CNS penetration [80, 81]. Despite a relatively rapid clearance in

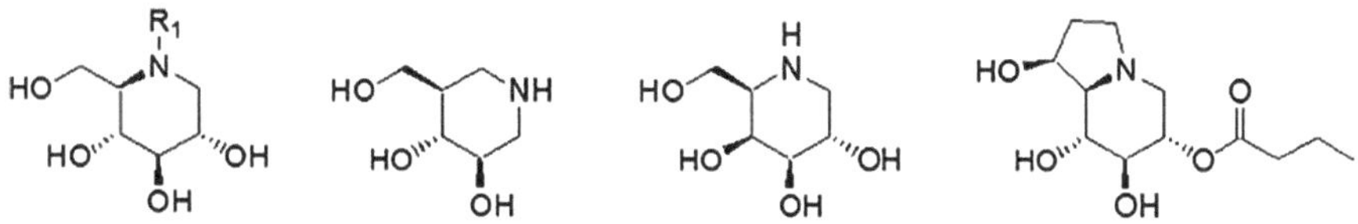

19 $R_1 = CH_2CH_2OH$
20 $R_1 = {}^nBu$
21
22
23

Fig. 9 Iminosugar marketed drugs and clinical candidates

Table 1 A summary of the clinical status for selected iminosugar therapeutics

Structure	Therapeutic indication	Therapeutic target	Clinical status	Company name
19	Type 2 diabetes	Intestinal α-1,4-glucosidases	Marketed	Bayer
20	Gaucher's disease	Glucosylceramide synthase	Marketed	Actelion
22	Fabry's disease	α-Galactosidase A (Chaperone monotherapy)	Phase 3	Amicus Therapeutics
		α-Galactosidase A (Chaperone-ERT co-administration)	Phase 2	
5	Pompe's disease	Acid α-glucosidase (Chaperone monotherapy)	Phase 2	
		Acid α-glucosidase (Chaperone-ERT co-administration)	Phase 2	
21	Gaucher's disease	Glucocerebrosidase (Chaperone monotherapy)	Phase 2	
6	Renal cell/colorectal cancer	Golgi α-mannosidase II	Phase 2	Glyco Design
23	HCV	Endoplasmic reticulum α-glucosidase	Phase 2	Migenix
	Dengue virus		Phase 1/2	
	HIV		Phase 2	Merrell Dow/Hoechst Marion Roussel

in vivo models, both the mouse and rat have proven adequate species to demonstrate efficacy. Such properties distinguish iminosugars from other small polar molecules and provide further advantage to their use as drugs.

Iminosugars were initially of interest in modulating the action of enzymes involved in carbohydrate synthesis and metabolism. The natural products deoxynojirimycin (5), swainsonine (6), castanospermine (7), and their closely related analogues have resulted in eight clinical candidates and two marketed drugs: miglitol (14; Glyset, Bayer [82]) for the treatment of Type II diabetes and miglustat (15; Zavesca, Actelion) for the treatment of lysosomal storage diseases (LSDs), specifically Gaucher's disease and Niemann Pick type C disease [83–85] (Fig. 9, Table 1).

Gaucher's disease is a genetic disorder whereby normal lysosomal processing of glucosylceramide is compromised resulting in excess build-up of material within the lysosomes. Zavesca was initially shown to exert its effect through inhibiting the synthesis of glucosylceramide, a strategy known as substrate reduction therapy. However, recent reports have highlighted that Zavesca may also exert its therapeutic effect by chaperoning the dysfunctional enzyme glucosylcerebrosidase from the endoplasmic reticulum to the lysosome thereby restoring the natural degradative process [86].

Pharmacological chaperone therapy is a small-molecule approach that aims to restore defective glycan degradation by allowing the dysfunctional enzyme to correctly fold and be trafficked to the lysosome where the enzyme becomes biologically active upon chaperone dissociation. Pharmacological chaperones selectively bind to their target proteins, increasing stability and enabling the protein to attain correct folding. The protein is then able to pass from the ER to the lysosome thereby increasing protein activity and cellular function and reducing cellular stress. As the two intracellular environments differ in relative pH a key aspect of this approach is to identify enzyme modulators that have a higher binding affinity in a neutral environment (ER) and a lower binding affinity in an acidic environment (lysosome).

The majority of chaperones identified to date are reversible competitive inhibitors and, consequently, the relationship between pH and binding affinity is important. This is considered in greater detail in Sect. 5.

As the principal glycomimetics iminosugars have shown great application as pharmacological chaperones in a number of diseases. Migalastat (22), Plicera (21) and Duvoglustat (5) have either recently completed or are undergoing clinical evaluation in Fabry's, Gaucher's and Pompe's diseases, respectively [87–93]. These compounds work by stabilising the dysfunctional enzymes associated with each disorder through binding to the active site. This allows the mutant enzyme to bypass the degradative pathways of the endoplasmic reticulum and progress to the lysosome where the iminosugar chaperone dissociates releasing the functionally active enzyme. Migalastat (22), Plicera (21) and Duvoglustat (5) are currently being evaluated in the clinic as standalone therapies or in combination with the marketed enzyme replacement therapies for the corresponding lysosmal storage disorder. One of the limiting factors to the effectiveness of enzyme replacement therapy is in the relative instability of the administered enzymes. The rationale of pharmacological chaperones binding to and stabilising defective enzymes also extends to the stabilisation of exogenous recombinant enzyme. Several other iminosugars have been clinically evaluated but failed to progress beyond the early stages. This includes swainsonine 6 (renal cell cancer), castanospermine pro-drug Celgosivir 23 (Hepatitis C) and fuconojirimycin derivative UT-231B (Hepatitis B) [94–97]. More recently mannosidase inhibitor kifunensine has been granted orphan status within Europe for the potential treatment of gamma-sarcoglycanopathy [98, 99]. Although clinical evaluation is yet to be carried out, preliminary data from in vivo models has been positive [99].

Despite this relative success, as a compound class, iminosugars have failed to realise their true therapeutic potential. Poor selectivity and/or limited efficacy for the few compounds to reach the clinic greatly hindered the advancement of the field. In addition, iminosugars have generally been labelled as synthetically challenging. Primarily this has been a consequence of the compound class consisting of stereochemically complex scaffolds of a highly polar nature. However, with an appropriate synthetic strategy and sagacious use of starting materials these problems can and have been overcome. This has been possible through an increasing array of affordable chiral pool molecules becoming available and significant advances in analytical methodology. Synthetic access, purification and high cost of goods are no longer sufficient reasons to avoid iminosugars.

4 Strategies for the Synthesis of Iminosugars

The purpose of this section is not to provide an all-encompassing and detailed treatise on the synthesis of iminosugars but rather provide an overview of the rationale and synthetic strategies employed in the design of iminosugar therapeutics. With five natural ring scaffolds (indolizidine, nortropane, piperidine, pyrrolidine and pyrrolizidine; Fig. 2) and numerous synthetic ring-forms reported in the literature (azetidines, azapanes, azaazulenes, quinolizidines, *iso*- and *aza*-analogues) covering myriad stereochemistries and derivatisations the reader is directed to several excellent reviews concerned with the specific aspects of iminosugar synthesis [100–114].

The development and exploitation rate for iminosugars has, to some extent, been impeded by perceived limitations of the chemical class. This has undoubtedly had an adverse impact on the extent to which the iminosugar field has been explored by industry to date. As chirally complex and polar molecules iminosugar syntheses may appear, to the uninitiated, overly detailed and atom inefficient. Syntheses frequently require more than three contiguous stereogenic centres to be introduced in the target molecule with a high degree of control. The ring scaffold needs to be obtained efficiently with protecting group use minimised or chosen appropriately to maximise atom efficiency. As, in the majority of structures, no chromophore is present, analysis and purification of final materials may also be perceived as prohibitively challenging although recent advances in this field have generally overcome these analytical limitations [115–117].

A number of strategies towards the synthesis of iminosugars have been utilised by both synthetic and medicinal chemists. These include the use of chirons, asymmetric methodology as well as chemoenzymatic approaches. Unsurprisingly, because of their mimicry early syntheses of iminosugars utilised carbohydrate starting materials [118–126]. As a chiral pool, carbohydrates have proven to be of significant benefit in the synthesis of iminosugars with all of the clinical candidates referenced in Sect. 3.5 utilising such materials in their manufacture. Key to this approach is a sound knowledge of monosaccharide chemistry, protecting group manipulations, a knowledge of readily available carbohydrate starting materials and the ability to relate key stereochemical motifs in the target to the starting material [127–129].

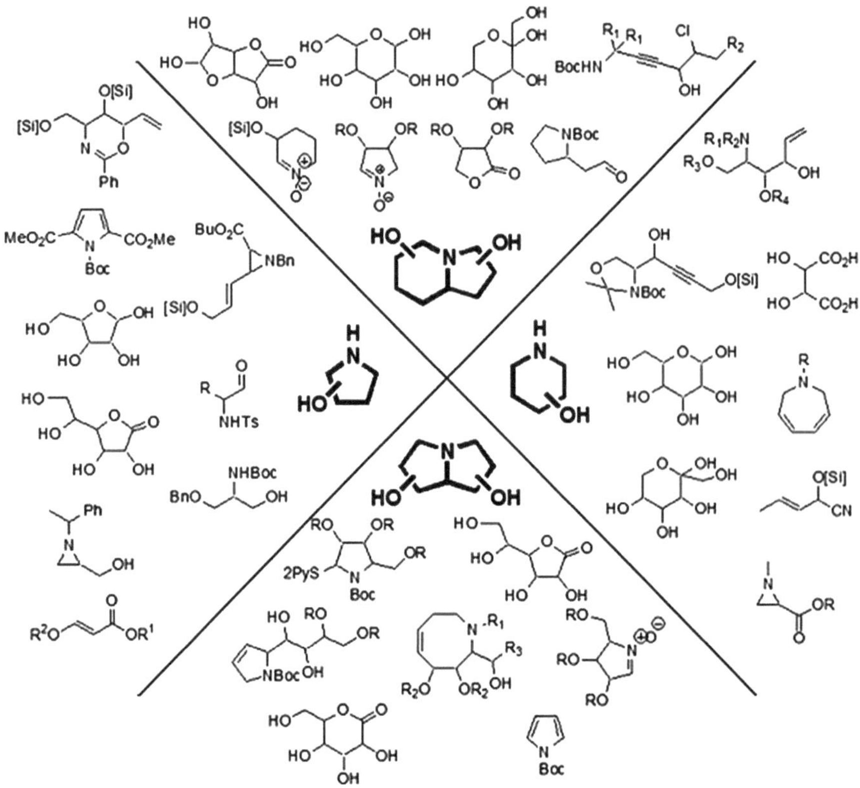

Fig. 10 Starting materials employed in the synthesis of the major iminosugar scaffolds

Stereochemical recognition is a key aspect of any chiral synthesis. The more chiral centres the more complex the synthesis tends to be. Naturally occurring iminosugars extend from the relatively simplistic (e.g. D-arabinose mimetic DAB-1) to the extremely complex (e.g. casuarine) and it is in the field of iminosugar synthesis where an understanding of carbohydrate chemistry has been of significant benefit. Pentose, hexose and heptose sugars and their derivatives are all readily available with cost of goods no longer an impediment to their use.

The approach of adopting carbohydrates or other chirons as starting materials introduces the endocyclic nitrogen through synthetic manipulation. This is frequently through the use of the azide group with ring closure mediated by reductive amination. Although a wide variety of target structures is increasingly accessible through these methods, such routes can be hampered by long synthetic sequences. An important objective for researchers in this area has been to simplify key synthetic steps, both as a matter of convenience and to significantly improve the commercial potential of this chemistry class. Recent syntheses have highlighted significant advances in this area (Fig. 10) with approaches employing amino acids and their derivatives, nitrogen containing heterocylces and nitrones all being successfully prosecuted [130–152]. Asymmetric and chemoenzymatic routes have also been extremely effective towards the synthesis of iminosugars.

5 Iminosugars: Drug-Like Properties, Concepts, Structure Design and Application

As discussed iminosugars principally exert their biological effect through mimicking the activities of carbohydrates. However, the activity of iminosugars is not restricted to interactions with these targets alone. Numerous literature reports highlight the utility of iminosugars to a range of targets. For example, iminosugar scaffolds have found utility in the design of inhibitors of matrix metalloproteases with potential for oncology [153–156]. The advantage of using an iminosugar in the design of these inhibitors is primarily one of improved solubility and bioavailability. Manipulation of the stereo-defined iminosugar fragment also enables a greater degree of target space to be explored in the pursuit of potent and selective inhibitors.

Nevertheless, the drug development of iminosugars has principally focused on carbohydrate targets, specifically inhibition of glycosyl hydrolases. As an example of how iminosugars have been applied to drug discovery, modulators of the enzyme glucocerebrosidase will be discussed.

5.1 Glucocerebrosidase: A Druggable Target

Glucosylceramide has a unique and abstruse role in mammalian cells. The generation of glucosylceramide from glucose and ceramide is the first pathway-committed step to the production of more complex glycosphingolipids such as lactosylceramide and the gangliosides. Although transient upregulation of glucosylceramide offers cellular protection and primes certain cells for proliferation a prolonged overabundance is detrimental as in the case of Gaucher's disease, the most common lysosomal storage disorder.

Gaucher's disease is caused by the deficient activity of the lysosomal hydrolase enzyme glucocerebrosidase (acid β-glucosidase, GCase, Enzyme Commission number 4.2.1.25) and resultant accumulation in the lysosme of glucosylceramide. Glucocerbrosidase is responsible for hydrolysing glucosylceramide to β-glucose and ceramide (Fig. 11).

Over 300 mutations in the GBA gene have been identified that adversely impact on the catalytic function, intracellular stability and trafficking of the enzyme. Patients present with a broad range of phenotype with the spectrum of disease correlating in part with residual enzyme activity [157]. The disease is classified into three subtypes based on age at onset and neurological manifestations [158]. Recently several studies

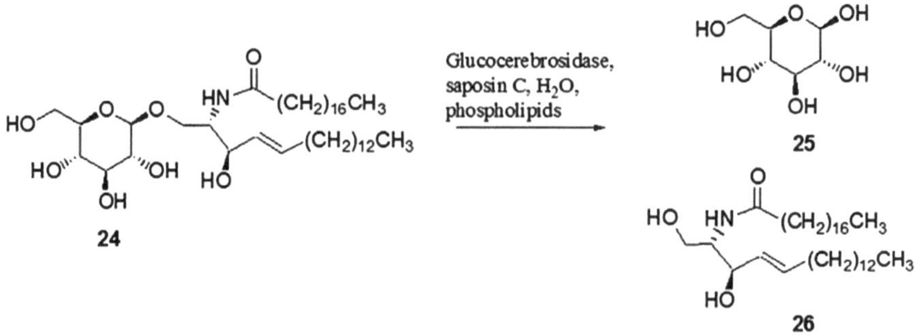

Fig. 11 Glucocerebrosidase schematic

have reported on the apparent association between mutations in the GBA gene and Parkinsonism [159–161]. An increased frequency of heterozygous GBA mutations has been observed in several cohorts of patients with Parkinsonism and other Lewy-body disorders. GBA mutation carriers have also been observed to exhibit diverse phenotypes presenting diffuse distribution patterns of Lewy bodies in the cerebral cortex. The apparent association between glucosylceramide accumulation and accumulation of synuclein has been rationalised through a bidirectional effect of α-synuclein and GCase forming a positive feedback loop that may lead to a self-propagating disease [162]. Improved targeting of GCase to lysosomes may represent a specific therapeutic approach for PD and other synucleinopathies.

Although enzyme replacement and substrate reduction therapies have improved the health of some affected individuals oral treatment with pharmacological chaperones may have wider therapeutic appeal. By restoring the native activity of the mutant glucocerebrosidase, chaperones are able to address the disease in the full gamut of afflicted tissue compartments as well as having the potential to restore the glucocerbrosidase-synuclein feedback loop implicated in Parkinson's disease. Increasing the activity of glucocerebrosidase in the brain with a small-molecule pharmacological chaperone may reduce the negative consequences of carrying GBA1 mutations that are not currently addressed with existing therapies.

5.2 *Iminosugar Chaperones for Therapeutic Intervention*

The principal purpose of a chaperone is to restore the natural degradative process by allowing the dysfunctional enzyme to correctly fold and be trafficked to the lysosome where the enzyme becomes biologically active upon chaperone dissociation. To enable this process the chaperone needs to selectively bind to the target mutant protein at the site of synthesis (endoplasmic reticulum) and then "guide" the enzyme to the site of activity (lysosome). Through selective binding to the dysfunctional enzyme the chaperone increases the enzymes stability enabling the protein to fold correctly and thereby bypass the degradative pathway. This allows

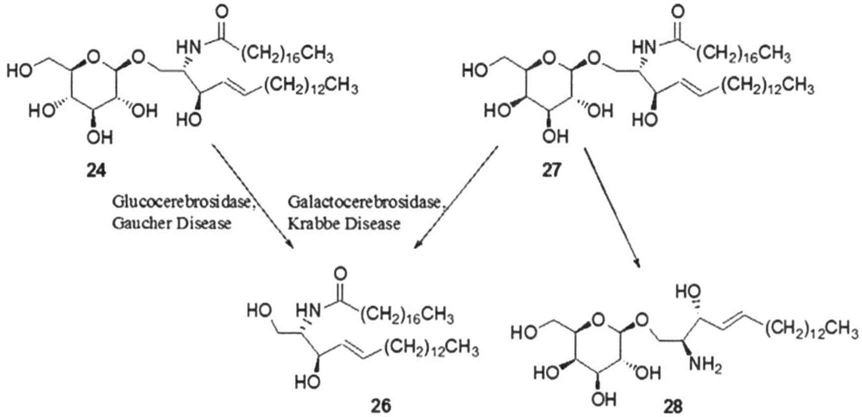

Fig. 12 Degradation of glycosyl ceramides

the mutant protein to be trafficked from the endoplasmic reticulum to the lysosome, thereby establishing protein activity and maintaining cellular function. A key aspect of this approach is to identify modulators of the target enzyme that have a high binding affinity in the neutral environment of the endoplasmic reticulum and a lower binding affinity in the acidic environment of the lysosome. Competitive inhibitors are preferred for this purpose. In light of this it is no surprise that iminosugars are ideally suited as pharmacological chaperones for the treatment of lysosomal storage diseases [163] (Table 1).

With respect to Gaucher's disease and glucocerebrosidase a number of compound classes have been evaluated as potential pharmacological chaperones including non-carbohydrate scaffolds identified through high-throughput-screening [164–168]. Although such compounds may appear to stabilise the enzyme and improve trafficking, increasing the turnover of glucosylceramide has not yet been demonstrated. This is not the case with iminosugars [169]. A significant advantage that iminosugars offer over non-carbohydrate and carbohydrate mimetics is the ability to not only mimic the natural ligand stereochemistry but also mimic transition states (Fig. 8). Additionally, target selectivity is key to the design of any therapeutic. This is of particular relevance to the design of inhibitors of enzymes involved in carbohydrate processing. Concerning the lysosomal storage disorders there are over 50 known diseases of which a significant proportion result from defective glycosidic enzymes. Krabbe's disease is caused by genetic defects in the lysosomal enzyme galactocerebrosidase, the enzyme responsible for the hydrolysis of galactosylceramide (27) to galactose and ceramide (26) (Fig. 12). Converse to Gaucher's disease it is the accumulation of the toxic galactosylceramide metabolite galactosylsphingosine (psychosine, 28) rather than the substrate of the defective enzyme that is causative [170]. Any inhibitor of glucosylceramidase employed in Gaucher's therapy may potentially inhibit galactosylceramidase inadvertently. Although the substrates for each enzyme share substantial structural similarity, the switch from a glucose moiety to its 4-epi congener allows the enzymes to discriminate between substrates. This has been demonstrated through crystallographic studies on both enzymes [171–173]. Although the active sites for each enzyme are similar, there are key differences notably in the active site interactions with the hydroxyl group at position 4. In the galactose-galactocerebrosidase structure this hydroxyl froms a hydrogen bond with a threonine reside that is responsible for the substrate specificity of the enzyme. Whereas a galactose configuration is congruent with recognition a glucose configuration would be incompatible with the galactocerebrosidase substrate-binding pocket as an equatorially configured hydroxyl would sterically clash with other residues within the site. From this perspective it can easily be understood how an appropriately configured and substituted iminosugar would have substantial advantage over other non-carbohydrate-based scaffolds.

The key residues involved in ligand recognition for glucocerebrosidase are depicted in Fig. 13 (adapted from [172]). Crudely this has been illustrated by segregating the active site into two regions for hydrophobic and polar interactions, respectively. Consideration of the structures obtained for *N*-nonyl deoxynojirimycin, 20 and 21 supports the use of iminosugars as chaperones for the enzyme. All inhibitors depicted are able to replicate the key interactions involved in substrate recognition most notably those concerned with the position 4 hydroxyl group. As expected all of these inhibitors present key attributes for glucose mimicry. As a direct glucose mimetic, analogues based on scaffold 5 have been evaluated extensively [174–178]. Although these compounds have demonstrated activity in vitro, historically, inhibitors based on this scaffold have been associated with signficant side effects resulting from limited selectivity. One key observation from these studies is that mimicry of both the carbohydrate and ceramide portions of the substrate may be necessary to maximise inhibition. It should be iterated that potent inhibition does not necessarily translate to greater chaperoning potential.

Glucose mimetic isofagomine (21) is the iminosugar that has, to date, presented the greatest therapeutic potential for the treatment of Gaucher's disease utilising pharmacological chaperone approaches. It was identified as a potential lead from a collection of natural and synthetic iminosugars screened in a patient-derived cell line [179] and is ideally suited as a chaperone for the glucocerebrosidase enzyme. As a 1-azasugar, an iminosugar in which the ring nitrogen is positioned at the anomeric carbon, 21 can be considered an excellent transition state analogue for a β-glucosidase (Fig. 8). It is a competitive inhibitor that exhibits preferential inhibition at the neutral pH of the endoplasmic reticulum relative to the acidic environment of the lysosome. In binding with the enzyme, all key inetractions with active site residues are retained despite only three hydroxyl groups being present. Additionally, whereas structures based on scaffold 5 are associated with non-specific inhibition no measurable effect on other related carbohydrate processing enzymes has been reported so far for 21.

Strategies to maximise the therapeutic potential of 21 have followed those utilised for chaperones based on the deoxynojirimycin scaffold 5. Primarily this has involved capturing the ceramide moiety through lipophilic addition to the ring (Fig. 14, Table 2). Approaches employing *N*-alkylation (compounds 29 and 30)

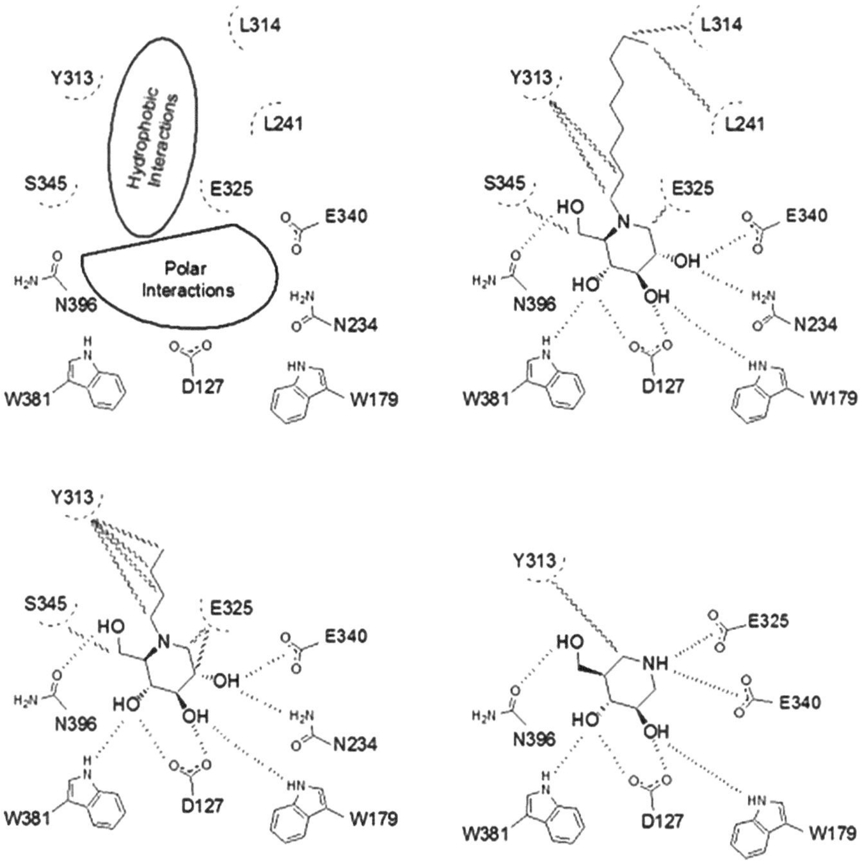

Fig. 13 Glucocerebrosidase active site architecture. Overview of enzyme active site (*top left*) and depiction of inhibitor-active site interactions for *N*-nonyl-deoxynojirimycin (*top right*) 20 (*bottom left*) and 21 (*bottom right*). Adapted from [172]

generated, as expected, weak inhibitors with reductions in activity greater than 3 orders of magnitude relative to 21 [180]. Similarly, adamantyl analogues 31–33 also exhibited significantly reduced inhibition relative to 21 [181]. The adamantyl moiety was selected to more closely resemble ceramide and despite reduced glucocerebrosidase inhibition all compounds were reported as increasing enzymatic activity in two patient cell lines. Re-positioning of the substitution so that the additional functionality is ideally suited to utilise the hydrophobic interactions within the binding pocket generated more potent inhibitors than the *N*-alkyl congeners [180, 182]. Inhibitors 35–38 were between 13- and 93-fold more potent than parent inhibitor 21 with increasing alkyl chain correlating with increased potency. Analogues 39–41 were also reported as competitive inhibitors of the enzyme. All inhibitors stabilised glucocerebrosidase against thermal denaturation with pharmacologically relevant levels of chaperoning activity observed in all cases,

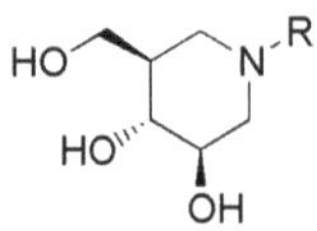

29 R = butyl
30 R = nonyl

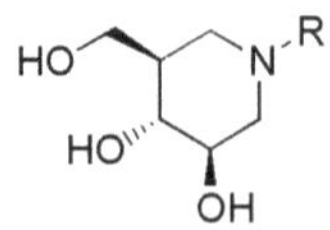

31 R = $(CH_2)_3CONH$(Adamantyl)
32 R = $(CH_2)_4CONH$(Adamantyl)
33 R = $(CH_2)_5CONH$(Adamantyl)

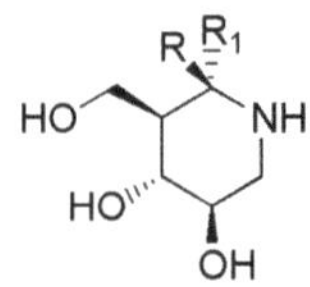

34 R = H, R_1 = butyl
35 R = H, R_1 = hexyl
36 R = H, R_1 = heptyl
37 R = H, R_1 = octyl
38 R = H, R_1 = nonyl
39 R = H, R_1 = propyl
40 R = propyl, R_1 = propyl
41 R = H, R_1 = hydroxypropyl

Fig. 14 Second generation glucocerebrosidase chaperones

Table 2 Selected second generation glucocerebrosidase chaperones and inhibition constants

Compound ID	Glucocerebrosdiase IC_{50}/K_i (μm)	Compound ID	Glucocerebrosdiase IC_{50}/K_i (μm)
29	44,000	**35**	4.2
30	>100,000	**36**	1.8
31	18,000	**37**	0.8
32	11,000	**38**	0.6
33	94,000	**39**	610
34	160	**40**	600

particularly for the propyl derivative. However, the incorporation of large alkyl chains may increase the potential for inhibition rather than enhancement and may adversely affect the drug-like properties of the molecule, both in terms of solubility, non-specific binding, metabolic liability and toxicity.

Although none of the compounds reported have, to date, progressed beyond 21 with respect to clinical evaluation they serve as exemplifiers for the approach of using iminosugars for therapeutic benefit and highlight their utility to drug discovery. This is demonstrated further by consideration of fluorinated analogues of 21 that may offer greater therapeutic utility [183, 184]. Previous studies have demonstrated that treatment with 21 significantly increased glucocerebrosidase activity in disease-relevant tissues including liver and spleen [90]. However, 21 has limited CNS exposure and although activity in synucleinopathy models has been reported this reduced exposure may limit its utility in addressing CNS manifestations and neurodegenerative disease. For a pharmacological chaperone targeting glucocerebrosidase a significant improvement in drug-like properties including CNS penetration would be necessary for wider application to neuronopathic Gaucher disease and Parkinson's disease. One established strategy in drug discovery and development for maximising CNS penetration is the strategic use of fluorine substitution. The small size and high electronegativity of fluorine are among properties of this element that lend special advantages such as blood brain barrier penetration. Fluorinated analogues of 21 may greatly improve the therapeutic potential without compromising the chaperoning capacity. Bioavailability, CNS penetration and clearance are all enhanced and may result in associated improvements in CNS glucocerebrosidase enhancement and enhanced therapies for both Gaucher disease and Parkinson's disease in Gaucher gene carriers.

6 Summary, Conclusions, Outlook

The pioneering work of Dwek, Freeze, Varki and others in realising the importance of glycobiology in health and disease is becoming increasingly apparent [12, 27, 185–190]. The number of druggable targets emerging from these early reports is substantial and growing, covering a therapeutically broad and diverse area. Many, if not all, of these targets can be addressed by employing carbohydrate mimetics the most notable of which are the iminosugars. Whereas classical small molecule approaches have a strong history in drug development, iminosugars are relative newcomers. Early clinical failures and misguided concerns regarding ease of synthesis and drug-likeness have greatly limited the level to which this field has been explored. Synthetic and analytical issues have been addressed and are no longer an impediment to progress. Combined with a growing understanding of the impact that glycobiology has on disease and development a resurgence of inerest in iminosugars has emerged. As a chemical class iminosugars offer many advantages as potential drug candidates. They occupy a different area of chemical space to compounds typically found in pharmaceutical screening libraries and "new chemical space" is a concept much prized and sought after. As stable and bioavailable glycomimetics iminosugars provide substantial opportunity to access carbohydrate-related targets, an area increasingly desired. With two marketed drugs and several in clinical development the potential for iminosugars to deliver drug molecules is already established. The full exploitation of this unique compound class represents a major prospect for drug discovery in the near and medium term with these successes providing substantial evidence for the opportunities that now exist for innovation and clinical discovery in iminosugars.

References

1. Gupta G, Surolia A (2012) Glycomics: an overview of the complex glycocode. Adv Exp Med Biol 749:1–13
2. Furukawa K et al (2012) Fine tuning of cell signals by glycosylation. J Biochem 151(6):573–578
3. Zhang L, Ten Hagen KG (2011) The cellular microenvironment and cell adhesion: a role for O-glycosylation. Biochem Soc Trans 39(1):378–382
4. Neelamegham S, Liu G (2011) Systems glycobiology: biochemical reaction networks regulating glycan structure and function. Glycobiology 21(12):1541–1553
5. Gabius HJ et al (2011) From lectin structure to functional glycomics: principles of the sugar code. Trends Biochem Sci 36(6):298–313

6. Tian E, Ten Hagen KG (2009) Recent insights into the biological roles of mucin-type O-glycosylation. Glycoconj J 26(3):325–334
7. Cummings RD (2009) The repertoire of glycan determinants in the human glycome. Mol Biosyst 5(10):1087–1104
8. Varki A (2008) Sialic acids in human health and disease. Trends Mol Med 14(8):351–360
9. Alavi A, Axford JS (2008) Sweet and sour: the impact of sugars on disease. Rheumatology (Oxford) 47(6):760–770
10. Gabius HJ (2009) The sugar code: fundamentals of glycosciences. Wiley, Weinheim
11. Moran AP, Holst O, Brennan PJ, von Itzstein M (2009) Microbial glycobiology. Academic, London
12. Varki A, Cummings RD, Esko JD, Freeze HH, Stanley P, Bertozzi CR, Hart GW, Etzler ME (2008) Essentials of glycobiology. Cold Spring Harbor Laboratory Press, Cold Spring Harbor
13. Rabinovich G, Cobb B, van Kooyk Y (2012) Glycobiology of the immune response. Wiley, Weinheim
14. Wang B, Boons G-J (2011) Carbohydrate recognition: biological problems, methods, and applications. Wiley, Hoboken
15. Fukuda M, Rutishauser U, Schnaar R (2005) Neuroglycobiology. Oxford University Press, London
16. Fukuda M (2012) Recent progress in carbohydrate biosynthesis and function in relation to tumor biology. Biol Pharm Bull 35(10):1622–1625
17. Ghazarian H, Idoni B, Oppenheimer SB (2011) A glycobiology review: carbohydrates, lectins and implications in cancer therapeutics. Acta Histochem 113(3):236–247
18. Cazet A et al (2010) Tumour-associated carbohydrate antigens in breast cancer. Breast Cancer Res 12(3):204
19. Abbott KL (2010) Glycomic analysis of ovarian cancer: past, present, and future. Cancer Biomark 8(4–5):273–280
20. Saldova R et al (2008) Glycosylation changes on serum glycoproteins in ovarian cancer may contribute to disease pathogenesis. Dis Markers 25(4–5):219–232
21. Ohyama C (2008) Glycosylation in bladder cancer. Int J Clin Oncol 13(4):308–313
22. Chandra S et al (2010) Glycobiology of the *Leishmania parasite* and emerging targets for antileishmanial drug discovery. Expert Opin Ther Targets 14(7):739–757
23. Lefebvre T et al (2010) Dysregulation of the nutrient/stress sensor O-GlcNAcylation is involved in the etiology of cardiovascular disorders, type-2 diabetes and Alzheimer's disease. Biochim Biophys Acta 1800(2):67–79
24. Barone R et al (2012) Glycomics of pediatric and adulthood diseases of the central nervous system. J Proteomics 75(17):5123–5139
25. Boomkamp SD, Butters TD (2008) Glycosphingolipid disorders of the brain. Subcell Biochem 49:441–467
26. Sugahara K, Mikami T (2007) Chondroitin/dermatan sulfate in the central nervous system. Curr Opin Struct Biol 17(5):536–545
27. Freeze HH (2007) Congenital disorders of glycosylation: CDG-I, CDG-II, and beyond. Curr Mol Med 7(4):389–396
28. Endo T (2007) Dystroglycan glycosylation and its role in alpha-dystroglycanopathies. Acta Myol 26(3):165–170
29. Nash RJ et al (2011) Iminosugars as therapeutic agents: recent advances and promising trends. Future Med Chem 3(12):1513–1521
30. Nash RJ (2011) Advances in pharmaceutical applications of iminosugars. Spec Publ R Soc Chem 320:129–139 (Functional Molecules from Natural Sources)
31. Winchester BG (2009) Iminosugars: from botanical curiosities to licensed drugs. Tetrahedron Asymmetry 20(6–8):645–651
32. Asano N (2007) Naturally occurring iminosugars and related alkaloids: structure, activity and applications. In: Compain P, Martin OR (eds) Iminosugars. Wiley, West Sussex, pp 7–24
33. Stutz AE, Wrodnigg TM (2011) Imino sugars and glycosyl hydrolases: historical context, current aspects, emerging trends. Adv Carbohydr Chem Biochem 66:187–298

34. Horne G et al (2011) Iminosugars past, present and future: medicines for tomorrow. Drug Discov Today 16(3/4):107–118
35. Horne G, Wilson FX (2011) Therapeutic applications of iminosugars: current perspectives and future opportunities. Prog Med Chem 50:135–176
36. Nishikawa T, Ishida N (1965) A new antibiotic R-468 active against drug-resistant Shigella. J Antibiot (Tokyo) 18:132–133
37. Ishida N et al (1967) Nojirimycin, a new antibiotic. II. Isolation, characterization and biological activity. J Antibiot (Tokyo) 20(2):66–71
38. Ishida N et al (1967) Nojirimycin, a new antibiotic. I. Taxonomy and fermentation. J Antibiot (Tokyo) 20(2):62–65
39. Inouye S, Tsuruoka T, Nida T (1966) The structure of nojirimycin, a piperidinose sugar antibiotic. J Antibiot (Tokyo) 19(6):288–292
40. Niwa T et al (1984) Novel glycosidase inhibitors, nojirimycin B and D-mannonic-delta-lactam. Isolation, structure determination and biological property. J Antibiot (Tokyo) 37(12):1579–1586
41. Miyake Y, Ebata M (1987) Galactostatin, a new beta-galactosidase inhibitor from *Streptomyces lydicus*. J Antibiot (Tokyo) 40(1):122–123
42. Colegate SM, Dorling PR, Huxtable CR (1979) A spectroscopic investigation of swainsonine: an α-mannosidase inhibitor isolated from *Swainsona canescens*. Aust J Chem 32(10):2257–2264
43. Hohenschutz LD et al (1981) Castanospermine, a 1,6,7,8-tetrahydroxyoctahydroindolizine alkaloid, from seeds of *Castanospermum australe*. Phytochem (Elsevier) 20(4):811–814
44. Nash RJ et al (1988) Isolation from *Alexa leiopetala* and X-ray crystal structure of Alexine, (1r,2r,3r,7s,8s)-3-hydroxymethyl-1,2,7-trihydroxypyrrolizidine [(2r,3r,4r,5s,6s)-2-hydroxymethyl-1-azabicyclo[3.3.0]octan-3,4,6-triol], a unique pyrrolizidine alkaloid. Tetrahedron Lett 29(20):2487–2490
45. Molyneux RJ et al (1988) Australine, a novel pyrrolizidine alkaloid glucosidase inhibitor from *Castanospermum australe*. J Nat Prod 51(6):1198–1206
46. Nash RJ et al (1995) Casuarine: a very highly oxygenated pyrrolizidine alkaloid. Tetrahedron Lett 35(42):7849–7852
47. Kato A et al (1999) Polyhydroxylated pyrrolidine and pyrrolizidine alkaloids from hyacinthoides non-scripta and *Scilla campanulata*. Carbohydr Res 316(1–4):95–103
48. Tepfer D et al (1988) A plasmid of rhizobium meliloti 41 encodes catabolism of two compounds from root exudate of calystegium sepium. J Bacteriol 170(3):1153–1161
49. Michalik A et al (2010) Steviamine, a new indolizidine alkaloid from *Stevia rebaudiana*. Phytochem Lett 3(3):136–138
50. Zhu J-S et al (2013) Synthesis of eight stereoisomers of pochonicine: nanomolar inhibition of β-N-acetylhexosaminidases. J Org Chem 78(20):10298–10309
51. Aoyagi T et al (1992) Nagstatin, a new inhibitor of N-acetyl-β-D-glucosaminidase, produced by *Streptomyces amakusaensis* MG846-fF3. Taxonomy, production, isolation, physico-chemical properties and biological activities. J Antibiot (Tokyo) 45(9):1404–1408
52. Marlier M, Dardenne GA, Casimir J (1972) 4,5-Dihydroxy-L-pipecolic acid from *Calliandra haematocephala*. Phytochem (Elsevier) 11(8):2597–2599
53. Shibano M et al (2001) Two new pyrrolidine alkaloids, radicamines A and B, as inhibitors of α-glucosidase from Lobelia chinensis Lour. Chem Pharma Bull 49(10):1362–1365
54. Campbell SJ et al (2010) Visualizing the drug target landscape. Drug Discov Today 15:3–15
55. Hajduk PJ, Huth JR, Fesik SW (2005) Druggability indices for protein targets derived from NMR-based screening data. J Med Chem 48:2518–2525
56. Halgren TA (2009) Identifying and characterizing binding sites and assessing druggability. J Chem Inf Model 49:377–389
57. Schmidtke P, Barril X (2010) Understanding and predicting druggability. A high-throughput method for detection of drug binding sites. J Med Chem 53:5858–5867
58. Sheridan RP et al (2010) Drug-like density: a method of quantifying the "bindability" of a protein target based on a very large set of pockets and druglike ligands from the Protein Data Bank. J Chem Inf Model 50:2029–2040

59. Lipinski CA (2004) Lead- and drug-like compounds: the rule-of-five revolution. Drug Discov Today Technol 2004:337–341
60. Lipinski CA et al (1997) Experimental and computational approaches to estimate solubility and permeability in drug discovery and development settings. Adv Drug Del Rev 23:3–25
61. Proudfoot JR (2005) The evolution of synthetic oral drug properties. Bioorg Med Chem Lett 15:1087–1090
62. Vieth M et al (2004) Characteristic physical properties and structural fragments of marketed oral drugs. J Med Chem 47:224–232
63. Wenlock MC et al (2003) A comparison of physiochemical property profiles of development and marketed oral drugs. J Med Chem 46:1250–1256
64. Lipinski CA, Hopkins A (2004) Navigating chemical space for biology and medicine. Nature 432:855–861
65. Gorse AD (2006) Diversity in medicinal chemistry space. Curr Top Med Chem 6(1):3–18
66. Ruddigkeit L et al (2012) Enumeration of 166 billion organic small molecules in the chemical universe database GDB-17. J Chem Inf Model 52(11):2864–2875
67. Reymond JL, Awale M (2012) Exploring chemical space for drug discovery using the chemical universe database. ACS Chem Neurosci 3(19):649–657
68. Werz DB et al (2007) Exploring the structural diversity of mammalian carbohydrates ("glycospace") by statistical databank analysis. ACS Chem Biol 2(10):685–691
69. Angata T, Varki A (2002) Chemical diversity in the sialic acids and related alpha-keto acids: an evolutionary perspective. Chem Rev 102(2):439–469
70. Herget S et al (2008) Statistical analysis of the bacterial carbohydrate structure data base (BCSDB): characteristics and diversity of bacterial carbohydrates in comparison with mammalian glycans. BMC Struct Biol 8:35
71. O'Brien JS (1972) Sanfilippo syndrome. Profound deficiency of α-acetylglucosaminidase activity in organs and skin fibroblasts from type-B patients. Proc Natl Acad Sci U S A 69(7):1720–1722
72. Sandhoff K, Andreae WA, Jatzkewitz H (1968) Deficient hexosaminidase activity in an exceptional case of Tay-Sachs disease with additional storage of kidney globoside in visceral organs. Life Sci 7(6):283–288
73. Hart GW et al (2011) Cross talk between O-GlcNAcylation and phosphorylation: roles in signaling, transcription, and chronic disease. Annu Rev Biochem 80:825–858
74. Darley-Usmar VM, Ball LE, Chatham JC (2012) Protein O-linked β-N-acetylglucosamine: a novel effector of cardiomyocyte metabolism and function. J Mol Cell Cardiol 52(3):538–549
75. Zachara NE (2012) The roles of O-linked β-N-acetylglucosamine in cardiovascular physiology and disease. Am J Physiol Heart Circ Physiol 302(10):H1905–H1918
76. van Diggelen OP et al (1988) Alpha-N-acetylgalactosaminidase deficiency: a new lysosomal storage disorder. J Inherit Metab Dis 11(4):349–357
77. Dorfmueller HC et al (2010) Cell-penetrant, nanomolar O-GlcNAcase inhibitors selective against lysosomal hexosaminidases. Chem Biol 17:1250–1255
78. Heightman TD, Vasella AT (1999) Recent insights into inhibition, structure, and mechanism of configuration-retaining glycosidases. Angew Chem Int Ed 38(6):750–770
79. Douroumis D, Fahr A (2013) Drug delivery strategies for poorly water-soluble drugs. Wiley, Oxford
80. Lachmann RH (2006) Miglustat: substrate reduction therapy for glycosphingolipid lysosomal storage disorders. Drugs Today (Barc) 42:29–38
81. Benito JM, García Fernández JM, Ortiz Mellet C (2011) Pharmacological chaperone therapy for Gaucher disease: a patent review. Expert Opin Ther Pat 21(6):885–903
82. Campbell LK, Baker DE, Campbell RK (2000) Miglitol: assessment of its role in the treatment of patients with diabetes mellitus. Ann Pharmacother 34(11):1291–1301
83. Dulsat C (2009) Gaucher's disease. Drugs Future 34:147–149
84. Wraith JE, Imrie J (2009) New therapies in the management of Niemann-Pick type C disease: clinical utility of miglustat. Ther Clin Risk Manag 5:877–887

85. Sorbera LA, Castaner J, Bayes M (2003) Miglustat. Treatment of Gaucher's disease, ceramide glucosyltransferase inhibitor, a-glucosidase inhibitor. Drugs Future 28:229–236
86. Abian O et al (2011) Therapeutic strategies for Gaucher disease: miglustat (NB-DNJ) as a pharmacological chaperone for glucocerebrosidase and the different thermostability of velaglucerase alfa and imiglucerase. Mol Pharm 8(6):2390–2397
87. Parenti G et al (2007) Pharmacological enhancement of mutated alpha-glucosidase activity in fibroblasts from patients with Pompe disease. Mol Ther 15:508–514
88. Sugawara K et al (2009) Molecular interaction of imino sugars with human alpha-galactosidase: insight into the mechanism of complex formation and pharmacological chaperone action in Fabry disease. Mol Genet Metab 96:233–238
89. Khanna R et al (2010) The pharmacological chaperone 1-deoxygalactonojirimycin reduces tissue globotriaosylceramide levels in a mouse model of Fabry disease. Mol Ther 18(1):23–33
90. Khanna R et al (2010) The pharmacological chaperone isofagomine increases the activity of the Gaucher disease L444P mutant form of β-glucosidase. Febs J 277(7):1618–1638
91. Flanagan JJ et al (2009) The pharmacological chaperone 1-deoxynojirimycin increases the activity and lysosomal trafficking of multiple mutant forms of acid alpha-glucosidase. Hum Mutat 30(12):1683–1692
92. Khanna R et al (2012) The pharmacological chaperone AT2220 increases recombinant human acid α-glucosidase uptake and glycogen reduction in a mouse model of Pompe disease. PLoS One 7(7):e40776
93. Benjamin ER et al (2012) Co-administration with the pharmacological chaperone AT1001 increases recombinant human α-galactosidase a tissue uptake and improves substrate reduction in Fabry mice. Mol Ther 20(4):717–726
94. Durantel D (2009) Celgosivir, an α-glucosidase I inhibitor for the potential treatment of HCV infection. Curr Opin Investig Drugs 10(8):860–870
95. Mehta AS et al (2004) α-Galactosylceramide and novel synthetic glycolipids directly induce the innate host defense pathway and have direct activity against hepatitis B and C viruses. Antimicrob Agents Chemother 48:2085–2090
96. Goss PE et al (1994) A phase I study of swainsonine in patients with advanced malignancies. Cancer Res 54(6):1450–1457
97. Pavlović D et al (2003) The hepatitis C virus p7 protein forms an ion channel that is inhibited by long-alkyl-chain iminosugar derivatives. Proc Natl Acad Sci U S A 100(10):6104–6108
98. Soheili T et al (2012) Rescue of sarcoglycan mutations by inhibition of endoplasmic reticulum quality control is associated with minimal structural modifications. Hum Mutat 33(2):429–439
99. Bartoli M et al (2008) Mannosidase I inhibition rescues the human α-sarcoglycan R77C recurrent mutation. Hum Mol Genet 17(9):1214–1221
100. Takahata H (2012) Chiral synthesis of iminosugars. Heterocycles 85(6):1351–1376
101. Dragutan I, Dragutan V, Demonceau A (2012) Targeted drugs by olefin metathesis: piperidine-based iminosugars. RSC Adv 2(3):719–736
102. Dragutan I et al (2011) Metathesis access to monocyclic iminocyclitol-based therapeutic agents. Beilstein J Org Chem 7(81):699–716
103. Kim IS, Jung YH (2011) Recent advances in the total synthesis of indolizidine iminosugars. Heterocycles 83(11):2489–2507
104. Stecko S et al (2011) Synthesis of iminosugars via 1,3-dipolar cycloaddition reactions of nitrones to α, β-unsaturated sugar aldonolactones. C R Chim 14(1):102–125
105. Stocker BL et al (2010) Recent developments in the synthesis of pyrrolidine-containing iminosugars. Eur J Org Chem 9:1615–1637
106. Compain P, Chagnault V, Martin OR (2009) Tactics and strategies for the synthesis of iminosugar C-glycosides: a review. Tetrahedron Asymmetry 20(6–8):672–711
107. Clapes P, Sprenger GA, Joglar J (2009) Novel strategies in aldolase-catalyzed synthesis of imino-sugars. Mod Biocatal 299–311

108. Lopez MD, Cobo J, Nogueras M (2008) Building bicyclic polyhydroxylated alkaloids: an overview from 1995 to the present. Curr Org Chem 12(9):718–750
109. La Ferla B, Cipolla L, Nicotra F (2007) General strategies for the synthesis of imino-sugars and new approaches towards imino-sugar libraries. In: Campain P, Martin OR (eds) Iminosugars, from synthesis to terapeutic applications. Wiley, West Sussex, pp 25–61
110. Behr J-B, Plantier-Royon R (2006) Addition of organometallics to aldimines, aldoximes and aldononitriles: a key step towards the synthesis of azasugars. Recent Res Dev Org Chem 10:23–52
111. Cipolla L, La Ferla B, Gregori M (2006) Combinatorial approaches to iminosugars as glycosidase and glycosyltransferase inhibitors. Comb Chem High Throughput Screen 9(8):571–582
112. Ayad T, Genisson Y, Baltas M (2004) Chemical approaches towards synthesis of some naturally occurring iminosugars. Curr Org Chem 8(13):1211–1233
113. El Ashry ESH, El Nemr A (2005) Synthesis of naturally occurring nitrogen heterocycles from carbohydrates. Wiley-Blackwell, Oxford
114. El-Ashry SH, El Nemr A (2003) Synthesis of mono- and di-hydroxylated prolines and 2-hydroxymethylpyrrolidines from non-carbohydrate precursors. Carbohydr Res 338 (22):2265–2290
115. Kato A et al (2008) Iminosugars from Baphia nitida Lodd. Phytochem (Elsevier) 69:1261–1265
116. Nakagawa K et al (2010) Determination of iminosugars in mulberry leaves and silkworms using hydrophilic interaction chromatography-tandem mass spectrometry. Anal Biochem 404 (2):217–222
117. Nuengchamnong N et al (2007) Quantitative determination of 1-deoxynojirimycin in mulberry leaves using liquid chromatography-tandem mass spectrometry. J Pharm Biomed Anal 44(4):853–858
118. Fleet GWJ, Smith PW (1987) Methyl 2-azido-3-O-benzyl-2-deoxy-α-D-mannofuranoside as a divergent intermediate for the synthesis of polyhydroxylated piperidines and pyrrolidines. Synthesis of 2,5-dideoxy-2,5-imino-D-mannitol [2R,5R-bis(hydroxymethyl)-3R,4R-dihydroxypyrrolidine]. Tetrahedron Asymmetry 43(5):971–978
119. Kinast G, Schedel M (1981) A four-stage synthesis of 1-deoxynojirimycin with a biotransformation as the central reaction step. Angew Chem 93(9):799–800
120. Saeki H, Ohki E (1968) 5,6-Epimino-D-glucofuranose and synthesis of nojirimycin (5-amino-5-deoxy-D-glucose). Chem Pharm Bull 16(12):2477–2481
121. Bernotas RC, Ganem B (1984) Total syntheses of (+)-castanospermine and (+)-deoxynojirimycin. Tetrahedron Lett 25(2):165–168
122. Hamana H, Ikota N, Ganem B (1987) Chelate selectivity in chelation-controlled allylations. A new synthesis of castanospermine and other bioactive indolizidine alkaloids. J Org Chem 52(24):5492–5494
123. Hendry D, Hough L, Richardson AC (1988) Enantiospecific synthesis of polyhydroxylated indolizidines related to castanospermine: 1-deoxycastanospermine. Tetrahedron Asymmetry 44(19):6143–6152
124. Hendry D, Hough L, Richardson AC (1988) Enantiospecific synthesis of polyhydroxylated indolizidines related to castanospermine: (6R,7S,8aR)-6,7-dihydroxyindolizidine and (6R,7R,8S,8aR)-6,7,8-trihydroxyindolizidine. Tetrahedron Asymmetry 44(19):6153–6168
125. Suami T, Tadano K, Iimura Y (1984) Total synthesis of (−)-swainsonine, an α-mannosidase inhibitor isolated from *Swainsona canescens*. Chem Lett 4:513–516
126. Mezher HA, Hough L, Richardson AC (1984) A chiral synthesis of swainsonine from D-glucose. J Chem Soc Chem Commun (7):447–448
127. Bols M (1996) Carbohydrate building blocks. Wiley, Chichester
128. Hanessian S (1983) Total synthesis of natural products: "Chiron" approach. Pergamon Press, Oxford

129. Hanessian S, Giroux S, Merner BL (2013) Design and strategy in organic synthesis. Wiley, Weinheim
130. Donohoe TJ et al (2003) Flexibility in the partial reduction of 2,5-disubstituted pyrroles: application to the synthesis of DMDP. Org Lett 5(7):999–1002
131. Kim J-Y et al (2011) Efficient and stereoselective syntheses of DAB-1 and d-fagomine via chiral 1,3-oxazine. Tetrahedron 67(48):9426–9432
132. Choi HG et al (2013) An efficient synthesis of 1,4-dideoxy-1,4-imino-d- and l-arabinitol and 1,4-dideoxy-1,4-imino-d- and l-xylitol from chiral aziridines. Tetrahedron Lett 54(43):5775–5777
133. Singh S, Han H (2004) Stereodivergent total asymmetric synthesis of polyhydroxylated pyrrolidines via tandem allylic epoxidation and intramolecular cyclization reactions. Tetrahedron Lett 45(33):6349–6352
134. Restorp P, Fischer A, Somfai P (2006) Stereoselective synthesis of functionalized pyrrolidines via a [3 + 2]-annulation of N-Ts-α-amino aldehydes and 1,3-Bis(silyl)propenes. J Am Chem Soc 128(39):12646–12647
135. Kondo Y et al (2012) Enantioselective construction of a polyhydroxylated pyrrolidine skeleton from 3-vinylaziridine-2-carboxylates: synthesis of (+)-DMDP and a potential common intermediate for (+)-hyacinthacine A1 and (+)-1-epi-australine. J Org Chem 77(18):7988–7999
136. Izquierdo I et al (2010) Total synthesis of natural (+)-hyacinthacine A6 and non-natural (+)-7a-epi-hyacinthacine A1 and (+)-5,7a-diepi-hyacinthacine A6. Tetrahedron 66(21):3788–3794
137. Izquierdo I et al (2008) Synthesis of (+)-1-epi-castanospermine from L-sorbose. Tetrahedron 64(34):7910–7913
138. Wang N, Zhang L-H, Ye X-S (2010) A new synthetic access to bicyclic polyhydroxylated alkaloid analogues from pyranosides. Org Biomol Chem 8(11):2639–2649
139. Kalamkar NB, Puranik VG, Dhavale DD (2011) Synthesis of C1– and C8a-epimers of (+)-castanospermine from D-glucose derived γ, δ-epoxyazide: intramolecular 5-endo epoxide opening approach. Tetrahedron 67(15):2773–2778
140. Kalamkar NB, Dhavale DD (2011) Chiron approach strategy to the bicyclic oxazolidinylpiperidine: a building block for preparing mono- and bi-cyclic imino-sugars. Tetrahedron Lett 52(48):6363–6365
141. Dhand V et al (2013) A short, organocatalytic formal synthesis of (−)-swainsonine and related alkaloids. Org Lett 15(8):1914–1917
142. Zhang H-K et al (2012) A flexible enantioselective approach to 3,4-dihydroxyprolinol derivatives by SmI2-mediated reductive coupling of chiral nitrone with ketones/aldehydes. Tetrahedron 68(33):6656–6664
143. Archibald G et al (2012) A divergent approach to 3-piperidinols: a concise syntheses of (+)-swainsonine and access to the 1-substituted quinolizidine skeleton. J Org Chem 77(18):7968–7980
144. Yun H et al (2012) Asymmetric syntheses of 1-deoxy-6,8a-di-epi-castanospermine and 1-deoxy-6-epi-castanospermine. J Org Chem 77(12):5389–5393
145. Ameijde JV et al (2006) Isolation synthesis and glycosidase inhibition profile of 3-epi-casuarine. Tetrahedron Asymmetry 17(18):2702–2712
146. Ritthiwigrom T, Willis AC, Pyne SG (2010) Total synthesis of uniflorine a, casuarine, australine, 3-epi-australine, and 3,7-Di-epi-australine from a common precursor. J Org Chem 75(3):815–824
147. Liu X-K et al (2011) SmI2-mediated radical cross-couplings of α-hydroxylated Aza-hemiacetals and N, S-acetals with α, β-unsaturated compounds. Asymmetric synthesis of (+)-hyacinthacine A2, (−)-uniflorine A, and (+)-7-epi-casuarine. J Org Chem 76(12):4952–4963
148. Ribes C et al (2007) Stereoselective synthesis of the glycosidase inhibitor australine through a one-pot, double-cyclization strategy. Org Lett 9(1):77–80

149. Davies SG et al (2013) Asymmetric syntheses of (−)-1-deoxymannojirimycin and (+)-1-deoxyallonojirimycin via a ring-expansion approach. Org Lett 15(8):2042–2045
150. Jenkinson SF et al (2011) Looking-glass synergistic pharmacological chaperones: DGJ and L-DGJ from the enantiomers of tagatose. Org Lett 13(15):4064–4067
151. Donohoe TJ et al (2008) Flexible strategy for the synthesis of pyrrolizidine alkaloids. Org Lett 10(16):3615–3618
152. Gilles P, Py S (2012) SmI2-mediated cross-coupling of nitrones with β-silyl acrylates: synthesis of (+)-australine. Org Lett 14(4):1042–1045
153. Moriyama H et al (2003) Structure–activity relationships of azasugar-based MMP/ADAM inhibitors. Bioorg Med Chem Lett 13(16):2737–2740
154. Moriyama H et al (2003) Design, synthesis and evaluation of novel azasugar-based MMP/ADAM inhibitors. Bioorg Med Chem Lett 13(16):2741–2744
155. Moriyama H et al (2004) Azasugar-based MMP/ADAM inhibitors as antipsoriatic agents. J Med Chem 47:1930–1938
156. Chikaraishi Y et al (2009) CB-12181, a new azasugar-based matrix metalloproteinase/tumor necrosis factor-alpha converting enzyme inhibitor, inhibits vascular endothelial growth factor-induced angiogenesis in vitro and retinal neovascularization in vivo. Curr Neurovasc Res 6(3):140–147
157. Cox TM, Schofield JP (1997) Gaucher's disease: clinical features and natural history. Baillieres Clin Haematol 10:657–689
158. Velayati A, Yu WH, Sidransky E (2010) The role of glucocerebrosidase mutations in Parkinson disease and Lewy body disorders. Curr Neurol Neurosci Rep 10:190–198
159. Lwin A et al (2004) Glucocerebrosidase mutations in subjects with parkinsonism. Mol Genet Metab 81:70–73
160. Neumann J et al (2009) Glucocerebrosidase mutations in clinical and pathologically proven Parkinson's disease. Brain 132:1783–1794
161. Setó-Salvia N et al (2011) Glucocerebrosidase mutations confer a greater risk of dementia during Parkinson's disease course. Mov Disord 27:393–399
162. Mazzulli JR et al (2011) Gaucher disease glucocerebrosidase and α-synuclein form a bidirectional pathogenic loop in synucleinopathies. Cell 146(1):37–52
163. Boyd RE et al (2013) Pharmacological chaperones as therapeutics for lysosomal storage diseases. J Med Chem 56(7):2705–2725
164. Maegawa GHB et al (2009) Identification and characterization of ambroxol as an enzyme enhancement agent for Gaucher disease. J Biol Chem 284:23502–23516
165. Tropak MB et al (2008) Identification of pharmacological chaperones for Gaucher disease and characterization of their effects on β-glucocerebrosidase by hydrogen/deuterium exchange mass spectrometry. ChemBioChem 9:2650–2662
166. Zheng W et al (2007) Three classes of glucocerebrosidase inhibitors identified by quantitative high-throughput screening are chaperone leads for Gaucher disease. Proc Natl Acad Sci U S A 104:13192–13197
167. Marugan JJ et al (2011) Evaluation of quinazoline analogues as glucocerebrosidase inhibitors with chaperone activity. J Med Chem 54:1033–1058
168. Patnaik S, Marugan JJ (2012) Discovery, SAR, and biological evaluation of non-inhibitory small molecule chaperones of glucocerebrosidase. J Med Chem 55:5734–5748
169. Fan J-Q (2007) Iminosugars as active-site-specific chaperones for the treatment of lysosomal storage disorders. In: Compain P, Martin OR (eds) Iminosugars: from synthesis to therapeutic applications. Wiley, West Sussex
170. Svennerholm L, Vanier MT, Månsson JE (1980) Krabbe disease: a galactosylsphingosine (psychosine) lipidosis. J Lipid Res 21(1):53–64
171. Lieberman Raquel L et al (2007) Structure of acid beta-glucosidase with pharmacological chaperone provides insight into Gaucher disease. Nat Chem Biol 3(2):101–107

172. Brumshtein B et al (2007) Crystal structures of complexes of N-butyl- and N-nonyl-deoxynojirimycin bound to acid β-glucosidase: insights into the mechanism of chemical chaperone action in Gaucher disease. J Biol Chem 282(39):29052–29058
173. Deane JE et al (2011) Insights into Krabbe disease from structures of galactocerebrosidase. Proc Natl Acad Sci U S A 108(37):15169–15173
174. Diot JD et al (2011) Amphiphilic 1-deoxynojirimycin derivatives through click strategies for chemical chaperon in N370S Gaucher cells. J Org Chem 76(19):7757–7768
175. Yu L et al (2006) α-1-C-octyl-1-deoxynojirimycin as a pharmacological chaperone for Gaucher disease. Bioorg Med Chem 14(23):7736–7744
176. Sawkar AR et al (2002) Chemical chaperones increase the cellular activity of N370S β-glucosidase: a therapeutic strategy for Gaucher disease. Proc Natl Acad Sci U S A 99:15428–15433
177. Fröhlich RFG et al (2010) 1-Deoxynojirimycins with dansyl capped N substituents as probes for morbus Gaucher affected cell lines. Carbohydr Res 345:1371–1376
178. Luan Z et al (2010) A fluorescent sp2-iminosugar with pharmacological chaperone activity for Gaucher disease: synthesis and intracellular distribution studies. ChemBioChem 11(17):2453–2464
179. Chang HH et al (2006) Hydrophilic iminosugar active-site-specific chaperones increase residual glucocerebrosidase activity in fibroblasts from Gaucher patients. FEBS J 273(17):4082–4092
180. Zhu X et al (2005) Rational design and synthesis of highly potent beta-glucocerebrosidase inhibitors. Angew Chem Int Ed Engl 44(45):7450–7453
181. Yu Z et al (2007) Isofagomine- and 2,5-anhydro-2,5-imino-D-glucitol-based glucocerebrosidase pharmacological chaperones for Gaucher disease intervention. J Med Chem 50(1):94–100
182. Hill T et al (2011) Synthesis, kinetic evaluation and cell-based analysis of C-alkylated isofagomines as chaperones of β-glucocerebrosidase. ChemBioChem 12(14):2151–2154
183. Boyd R, Lee G, Rybczynski P (2011) Novel compositions for preventing and/or treating lysosomal storage disorders using piperidine diol derivatives. Amicus Therapeutics, Inc., USA. Application: US. p 34
184. Boyd R, Lee G, Rybczynski P (2011) Novel compositions for preventing and/or treating degenerative disorders of the central nervous system. Amicus Therapeutics, Inc., USA. Application: US. p 38
185. Rademacher TW, Parekh RB, Dwek RA (1988) Glycobiology. Annu Rev Biochem 57:785–838
186. Varki A (1993) Biological roles of oligosaccharides: all of the theories are correct. Glycobiology 3:97–130
187. Dwek RA (1996) Glycobiology: toward understanding the function of sugars. Chem Rev 96:683–720
188. Spiro RG (2002) Protein glycosylation: nature, distribution, enzymatic formation, and disease implications of glycopeptide bonds. Glycobiology 12:43R–56R
189. Ohtsubo K, Marth JD (2006) Glycosylation in cellular mechanisms of health and disease. Cell 126:855–867
190. Freeze HH (2006) Genetic defects in the human glycome. Nat Rev Genet 7:537–551

Top Med Chem (2014) 12: 53–72
DOI: 10.1007/7355_2014_42

Published online: 5 June 2014

Computational Docking as a Tool for the Rational Design of Carbohydrate-Based Drugs

Martin Frank

Abstract Docking methods are a valuable tool for the prediction of carbohydrate binding sites and the design of carbohydrate-based drugs. However, there are also significant limitations and care needs to be taken when evaluating the docking results. In this chapter the challenges, limitations, and possible pitfalls in docking of carbohydrates are described. Practical examples explain the use of docking methods for the rational design of carbohydrate-based inhibitors as well as the prediction of carbohydrate binding sites.

Keywords Computational chemistry, Docking, Structure-based drug design

Contents

Abbreviations

3D	Three-dimensional
CAT	Conformational analysis tools
CPU	Central processing unit
GA-LS	Genetic algorithm – local search

M. Frank (✉)
Biognos AB, Generatorsgatan 1, 41705 Göteborg, Sweden
e-mail: martin@glycosciences.org

GPU	Graphical processing unit
LIE	Linear interaction energy
MD	Molecular dynamics
MM/GBSA	Molecular mechanics/generalized born surface area
RMSD	Root mean square deviation
TI	Thermodynamic integration

1 Introduction

The application of computational methods has become an integral part in modern drug design workflows. Bioinformatic methods are used routinely to understand the evolutionary relationship between protein sequences and – in case no experimentally determined structure is available – to aid in finding structural templates for homology modeling of three-dimensional (3D) structures of protein targets [1, 2]. Molecular modeling methods are required frequently to interpret experimental data in NMR and used to build models in X-ray crystallography [3–5]. Docking methods can be applied to predict the binding sites and binding mode(s) of small molecules on protein surfaces [6]. Structure-based virtual screening of chemical compound libraries is a cost-efficient alternative to high-throughput screening campaigns for finding active molecules that can serve as new leads for drug design [7, 8]. Chemoinformatic methods can be used for ligand-based screening (and design) of small molecule libraries containing millions of chemical compounds [9]. Finally, current research in life sciences would 'collapse' without the hundreds of databases that store (and make publicly accessible) the wealth of experimental data (sequences and 3D structures of biomolecules, affinity data, physical chemical properties of small molecules, and many more) that is generated daily worldwide [10–12].

All of the methods mentioned above are routinely applied also in glycobiology and glycomics. However, carbohydrate building blocks occur in small molecules (e.g., substituted monosaccharides, glycomimetics) as well as in large macromolecular tree-like structures (polysaccharides, glycans) which render these molecules difficult to be handled in a consistent way by the established bio- and chemoinformatics algorithms [13]. Therefore it was necessary to establish glycoinformatics (standards, algorithms, and databases) at the interface between bioinformatics and chemoinformatics (reviews and books on the topic are found in, for example, [14–19]).

Molecular modeling of carbohydrates is challenging due to the intrinsic flexibility of glycosidic linkages [20] and the requirement for special carbohydrate force fields that are compatible with existing force fields used for proteins or small molecules [21, 22]. Three-dimensional structures of carbohydrates are best described as conformational ensembles of interchanging conformational states and the method of choice to simulate carbohydrates in their biological context (glycoprotein, protein–carbohydrate complex, glycolipid embedded in a membrane) is molecular dynamics (MD) simulation in explicit solvent. The recent development of algorithms to use GPUs for routine MD simulations [23] has made it possible to study carbohydrate structure on the microsecond timescale [24]. MD simulation methods can also be used to estimate the free energies of binding based on thermodynamic integration (TI) [25, 26], linear interaction energy (LIE) [27], or MM/GBSA [28] methods. However these methods require significant expert knowledge for proper setup and analysis, are computationally demanding, and have also severe other limitations for practical use in a drug design project. Docking methods are less accurate, but very fast and can be performed routinely also by 'bench-chemists' using graphical software interfaces [29]. Docking methods have been applied successfully in the area of carbohydrate-based inhibitor development in several recent projects, some of which are described in [30–36].

2 The Application of Docking Methods in Carbohydrate-Based Drug Design

2.1 Basics, Limitations, and Pitfalls

Automated docking methods were introduced already in the early 1980s [37] but became popular in the 1990s fuelled by the increasing number of available X-ray crystal structures and several publications demonstrating that novel bioactive molecules can be identified using docking. The expectations were high at the beginning but the success was very limited, which led to an opinion that computerized 'structure-based drug design' is not practical. Uncritical application of virtual screening methods may still result nowadays in inappropriate compound selection [8]. The search for new drugs, especially in lead optimization, is an evolutionary process that is only likely to be successful if computational methods merge with classical medicinal chemistry knowledge [38]. As a matter of fact, there is probably still nowadays not a single drug that was discovered entirely by structure-based methods; however, there are a significant number of drugs on the market where structure-based methods were critical for the development in the early phase [39]. It has to be realized that structure-based methods can contribute only to the discovery and early optimization of leads in drug design. Many properties of a compound that are important to make it a 'good drug' are outside the scope or predictive possibilities of structure-based computational methods.

The aims in molecular docking are the accurate prediction of the binding mode (conformation and orientation) of a ligand in the binding site of a receptor of known three-dimensional structure and the correct prediction of the binding affinity of the ligand. This includes also the correct ranking of the binding affinities of different ligands to the same binding site. There are excellent reviews and protocols on molecular docking (e.g., [6, 40–44]); therefore, the method is described here only very briefly.

In principle, the docking process consists of two parts, which both have their special challenges and limitations: posing and scoring. The docking algorithm tries to place (pose) the ligand into the active site by sampling all conformational, rotational, and translational degrees of freedom. At each trial a 'score' that aims to approximate binding affinity is calculated and is used to rank all the generated poses of a ligand. The fundamental assumption is that the pose with the best score is the 'correct' binding mode. The difficulty is that sampling the conformational space systematically with a resolution that is high enough to identify the conformation that best matches the receptor structure requires an enormous number of ligand poses to be generated and scored even for simple molecules. The search would take too long to be useful for a practical application. Therefore stochastic or genetic search algorithms are generally used to explore the search space. The search stops when a user-defined number of trials have been carried out, or some other termination criteria are met. It is very important that the termination criteria are chosen appropriately for the search problem. If, for example, the number of trials is set to 10^5, but 10^7 are appropriate for the size of the search space, then there is a high risk that the correct pose is not found, simply because it may not be in the set of trial structures generated. In other words: poses that have not been generated cannot be scored! One way to check for convergence of a docking is to check whether the best scored pose has been found more than once.

Another consideration is the size of the search space used to represent the binding site in a receptor (e.g., the grid box in AutoDock [45]). Many crystal structures contain a small ligand that is typically used in docking setup to define the center and size of the binding site. Limiting the search space only to the direct neighborhood of this ligand may give excellent results when re-docking the same ligand; however, larger ligands from a compound library may not fit at all into the search space or only in an artificial orientation. Therefore it is required to take into account the size of the largest ligand that will be docked when defining the search space.

Docking scores are calculated using a scoring function [46–48], which may be based on either physical interaction terms, such as van-der-Waals or electrostatic interaction or 'knowledge-based' terms derived, e.g., from inter-atomic contact frequencies in crystal structures of protein–ligand complexes. Currently there is no single scoring function available that accurately predicts the binding affinity of a diverse set of small molecules to any protein [49]. A high accuracy can be obtained by training a scoring function to a class of small molecules interacting with a particular binding site. However, this requires experimental affinity values, which are not always available.

The calculation time required for a docking is also an important practical aspect. For docking of a single molecule a CPU time in the range of minutes to hours would be perfectly acceptable; however in a virtual screening project, the calculation times have to be below one second. In this case the docking of 10^6 compounds would require still about 10 days of CPU time. In order to speed up the docking process, different scoring functions may be used in different stages. For pose searching ('docking') a simple scoring function ('docking score') that can be evaluated rapidly can be used, while a more sophisticated scoring function will be used for scoring of the final pose(s). Typically a defined number of 'best' docked poses are output in each docking run, which are then re-ranked using the final scoring function. However, this assumes that the pose with the best final score is among the 'best' poses that are kept using the simple docking score as criterion.

It is also possible to use a docking software only for the generation of a larger set of reasonable poses and use external more sophisticated scoring functions [50] or other methods such as MM/GBSA simulations for the final ranking of the poses (re-scoring) [51]. Additionally, a molecular library containing a larger number of docked compounds can be (post)processed using 'filters.' In this way compounds with 'unwanted properties' can be removed from the list, thereby lowering the number of compounds that need to be chemically synthesized and tested experimentally, e.g., in a biological assay.

The receptor is usually kept rigid during a docking run in order to speed up the calculations. This is problematic in cases where the available crystal structure represents a conformation that is not shaped properly for optimal binding of the ligand. In this case receptor flexibility needs to be included in the docking protocol [52]. Another current challenge in docking is that water molecules frequently mediate the binding of a ligand. It has been realized that neglecting these 'bridging waters' has a great impact on the docking accuracy; however, no generally applied method has emerged yet that takes into account water molecules in docking [53–55]. In the scientific literature there are many new or improved protocols and scoring functions published every year, which highlight that there are still many developments and improvements possible. Although many of the published methods are very sophisticated, if they are not made available to the scientific community as a ready-to-use software product they will be of very limited value. In reality, a user is limited to the software available. Each docking software has not only different 'useful' features and advantages but also different disadvantages. There are many docking software available (http://www.click2drug.org), but some of the most frequently used for docking of carbohydrates are AutoDock [45], Vina [56], and Glide [57]. The first two programs are free of charge, Glide is a commercial program distributed by Schrödinger, Inc.

2.2 *What Makes Docking of Carbohydrates Challenging?*

In principle small carbohydrates (e.g., mono-, di-, or trisaccharides) can be docked like any other organic molecule. Unfortunately, the simple force fields used in most docking software to calculate the conformational energy of a ligand are generally not optimized for carbohydrates. Therefore it can happen that poses with very good interaction energy scores have conformations that are not very reasonable for carbohydrates. This affects typically the values for the glycosidic linkage torsions. A practical approach to deal with this problem is the use of energy penalty functions for rescoring of the poses generated [58]: Quantum chemical conformational energy profiles, calculated for the φ and ψ torsions in a variety of disaccharides models, are used to assign an energy value to each glycosidic torsion in a docked pose. If a torsion value deviates from the minimum value, this results in a positive conformational energy value ('energy penalty'), which is summed up for all φ and ψ torsions and added to the total docking energy. In this way poses that have 'wrong' torsion values get a more positive (unfavorable) energy score and are therefore moved further down the sorted ranking list of poses.

Difficulties may also arise when ring flexibility needs to be taken into account in docking (as it required, e.g., for docking of furanoses or heparin fragments). Many docking software (e.g., AutoDock and Vina) cannot change the conformations of rings during docking. One solution may be to generate a set of starting structures with different ring conformations and dock them individually.

Docking of larger carbohydrates (N-glycans, polysaccharides) faces two limiting factors: First of all, the number of rotatable bonds may exceed the maximum that, e.g. AutoDock and Vina, can handle; therefore, some of the rotatable bonds need to be kept rigid during docking, which can mean that the conformation required for binding cannot be adopted and therefore will not be found. Secondly, in the biological recognition event only a subset of the carbohydrate residues present in a larger molecule are usually involved in binding. Unfortunately, the standard docking methods are not very well suited to predict binding of sub-structures, since the aim of the algorithms is to generate a pose with maximum interaction energy, which usually means that as many residues as possible will be put in contact with the protein surface (since many weak interactions may give a better score than a few strong interactions). A solution might be to dock fragments of variable size to the protein surface and to check which fragment results in the best (normalized) score.

However, if the binding mode of a fragment of the larger natural carbohydrate ligand has been determined by X-ray crystallography, there are several possibilities to use this experimental data to guide the structure building of the complete complex: In principle it is possible to generate thousands of poses of the complete carbohydrate using standard docking software and then use the coordinates of the fragment as a filter to extract poses that are fitting to the template. Alternatively, the fragment can also be used as an anchor point for positioning of a library of pre-build conformations of the complete carbohydrate. The preliminary complex(es) can subsequently be further refined using MD simulations in explicit solvent [59–61].

It is important to know that the empirical free energy models that are used to calculate the docking scores can be trained also for use with carbohydrates, but the standard errors still remain in the range of 1.1–1.4 kcal/mol [62, 63]. Energy (score) differences less than 1 kcal/mol might therefore be meaningless. Since not every scientist would like to invest a significant amount of time to develop a tailored scoring function to improve docking accuracy, it is of great interest to know how well standard docking programs do perform ‘out-of-the-box’ in carbohydrate docking. This has been evaluated in a variety of reviews [64–66]. The bottom line is that none of the docking programs is perfect, the performance depends to some extent on the molecular system studied and in many cases the top ranked pose was not the ‘best pose’ (the pose with the lowest RMSD to the bound ligand in the crystal structure). It has to be noted that studies that use crystal structures of protein–carbohydrate complexes and evaluate how well a ligand can be re-docked to the binding site may not simulate a realistic scenario in a research project (although such tests are important of course). In reality the available protein (crystal) structures may need to undergo a conformational change in order to accommodate the ligand properly (induced fit) [67]. The inclusion of induced fit effects in docking is difficult and only few of the docking programs can change the conformation of receptor and ligand while the docking is performed. Alternatively multiple receptor conformations can be used as input for several independent docking runs (ensemble docking) [68]. In general it can be expected that taking into account receptor flexibility will increase the calculation time significantly. It was also found that the flexible receptor approach was less accurate in reproducing binding modes from known antibody–carbohydrate complexes [66].

In many experimentally determined protein–carbohydrate complexes there are one or more (bridging) water molecules that mediate binding. Since water molecules are usually removed from the receptor structure prior to docking, this can be one of the reasons why docking programs fail to reproduce the binding mode of carbohydrates in some cases. Despite this knowledge, there is still no obvious general solution how to include the effect of bridging waters in docking algorithms, but there are promising new developments [55, 69–72]. It would be possible to leave the 'relevant' water molecules in the receptor structure and it is obvious that this will improve the re-docking accuracy of the crystallographic ligand. However, water molecules are not always well resolved in crystal structures and it may be difficult to determine the positions of the relevant waters computationally without the ligand of interest in place. Additionally, the number of relevant waters may change from ligand to ligand and therefore it may be necessary to switch 'on' and 'off' some of the water molecules during docking so that a ligand can bind properly [54]. In general, the tight binding of water molecules on protein surfaces is entropically unfavorable. A popular strategy in rational drug design is therefore to add a functional group to a ligand (e.g., an OH group) that will occupy the space of an ordered water molecule upon ligand binding and release it into the bulk solvent [73]. Consequently, the detection of ordered water molecules has gained significant attention in recent years [54, 73–75] and future developments in this area will be beneficial also for the accurate docking of carbohydrates.

The above highlighted aspects have important consequences for the rational design of carbohydrate-based drugs using docking methods: a 'black-box docking' of compounds containing carbohydrates is currently likely to fail, since there is no guarantee that the poses with the best score are the 'correct' poses. Therefore all poses need to be carefully evaluated by an experienced carbohydrate modeler, either manually on the computer screen or using (usually in-house developed) specialized post-processing software. Additionally, since the success of a docking calculation also depends on the molecular system studied, it is in general also a good idea to test whether the docking software available is suitable for the planned project.

2.3 Docking of Modified Carbohydrates: A Practical Guide

Siglecs are a family of sialic-acid-binding immunoglobulin-like lectins that are thought to promote cell–cell interactions and regulate the functions of cells in the

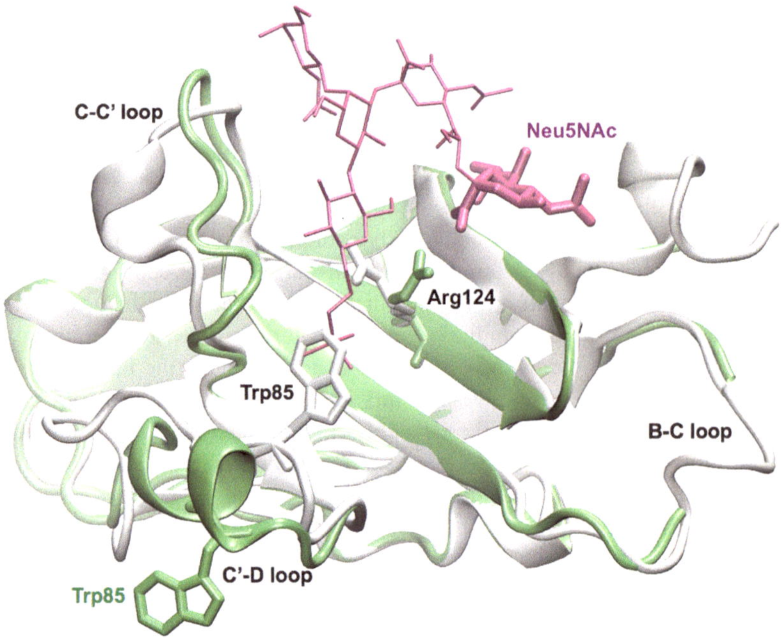

Fig. 1 Crystal structures of Siglec-7: *apo* structure (*white*), complex with a synthetic oligosaccharide containing GT1b (*green*, ligand in *purple*). The dynamic loops and the binding site of Neu5NAc are indicated

innate and adaptive immune systems through glycan recognition [76]. Siglec-7 [77] is an inhibitory receptor expressed on NK cells and a challenging target for the design of inhibitors [78, 79]. A classical strategy for inhibitor design would be to maintain the Neu5NAc moiety as a core and aim at increasing the binding affinity by means of attaching additional chemical groups to the various positions available in Neu5NAc. For a more reliable outcome of the computational predictions it is required that the Neu5Ac binding site is known from crystal structures, although it is possible in some cases to predict carbohydrate binding sites (see next section).

Over the years several crystal structures of Siglec-7 have become available, demonstrating that the sialic-acid-binding V-set domain is highly dynamic [78, 80–82].

Two of the six currently available crystal structures are shown in Fig. 1, an *apo* structure (pdb code 1O7V) and the structure of a complex with a synthetic oligosaccharide containing GT1b (pdb code 2HRL). The structure 2HRL has the terminal Neu5NAc in a position that is very similar to the binding modes found for other Siglecs. Therefore it can be assumed that this binding mode determines the specificity and has to be regarded as a reference for judging the poses of modified sialic-acids generated by docking. Neu5NAc interacts with side chains of Arg124, Trp132, and Tyr26 and forms hydrogen bonds with the backbone atoms of Lys131 and Asn133. Additionally there is a water bridge between Neu5NAc(O4) and Asn129(O). Mutation studies revealed that Arg124 plays a key role for binding by establishing a salt-bridge to the carboxylate group of the terminal Neu5NAc moiety.

With respect to the application of docking methods it should be noted that in the *apo* structure Arg124 is not in an orientation suitable for establishing the salt-bridge; therefore, docking of Neu5Ac to the *apo* structure would not result in the correct pose for Neu5Ac. The two structures also differ significantly in the conformations of the C–C′ and C′–D loops. Most importantly the trimethylsilyl group at the reducing end of the 'GT1b ligand' is located in a cavity that does not exist in the *apo* structure. Since it is unlikely that the ligand has pushed out Trp85 to form the cavity, it can be assumed that the cavity also is formed/is present in a fraction of the structures in the conformational ensemble of Siglec-7 present in solution. Therefore Siglec-7 needs to be seen as a highly dynamic receptor with cavities formed and disappearing and it is unclear at this stage what other conformations of the receptor exist in solution and which of them are relevant for inhibitor design. Closer inspection of the 2HRL structure also reveals that OH groups 2, 4, 7, and 9 of Neu5Ac are accessible for attachment of additional chemical groups. Additionally the *N*-acetyl group at position 5 could be modified. The goal is now to find chemical groups that fit sterically and would be able to establish additional favorable interactions. In principle, this could be automated by using dedicated computer software and a chemical fragment library, but not all combinations that are theoretically fitting are synthetically feasible. Alternatively a synthetic or medicinal chemist could suggest a set of synthetically feasible compounds that could be checked *in silico* with respect to their 3D fit or predicted affinity prior to a costly chemical synthesis and biological testing.

The most basic requirement for the application of docking methods for the project outlined is that there is docking software (and a scoring function) available that is able to reproduce reliably the binding mode of Neu5Ac in 2HTY. 200 GA-LS runs were therefore performed using AutoDock 3.05 and AutoDock 4.2.5.1 (grid size = 64 × 64 × 64, grid resolution 0.375, 10^7 energy evaluations). Kollman charges were used for the receptor atoms in AutoDock 3.05. Otherwise Gasteiger charges were assigned to the receptor and ligand atoms using AutoDock Tools. The docked poses are shown in Fig. 2a, b. AutoDock Vina 1.1.2 was also tested (box dimensions 30 × 30 × 30 Å^3, exhaustiveness = 100) using the input structures generated for AutoDock 4. The maximum number of poses that Vina outputs per run is 20. In principle it is also possible to generate 200 poses by running Vina 10 times with a different random seed number, but the correct pose was contained already in the set generated by the first run (Fig. 2c). AutoDock 3.05 reproduced the Neu5NAc binding mode very well, the pose with the best score has an RMSD of about 1 Å. For Vina the correct pose has rank four, but all the 20 poses have very similar scores in the range of 0.5 kcal/mol. The pose with the best score generated by AutoDock 4.2.1 is shown in Fig. 3d. Although the pose is close to the Neu5Ac binding site, it does not reproduce the correct binding mode. The pose is reasonable – in principle – with respect to the intermolecular interaction pattern; however, the conformation of the Neu5NAc is very unrealistic. Cleary AutoDock 4 (and to some extent also Vina) seems to have difficulties reproducing the experimentally known conformation of Neu5NAc; therefore, the program is probably not very well suited for this system. However the results of AutoDock 3.05 are very satisfying.

Next a Neu5NAc modified in 9-position – the sialoside inhibitor 'oxamido-Neu5NAc' [methyl α-9-(amino-oxalyl-amino)-9-deoxy-Neu5NAc] – was docked

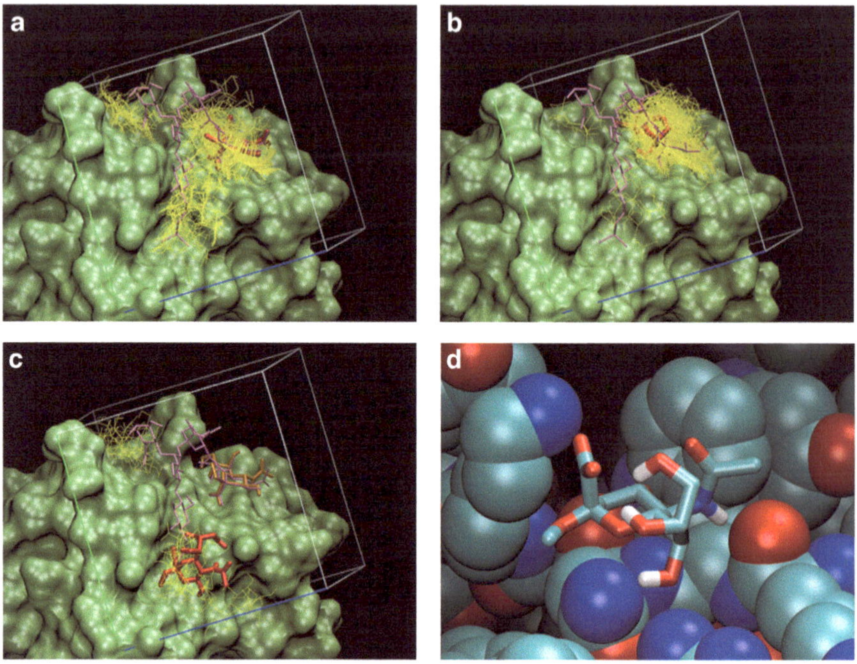

Fig. 2 Docking of Neu5NAc to Siglec-7. The GT1b ligand is shown in *purple* as a reference. The poses generated are shown in *yellow* and the pose with the best score is shown in *red*. (**a**) AutoDock 3; (**b**) AutoDock 4; (**c**) AutoDock Vina (the correct pose has rank four and is shown in *orange*); (**d**) Pose with best score from AutoDock 4 which has an unrealistic conformation of Neu5NAc (carbon = *cyan*, oxygen = *red*, nitrogen = *blue*)

to 2HRL using AutoDock 3.05 and the best pose is shown in Fig. 3a. It is predicted that the oxamido group would extend the glycerol side chain of Neu5NAc straightforward and be located between Lys135 and Trp132. However the crystal structure (pdb code 2G5R) reveals that the side chain of Lys135 has changed the orientation and there is a slight kink between the glycerol and the oxamido group (Fig. 3b). Based on the inhibition constants predicted by AutoDock the oxamido-Neu5Ac compound should have about 6-fold greater inhibitory activity than Neu5Ac. In hapten inhibition assays it was found that it is at least 3-fold [78].

In summary docking methods can predict binding modes of modified sialic-acids and also can make reasonable predictions of the relative binding affinity. However it is very important to realize that the results have to be evaluated carefully. Not all poses with a high score are realistic and also the differences in the scores between 'correct' and 'wrong' poses are often within the error range of scoring functions. It needs to be emphasized again that it currently requires knowledge about carbohydrate conformation in order to recognize and filter out unlikely conformations since current docking software has significant shortcomings in this respect. If docking methods are applied for screening of small libraries of modified carbohydrates one should not only output the pose with the best score for each compound

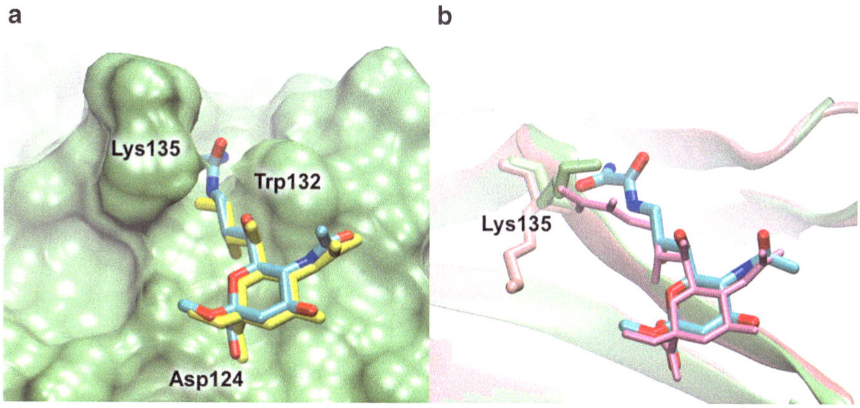

Fig. 3 (**a**) Docking of oxamido-Neu5NAc to receptor structure 2HRL (*green*). The correct Neu5NAc binding mode is shown as reference in *yellow*. (**b**) Redocking of oxamido-Neu5NAc to receptor structure 2G5R (*pink*). The ligand of the crystal structure is shown in *purple*. The docked structures are colored by element (carbon = *cyan*, oxygen = *red*, nitrogen = *blue*)

to the results list. It is definitely required to output a larger number of poses and perform post-processing of the results; e.g. remove all poses that do not have the Neu5Ac in the correct binding mode (which could be done, for example, using Conformational Analysis Tools (www.md-simulations.de/CAT/)).

2.4 *Predicting Carbohydrate Binding Sites: A Practical Example*

If a crystal structure of the carbohydrate-binding protein is available but the binding site of a carbohydrate is unknown, ‘blind docking’ [83] approaches might be useful to propose possible binding sites, which could be confirmed by side-directed mutagenesis experiments. In principle, the method is similar to a normal docking run, except that the whole protein surface is used as search space. Since protein surfaces are usually very large, this requires adjustments with respect to the exhaustiveness of the search. In case of AutoDock, which is frequently used for blind docking, this can be done by increasing the number of energy evaluations and/or the number of docking runs significantly.

An illustrative example that highlights the possibilities is presented here with the ‘blind docking’ of oseltamivir (‘Tamiflu’) to the tetrameric structure of influenza N1 neuraminidase (pdb code 2HTY) [84] using AutoDock 3.05 [85]. Scientific background information on drug design against influenza virus can be found in several excellent reviews [86–88].

The docking procedure begins with the preparation of the receptor. The crystal structure 2HTY can be downloaded from the Protein Data Bank (www.pdb.org)

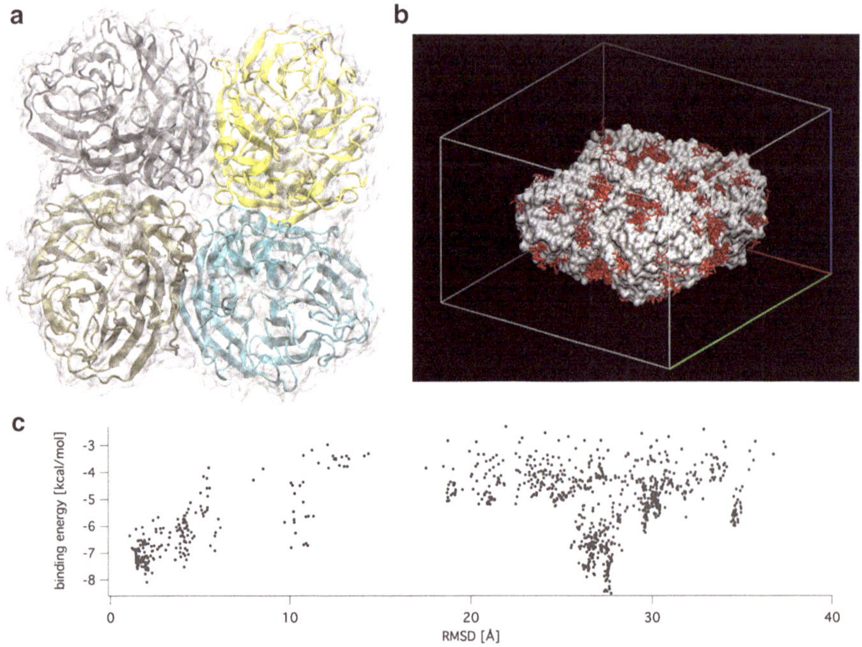

Fig. 4 'Blind docking' of oseltamivir to the influenza N1 neuramidase tetramer. (**a**) Overview of the crystal structure. The amino acid chains are shown as *cartoons*, the protein surface is indicated in (transparent) *white*. (**b**) Grid box defining the search space for 'blind docking.' The protein surface is shown in *white* and the 1000 docked poses are shown in *red*. (C) Free energy of binding calculated by AutoDock for each pose. The RMSD values are calculated relative to the nearest oseltamivir ligand taken from PDB entry 2HU4

[89, 90]. Since it contains two tetramer units, one of them needs to be extracted (Fig. 4a). This can be done by processing the PDB file with either a simple text editor or a dedicated molecular modeling software. The initial coordinates of small molecules (e.g., oseltamivir) can be built using stand-alone molecular modeling software, molecular builders on the Internet (e.g., http://www.molecular-networks.com/online_demos/corina_demo) or downloaded from databases such as PDBeChem (http://www.ebi.ac.uk/pdbe-srv/pdbechem/), the ligand database of the PDB [91]. Here, oseltamivir was extracted from PDB entry 2HU4. Depending on the source of the coordinates it may be required to optimize the initial structure using a force field. For this task a molecular modeling software is required. Finally, AutoDockTools is very useful for the setup of AutoDock runs [45, 92]. There are many tutorials available on the Internet describing how to perform AutoDock dockings step-by-step.

The size of the search space is defined by the affinity grids, which are calculated with the AutoGrid program. The number of grid points is limited to 128 in each dimension. Using the default resolution in AutoDock, which is 0.375 Å, the maximum box size possible would be approx. $47 \times 47 \times 47$ Å^3. However, the

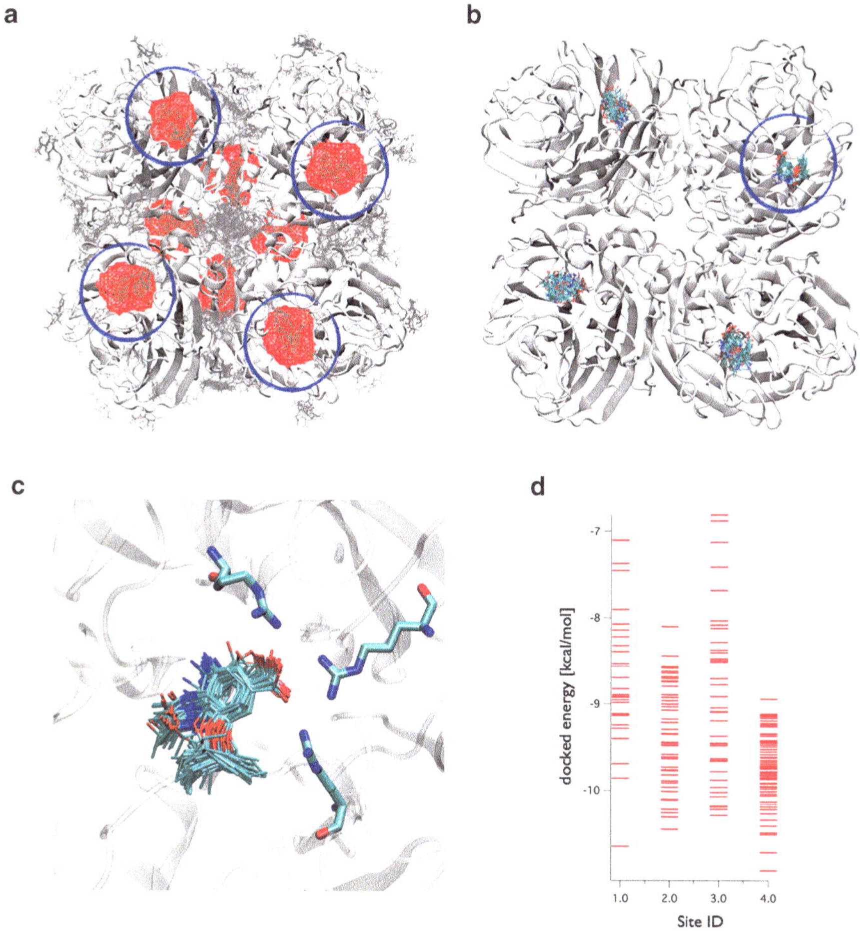

Fig. 5 'Blind docking' of oseltamivir to the influenza N1 neuramidase tetramer. (**a**) Predicted high affinity sites. The four sites of interest are *encircled*. (**b**) Extracted poses located in the four sites of interest. The site with the ordered poses (site 4) is *encircled*. (**c**) Close view of site 4. (**d**) Distribution of the docked energy in the four sites

diameter of the N1 tetramer is about 105 Å; therefore, the resolution needs to be reduced significantly. In this case a grid with 128x128x80 points and a resolution of 0.9 Å were used (Fig. 4b). One thousand Lamarckian genetic algorithm-local search (GA-LS) runs (with 10^7 energy evaluations each) were performed which takes about 23 CPU hours on an Intel Xeon processor (X5690 3.47GHz). The docking was distributed over 10 CPUs each performing 100 GA-LS runs in order to shorten the required wall-clock time. The log files were merged and processed with CAT. The 1,000 docked structures (poses) and their estimated free energy of binding are shown in Fig. 4b, c.

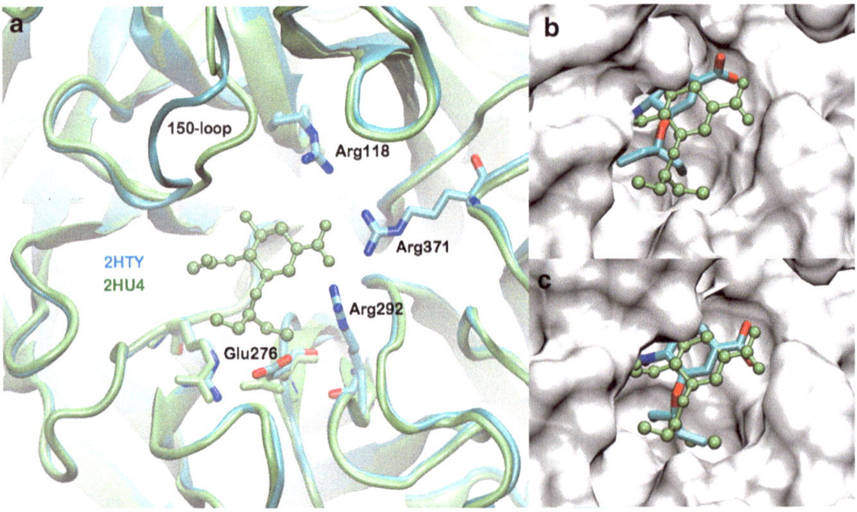

Fig. 6 (**a**) Crystal structures of N1 neuraminidase with (*green*) and without (*cyan*) oseltamivir in the binding site (pdb codes 2HTYand 2HU4). (**b**) Best docked pose in 2HTY. (**c**) Best docked pose in 2HU4

The poses are distributed over the complete surface in patches (Fig. 4b). One possibility would be to cluster the poses based on RMSD, but an alternative is to use 'docking score' iso-surfaces – calculated with CAT – to detect efficiently binding sites with high affinity [93] (Fig. 5a). Eight sites are clearly visible, but four are located in a region that is not accessible in the full virus, therefore only the four remaining (encircled in Fig. 5a) have to be considered. About 20% of the docked poses are located within these four sites and are extracted using the 'hotspot' filter of CAT (Fig. 5b). The poses in one of the sites (ID = 4) are highly ordered and close inspection reveals that the carboxylate group interacts with three arginines (Fig. 5c). The docked energy in this site spans a range of 2 kcal/mol [−10.9, −8.9] (Fig. 5d), which is within the expected error range of a docking score. That the docked poses in the other sites (1 to 3) – despite being essentially identical – are not as ordered as in site 4 may be explained by different shapes of the affinity grids due to the low resolution.

It has to be noted that 2HTY is an *apo* protein structure, which was derived without a ligand present in the binding site. Fortunately there is also a crystal structure available with oseltamivir in the binding site (pdb code 2HU4, Fig. 6a) which can be used to validate the predictions. It can be seen that the *apo* and the ligand bound conformation of the protein differ quite significantly in some regions (e.g., 150-loop and Glu276) and it is therefore remarkable that the 'blind docking' predicted all four sites and also the orientation of the ligand in one site with reasonable accuracy. In general, available crystal structures are often only a subset of a biologically functional molecular assembly. The pose with the overall best score is located in one of the sites that are most likely not accessible in the complete virus. Therefore taking into account only the pose with the highest score would have been misleading. Consequently in a 'blind docking' all the predicted binding sites need to be critically evaluated in terms of accessibility of the site. It also has to be also kept in mind that the available receptor structure might not be in a conformation perfectly shaped to accommodate the ligand. Although this is the case for 2HTY, the ligand with the best 'site 4'-score (shown in Fig. 6b) is similar to the position of the same ligand in 2HU4. The difference can be explained by the different orientation of Glu276. A 'blind docking' of oseltamivir to a N1 tetramer taken from the PDB structure 2HU4 predicted all four sites with highest scores (data not shown) and the pose with the overall best score reproduced the pose from the crystal structure with high accuracy (Fig. 6c).

In principle it is possible to take into account also receptor flexibility in 'blind docking'; however, the calculation times become very long [93]. It also has to be realized that normally there is no structure available that can be used for validation (as 2HU4 in the example presented) and in principle it cannot always be assumed that the site with the lowest score is the correct one. Therefore experimental validation (e.g., with side directed mutagenesis) of each predicted site is required.

3 Summary, Conclusions, Outlook

Docking methods are a valuable tool for the prediction of carbohydrate binding sites and the design of carbohydrate-based drugs. However, there are different limitations in the application of docking methods. One of the major fundamental shortcomings is that scoring functions are either not very accurate or not generally applicable. Another important aspect is that – even if the perfect scoring function would exist – one should be aware that the 'correct pose' might not have been found, simply because the search space has not been sufficiently explored due to calculation time restraints. With respect to docking of carbohydrates special attention should be paid to checking the conformation of the docked poses. However, despite all this limitations: If the method is applied correctly and the results are evaluated critically by an experienced computational chemist and finally 'scored' with medicinal chemists knowledge, docking methods are a very useful tool for the early phases of a drug design project.

Currently the applications of the method are mainly focused on the docking of a few carbohydrate compounds. Virtual screening approaches are difficult due to the lack of significantly large compound libraries of (modified) carbohydrates. The field will significantly benefit from new developments to include 'bridging waters' and receptor flexibility as well as from new scoring functions that take into account more accurately the conformational energy of carbohydrates.

References

1. Schmidt T, Bergner A, Schwede T (2013) Modelling three-dimensional protein structures for applications in drug design. Drug Discov Today 1–8. doi:10.1016/j.drudis.2013.10.027
2. Pavlopoulou A, Michalopoulos I (2011) State-of-the-art bioinformatics protein structure prediction tools (Review). Int J Mol Med 28:295–310. doi:10.3892/ijmm.2011.705
3. Forster MJ (2002) Molecular modelling in structural biology. Micron 33:365–384
4. Roldós V, Cañada FJ, Jiménez-Barbero J (2011) Carbohydrate–protein interactions: a 3D view by NMR. ChemBioChem 12:990–1005. doi:10.1002/cbic.201000705
5. van Gunsteren WF, Bakowies D, Baron R et al (2006) Biomolecular modeling: goals, problems, perspectives. Angew Chem Int Ed 45:4064–4092. doi:10.1002/anie.200502655
6. Huang S-Y, Zou X (2010) Advances and challenges in protein-ligand docking. IJMS 11: 3016–3034. doi:10.3390/ijms11083016
7. Cheng T, Li Q, Zhou Z et al (2012) Structure-based virtual screening for drug discovery: a problem-centric review. AAPS J 14:133–141. doi:10.1208/s12248-012-9322-0
8. Scior T, Bender A, Tresadern G et al (2012) Recognizing pitfalls in virtual screening: a critical review. J Chem Inf Model 52:867–881. doi:10.1021/ci200528d
9. Vogt M, Bajorath J (2012) Chemoinformatics: a view of the field and current trends in method development. Bioorg Med Chem 20:5317–5323. doi:10.1016/j.bmc.2012.03.030
10. Chen C, Huang H, Wu CH (2011) Protein bioinformatics databases and resources. Methods Mol Biol 694:3–24. doi:10.1007/978-1-60761-977-2_1
11. Aoki-Kinoshita KF (2013) Using databases and web resources for glycomics research. Mol Cell Proteomics 12:1036–1045. doi:10.1074/mcp.R112.026252
12. Nicola G, Liu T, Gilson MK (2012) Public domain databases for medicinal chemistry. J Med Chem 55:6987–7002. doi:10.1021/jm300501t
13. von der Lieth C-W, Bohne-Lang A, Lohmann KK, Frank M (2004) Bioinformatics for glycomics: status, methods, requirements and perspectives. Briefings Bioinformat 5:164–178
14. Frank M, Schloissnig S (2010) Bioinformatics and molecular modeling in glycobiology. Cell Mol Life Sci 67:2749–2772. doi:10.1007/s00018-010-0352-4
15. Ranzinger R, Maass K, Lutteke T (2011) Bioinformatics databases and applications available for glycobiology and glycomics. Functional and Structural Proteomics of Glycoproteins 59–90. doi:10.1007/978-90-481-9355-4_3
16. von der Lieth C-W, Luetteke T, Frank M (2010) Bioinformatics for glycobiology and glycomics. Wiley, Chichester
17. Aoki-Kinoshita KF (2008) Using glycome databases for drug discovery. Expert Opin Drug Discov 3:877–890
18. Perez S, Mulloy B (2005) Prospects for glycoinformatics. Curr Opin Struct Biol 15:517–524. doi:10.1016/j.sbi.2005.08.005
19. Neumann D, Lehr C-M, Lenhof H-P, Kohlbacher O (2004) Computational modeling of the sugar–lectin interaction. Adv Drug Delivery Rev 56:437–457. doi:10.1016/j.addr.2003.10.019
20. Frank M, Lutteke T, von der Lieth CW (2007) GlycoMapsDB: a database of the accessible conformational space of glycosidic linkages. Nucleic Acids Res 35:287–290. doi:10.1093/nar/gkl907
21. Demarco ML, Woods RJ (2008) Structural glycobiology: a game of snakes and ladders. Glycobiology 18:426–440. doi:10.1093/glycob/cwn026
22. Foley BL, Tessier MB, Woods RJ (2011) Carbohydrate force fields. Wiley Interdisciplinary Reviews. Computational Mol Sci 2:652–697. doi:10.1002/wcms.89
23. Götz AW, Williamson MJ, Xu D et al (2012) Routine microsecond molecular dynamics simulations with AMBER on GPUs. 1. Generalized born. J Chem Theory Comput 8: 1542–1555. doi:10.1021/ct200909j
24. Sattelle BM, Almond A (2014) Carbohydrate research. Carbohydr Res 383:34–42. doi:10.1016/j.carres.2013.10.011

25. Lawrenz M, Baron R, Wang Y, McCammon JA (2012) Independent-trajectory thermodynamic integration: a practical guide to protein–drug binding free energy calculations using distributed computing. Methods Mol Biol 819:469–486. doi:10.1007/978-1-61779-465-0_27
26. Radmer RJ, Kollman PA (1997) Free energy calculation methods: a theoretical and empirical comparison of numerical errors and a new method qualitative estimates of free energy changes. J Comput Chem 18:902–919
27. Gutiérrez-de-Terán H, Åqvist J (2012) Linear interaction energy: method and applications in drug design. Methods Mol Biol 819:305–323. doi:10.1007/978-1-61779-465-0_20
28. Massova I, Kollman PA (2000) Combined molecular mechanical and continuum solvent approach (MM-PBSA/GBSA) to predict ligand binding. Perspect Drug Discovery Design 18:113–135
29. Ritchie TJ, McLay IM (2012) Should medicinal chemists do molecular modelling? Drug Discovery Today 17:534–537. doi:10.1016/j.drudis.2012.01.005
30. Schwardt O, Kelm S, Ernst B (2013) SIGLEC-4 (MAG) antagonists: from the natural carbohydrate epitope to glycomimetics. Topics Curr Chem. doi:10.1007/128_2013_498
31. Nycholat CM, Rademacher C, Kawasaki N, Paulson JC (2012) In silico-aided design of a glycan ligand of sialoadhesin for in vivo targeting of macrophages. J Am Chem Soc 134: 15696–15699. doi:10.1021/ja307501e
32. Landström J, Persson K, Rademacher C et al (2012) Small molecules containing hetero-bicyclic ring systems compete with UDP-Glc for binding to WaaG glycosyltransferase. Glycoconjugate J 29:491–502. doi:10.1007/s10719-012-9411-4
33. Bhunia A, Vivekanandan S, Eckert T et al (2010) Why structurally different cyclic peptides can be glycomimetics of the HNK-1 carbohydrate antigen. J Am Chem Soc 132:96–105. doi:10.1021/ja904334s
34. Rudrawar S, Dyason JC, Rameix-Welti M-A et al (2010) Novel sialic acid derivatives lock open the 150-loop of an influenza A virus group-1 sialidase. Nat Commun 1:113–117. doi:10.1038/ncomms1114
35. Neres J, Brewer ML, Ratier L et al (2009) Discovery of novel inhibitors of *Trypanosoma cruzi* trans-sialidase from in silico screening. Bioorg Med Chem Lett 19:589–596. doi:10.1016/j.bmcl.2008.12.065
36. Welch KT, Turner TA, Preast CE (2008) Rational design of novel glycomimetics: inhibitors of concanavalin A. Bioorg Med Chem Lett 18:6573–6575. doi:10.1016/j.bmcl.2008.09.095
37. Kuntz ID, Blaney JM, Oatley SJ et al (1982) A geometric approach to macromolecule-ligand interactions. J Mol Biol 161:269–288. doi:10.1016/0022-2836(82)90153-X
38. Kubinyi H (2003) Drug research: myths, hype and reality. Nat Rev Drug Discov 2:665–668. doi:10.1038/nrd1156
39. Kolb P, Ferreira RS, Irwin JJ, Shoichet BK (2009) Docking and chemoinformatic screens for new ligands and targets. Curr Opin Biotechnol 20:429–436. doi:10.1016/j.copbio.2009.08.003
40. Bortolato A, Fanton M, Mason JS, Moro S (2013) Molecular docking methodologies. Methods Mol Biol 924:339–360. doi:10.1007/978-1-62703-017-5_13
41. Sapay N, Nurisso A, Imberty A (2013) Simulation of carbohydrates, from molecular docking to dynamics in water. Methods Mol Biol 924:469–483. doi:10.1007/978-1-62703-017-5_18
42. Kitchen DB, Decornez H, Furr JR, Bajorath J (2004) Docking and scoring in virtual screening for drug discovery: methods and applications. Nat Rev Drug Discov 3:935–949. doi:10.1038/nrd1549
43. Krovat EM, T S, Langer T (2005) Recent advances in docking and scoring. Curr Comput Aided Drug Des 1:93–102.
44. Laederach A, Reilly PJ (2005) Modeling protein recognition of carbohydrates. Proteins 60: 591–597. doi:10.1002/prot.20545
45. Morris GM, Huey R, Lindstrom W et al (2009) AutoDock4 and AutoDockTools4: automated docking with selective receptor flexibility. J Comput Chem 30:2785–2791. doi:10.1002/jcc.21256

46. Huang S-Y, Grinter SZ, Zou X (2010) Scoring functions and their evaluation methods for protein–ligand docking: recent advances and future directions. Phys Chem Chem Phys 12: 12899. doi:10.1039/c0cp00151a
47. Warren GL, Andrews CW, Capelli A-M et al (2006) A critical assessment of docking programs and scoring functions. J Med Chem 49:5912–5931. doi:10.1021/jm050362n
48. Leach AR, Shoichet BK, Peishoff CE (2006) Prediction of protein–ligand interactions. Docking and scoring: successes and gaps. J Med Chem 49:5851–5855. doi:10.1021/jm060999m
49. Marsden PM, Puvanendrampillai D, Mitchell JBO, Glen RC (2004) Predicting protein-ligand binding affinities: a low scoring game? Org Biomol Chem 2:3267–3273. doi:10.1039/B409570G
50. Koppisetty CAK, Frank M, Kemp GJL, Nyholm P-G (2013) Computation of binding energies including their enthalpy and entropy components for protein–ligand complexes using support vector machines. J Chem Inf Model 53:2559–2570. doi:10.1021/ci400321r
51. Guimarães CRW (2012) MM-GB/SA rescoring of docking poses. Methods Mol Biol 819: 255–268. doi:10.1007/978-1-61779-465-0_17
52. Sinko W, Lindert S, McCammon JA (2012) Accounting for receptor flexibility and enhanced sampling methods in computer-aided drug design. Chem Biol Drug Design 81:41–49. doi:10.1111/cbdd.12051
53. Roberts BC, Mancera RL (2008) Ligand-protein docking with water molecules. J Chem Inf Model 48:397–408. doi:10.1021/ci700285e
54. Huang N, Shoichet BK (2008) Exploiting ordered waters in molecular docking. J Med Chem 51:4862–4865. doi:10.1021/jm8006239
55. van Dijk ADJ, Bonvin AMJJ (2006) Solvated docking: introducing water into the modelling of biomolecular complexes. Bioinformatics 22:2340–2347. doi:10.1093/bioinformatics/btl395
56. Trott O, Olson AJ (2010) AutoDock Vina: improving the speed and accuracy of docking with a new scoring function, efficient optimization, and multithreading. J Comput Chem 31:455–461
57. Friesner RA, Banks JL, Murphy RB et al (2004) Glide: a new approach for rapid, accurate docking and scoring. 1. Method and assessment of docking accuracy. J Med Chem 47:1739–1749. doi:10.1021/jm0306430
58. Nivedha AK, Makeneni S, Foley BL, et al (2013) Importance of ligand conformational energies in carbohydrate docking: Sorting the wheat from the chaff. J Comput Chem n/a–n/a. doi:10.1002/jcc.23517
59. Neu U, Allen S-AA, Blaum BS et al (2013) A structure-guided mutation in the major capsid protein retargets BK polyomavirus. PLoS Pathog 9:e1003688. doi:10.1371/journal.ppat.1003688.s003
60. Nasir W, Frank M, Koppisetty CAK et al (2012) Lewis histo-blood group α1,3/α1,4 fucose residues may both mediate binding to GII.4 noroviruses. Glycobiology 22:1163–1172. doi:10.1093/glycob/cws084
61. Schulz EC, Schwarzer D, Frank M et al (2010) Structural basis for the recognition and cleavage of polysialic acid by the bacteriophage K1F tailspike protein EndoNF. J Mol Biol 397:341–351. doi:10.1016/j.jmb.2010.01.028
62. Laederach A, Reilly PJ (2003) Specific empirical free energy function for automated docking of carbohydrates to proteins. J Comput Chem 24:1748–1757. doi:10.1002/jcc.10288
63. Hill AD, Reilly PJ (2008) A Gibbs free energy correlation for automated docking of carbohydrates. J Comput Chem 29:1131–1141. doi:10.1002/jcc.20873
64. Mishra SK, Adam J, WimmerovA M, Koča J (2012) In silico mutagenesis and docking study of Ralstonia solanacearum RSL lectin: performance of docking software to predict saccharide binding. J Chem Inf Model 52:1250–1261. doi:10.1021/ci200529n
65. Nurisso A, Kozmon S, Imberty A (2008) Comparison of docking methods for carbohydrate binding in calcium-dependent lectins and prediction of the carbohydrate binding mode to sea cucumber lectin CEL-III. Mol Simulation 34:469–479

66. Agostino M, Jene C, Boyle T et al (2009) Molecular docking of carbohydrate ligands to antibodies: structural validation against crystal structures. J Chem Inf Model 49:2749–2760. doi:10.1021/ci900388a
67. Birch L, Murray CW, Hartshorn MJ et al (2002) Sensitivity of molecular docking to induced fit effects in influenza virus neuraminidase. J Comput Aided Mol Des 16:855–869
68. Korb O, Olsson TSG, Bowden SJ et al (2012) Potential and limitations of ensemble docking. J Chem Inf Model 52:1262–1274. doi:10.1021/ci2005934
69. Repasky MP, Murphy RB, Banks JL et al (2012) Docking performance of the glide program as evaluated on the Astex and DUD datasets: a complete set of glide SP results and selected results for a new scoring function integrating WaterMap and glide. J Comput Aided Mol Des. doi:10.1007/s10822-012-9575-9
70. Gauto DF, Petruk AA, Modenutti CP et al (2012) Solvent structure improves docking prediction in lectin-carbohydrate complexes. Glycobiology 23:241–258. doi:10.1093/glycob/cws147
71. Ross GA, Morris GM, Biggin PC (2012) Rapid and accurate prediction and scoring of water molecules in protein binding sites. PLoS ONE 7:e32036. doi:10.1371/journal.pone.0032036.t006
72. García-Sosa AT, Mancera RL, Dean PM (2003) WaterScore: a novel method for distinguishing between bound and displaceable water molecules in the crystal structure of the binding site of protein-ligand complexes. J Mol Model 9:172–182. doi:10.1007/s00894-003-0129-x
73. de Beer SBA, Vermeulen NPE, Oostenbrink C (2010) The role of water molecules in computational drug design. Curr Top Med Chem 10:55–66
74. Wang L, Berne BJ, Friesner RA (2011) Ligand binding to protein-binding pockets with wet and dry regions. Proc Natl Acad Sci U S A 108:1326–1330. doi:10.1073/pnas.1016793108/-/DCSupplemental
75. Thilagavathi R, Mancera RL (2010) Ligand-protein cross-docking with water molecules. J Chem Inf Model 50:415–421. doi:10.1021/ci900345h
76. Crocker PR, Paulson JC, Varki A (2007) Siglecs and their roles in the immune system. Nat Rev Immunol 7:255–266. doi:10.1038/nri2056
77. Angata T, Varki A (2000) Siglec-7: a sialic acid-binding lectin of the immunoglobulin superfamily. Glycobiology 10:431–438
78. Attrill H, Takazawa H, Witt S et al (2006) The structure of siglec-7 in complex with sialosides: leads for rational structure-based inhibitor design. Biochem J 397:271–278. doi:10.1042/BJ20060103
79. Rillahan CD, Schwartz E, Rademacher C et al (2013) On-chip synthesis and screening of a Sialoside library yields a high affinity ligand for Siglec-7. ACS Chem Biol 8: 130424130301009. doi:10.1021/cb400125w
80. Alphey MS, Attrill H, Crocker PR, van Aalten DMF (2003) High Resolution Crystal Structures of Siglec-7. Insights into ligand specificity in the SIGLEC family. J Biol Chem 278:3372–3377. doi:10.1074/jbc.M210602200
81. Dimasi N, Moretta A, Moretta L et al (2004) Structure of the saccharide-binding domain of the human natural killer cell inhibitory receptor p75/AIRM1. Acta Crystallogr Sect D Biol Crystallogr 60:401–403. doi:10.1107/S0907444903028439
82. Attrill H, Imamura A, Sharma RS et al (2006) Siglec-7 undergoes a major conformational change when complexed with the α (2, 8)-disialylganglioside GT1b. J Biol Chem 281: 32774–32783
83. Hetényi C, van der Spoel D (2002) Efficient docking of peptides to proteins without prior knowledge of the binding site. Protein Sci 11:1729–1737. doi:10.1110/ps.0202302
84. Russell RJ, Haire LF, Stevens DJ et al (2006) The structure of H5N1 avian influenza neuraminidase suggests new opportunities for drug design. Nature 443:45–49. doi:10.1038/nature05114

85. Morris GM, Goodsell DS, Halliday RS et al (1998) Automated docking using a Lamarckian genetic algorithm and an empirical binding free energy function. J Comput Chem 19: 1639–1662
86. von Itzstein M (2007) The war against influenza: discovery and development of sialidase inhibitors. Nat Rev Drug Discov 6:967–974. doi:10.1038/nrd2400
87. Du J, Cross TA, Zhou H-X (2012) Recent progress in structure-based anti-influenza drug design. Drug Discovery Today 17:1111–1120. doi:10.1016/j.drudis.2012.06.002
88. von Itzstein M, Thomson R (2009) Anti-influenza drugs: the development of sialidase inhibitors. Antiviral Strategies 111–154
89. Berman H, Henrick K, Nakamura H (2003) Announcing the worldwide Protein Data Bank. Nat Struct Biol 10:980. doi:10.1038/nsb1203-980
90. Berman HM, Westbrook J, Feng Z et al (2000) The Protein Data Bank. Nucleic Acids Res 28: 235–242
91. Dimitropoulos D, Ionides J, Henrick K (2006) Using MSDchem to search the PDB ligand dictionary. Curr Protoc Bioinformat Chap 14, doi:10.1002/0471250953.bi1403s15
92. Morris GM, Huey R, Olson AJ (2008) Using AutoDock for ligand-receptor docking. Curr Protoc Bioinformat Chap. 8, doi: 10.1002/0471250953.bi0814s24
93. Voss C, Eyol E, Frank M et al (2006) Identification and characterization of riproximin, a new type II ribosome-inactivating protein with antineoplastic activity from *Ximenia americana*. FASEB J 20:1194–1196. doi:10.1096/fj.05-5231fje

Top Med Chem (2014) 12: 73–94
DOI: 10.1007/7355_2014_41

Published online: 4 October 2014

Discovery and Development of Selective Renal Sodium-Dependent Glucose Cotransporter 2 (SGLT2) Dapagliflozin for the Treatment of Type 2 Diabetes

Alan Braem, Prashant P. Deshpande, Bruce A. Ellsworth, and William N. Washburn

Abstract When blood flows through the renal capillaries, glucose is one of the many substances filtered by the kidney. However, glucose is subsequently recovered primarily by the sodium-dependent glucose transporter 2 (SGLT2) as the glomerular filtrate flows down the renal tubules. SGLT2 inhibitors inhibit this transporter leading to the loss of a significant fraction of the filtered glucose. The resulting glucosuria is of sufficient magnitude to reduce diabetes-related hyperglycemia and ameliorate-associated complications of diabetes. A systematic study was conducted to identify superior SGLT2 inhibitors based on a β-*1C*-arylglucoside with substituted diarylmethane moieties. Such compounds are potent and selective SGLT2 inhibitors with metabolic stability that promote glucosuria when administered in vivo. Through this investigation, the β-*1C*-arylglucoside dapagliflozin was identified as a potent and selective *h*SGLT2 inhibitor with an EC_{50} for *h*SGLT2 of 1.0 nM and 1,200-fold selectivity over *h*SGLT1. Dapagliflozin produced glucosuria in normal Sprague Dawley rats in a dose-dependent fashion. Moreover, a 0.1 mg/kg oral dose reduced blood glucose levels by as much as 55% in rats that had been made hyperglycemic by streptozotocin, a pancreatic toxin. These findings, combined with a favorable ADME profile and *vivo* data, led to nomination of dapagliflozin as a drug for the treatment of type 2 diabetes. The structural architecture of β-*1C*-arylglucosides and their amphiphilic nature presented significant obstacles to the synthesis of dapagliflozin and similar candidates for toxicological and clinical testing, prompting the development of a new, safe, efficient, and economical process for the synthesis of C-4′ and C-4 substituted β-*1C*-arylglucosides. A key element of the process was a remarkable discovery of novel crystalline complexes that enabled isolation and quality control.

A. Braem and P.P. Deshpande (✉)
Bristol-Myers Squibb Company, Chemical Development, New Brunswick, NJ 08903, USA
e-mail: prashant.deshpande@bms.com

B.A. Ellsworth and W.N. Washburn
Bristol-Myers Squibb Company, Discovery Chemistry, Hopewell, NJ 08534, USA

Keywords β-*1C*-arylglucosides, Dapagliflozin, Diarylmethanes, Phenylalanine co-crystal, SGLT2, Type 2 diabetes

Contents

Abbreviations

μg	Microgram
μM	Micromolar
Ac_2O	Acetic anhydride
AcOH	Acetic acid
ADME	Absorption, distribution, metabolism, and excretion
AMG	α-Methyl-D-glucopyranoside
AP	Area percent
API	Active pharmaceutical ingredient
$BF_3 \cdot Et_2O$	Boron trifluoride diethyl etherate
Bn	Benzyl
CH_3CN	Acetonitrile
CHO	Chinese hamster ovary
Cl	Clearance
C_{max}	Maximum concentration
dL	Deciliter
DMAP	4-(dimethylamino)pyridine
EC_{50}	Half maximal effective concentration
EtOH	Ethanol
Et_3SiH	Triethylsilane
F	Fraction absorbed/bioavailability
GC	Gas chromatography
GLUT	Glucose transporter
GRAS	Generally regarded as safe
h	Hour(s)
HbA1c	Glycated hemoglobin
HPLC	High performance liquid chromatography
iPr_3SiH	Triisopropylsilane
iPrOAc	Isopropyl acetate
i.v.	Intravenous
Kg	Kilogram
$KK\text{-}A^y$	Mouse of the KK-strain carrying the A^y mutation

L	Liter(s)
mg	Milligram
mg/kg	Milligram per kilogram
mL	Milliliter
mmHg	Millimeters of mercury
MeOH	Methanol
Me	Methyl
MSA	Methanesulfonic acid
n-BuLi	*n*-Butyllithium
nM	Nanomolar
nm/s	Nanometer per second
NaOH	Sodium hydroxide
NMM	*N*-Methyl morpholine
Pd–C	Palladium on carbon
PO	Per os
s	Second(s)
S-PG	S-Propylene glycol
SAR	Structure–activity relationships
SD	Sprague Dawley
SGLT	Sodium-dependent glucose cotransporter
STZ	Streptozotocin, a pancreatic toxin used to induce diabetes
TEA	Triethylamine
TEMPO	(2,2,6,6-Tetramethylpiperidin-1-yl)oxy
THF	Tetrahydrofuran
TMSCl	Chlorotrimethylsilane
T_{max}	Time to reach C_{max}
TMS	Tetramethylsilane
Tol	Toluene
TGA	Therapeutic goods administration
UKPDS	United Kingdom Prospective Diabetes Study
V_{ss}	Apparent volume of distribution at steady state

1 Introduction

The incidence of type 2 diabetes has become an increasing worldwide concern, as the number of patients suffering from type 2 diabetes is projected to increase from approximately 371 million people currently to 552 million by 2030 ([1], www.idf.org/diabetesatlas/5e/Update2012). Type 2 diabetes is characterized by hyperglycemia due to excessive hepatic glucose production, a deficiency in insulin secretion and/or peripheral insulin resistance. As the disease advances, hyperglycemia becomes the major risk factor for diabetes complications, including retinopathy, neuropathy, nephropathy, and macrovascular diseases [2–4]. Due to the progressive nature of the disease, combination therapy is usually necessary to achieve the target glycemic level, thereby necessitating development of alternative agents acting by novel mechanisms to control hyperglycemia [5, 6]. This need is underscored by the United Kingdom Prospective Diabetes Study (UKPDS) findings that at present only 25–50% of type 2 diabetics are effectively treated by current therapies [7, 8].

In healthy individuals, greater than 99% of the plasma glucose that is filtered in the kidney glomerulus is reabsorbed, resulting in less than 1% of the total filtered glucose being excreted in urine [9, 10]. This reabsorption process is mediated by two sodium-dependent glucose cotransporters: SGLT1, a low capacity, high affinity transporter expressed in gut, heart, and kidney [11, 12] and SGLT2, a high capacity, low affinity transporter that is expressed mainly in kidney [13, 14]. It is estimated that 90% of renal glucose reabsorption is facilitated by SGLT2 residing on the surface of the epithelial cells lining the S1 segment of the proximal tubule; the remaining 10% is likely mediated by SGLT1 localized on the more distal S3 segment of the proximal tubule [15–20]. Humans with SGLT1 gene mutations experience glucose-galactose malabsorption, resulting in frequent, watery diarrhea and dehydration when on a glucose diet, confirming that SGLT1 is the major glucose transporter in the small intestine. These individuals present with little or no glucosuria, suggesting that SGLT1 is not the major glucose transporter in the kidney [21, 22]. Persistent renal glucosuria is the sole reported phenotype of humans with SGLT2 gene mutations [23, 24].

Inhibition of SGLT2 has been proposed to aid in the normalization of plasma glucose levels in diabetics by preventing the glucose reabsorption process and promoting glucose excretion in urine [25]. Selective SGLT2 inhibitors would be desirable, since gastrointestinal side effects associated with SGLT1 inhibition would be minimized. This mechanism is expected to be associated with low risk of hypoglycemia, because there would be no interference with the normal counterregulatory mechanisms for glucose regulation.

A systematic study was conducted to identify superior SGLT2 inhibitors that are potent, selective over SGLT1, promote glucosuria when administered in vivo, and exhibit a pharmacokinetic profile compatible with once-daily administration.

2 Results and Discussion

2.1 Discovery

The natural product *O*-glucoside phlorizin (**1**, Fig. 1) is a well-documented, potent glucosuric agent that was subsequently shown to be a nonselective SGLT inhibitor [26]. The finding that chronic subcutaneous administration of **1** reduced plasma glucose levels of diabetic rodents supported this mechanistic approach [27–29]. However, phlorizin itself is not considered to be a suitable drug candidate because of its ability to inhibit SGLT1 and poor metabolic stability due to susceptibility to β-glucosidase-mediated cleavage resulting in release of the aglycone phloretin.

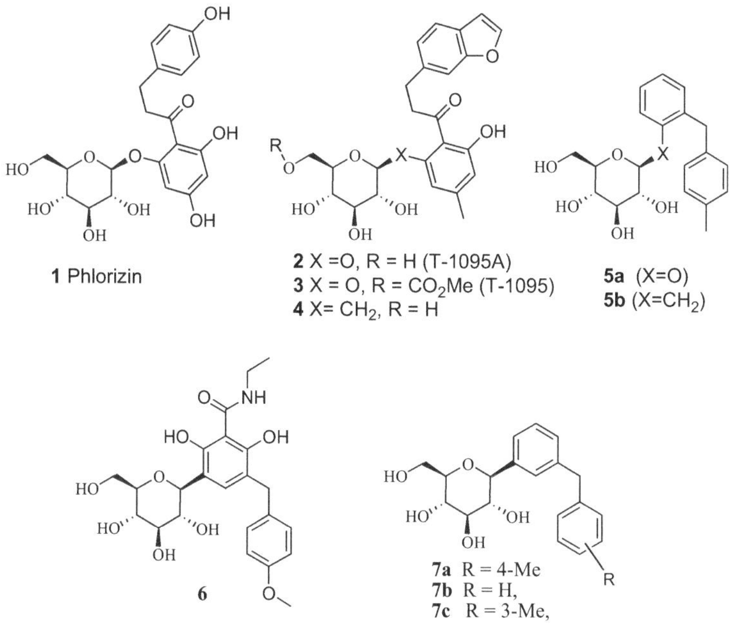

Fig. 1 Structures of some known SGLT inhibitors

In a series of papers, researchers at Tanabe disclosed the structure–activity relationships (SAR) of phlorizin analogs resulting in the identification of selective, potent SGLT2 inhibitors. However, in order to achieve a significant reduction of hyperglycemia with a concurrent increase of glucosuria, the metabolic instability of the *O*-glucoside linkage necessitated oral administration of their lead compound T-1095A (**2**, Fig. 1) to *KK-A*^{y} mice as the methyl carbonate pro-drug T1095 (**3**, Fig. 1) [25, 30–36]. Subsequently, Kissei disclosed two other series of *O*-glucosides containing SGLT2 inhibitors as potential treatment for type 2 diabetes which also required administration as pro-drugs [37–40].

In an attempt to increase the metabolic stability of the glucosyl linkage of *O*-arylglucoside **5a** (Fig. 1), we synthesized *C*-benzylglucoside **5b** (Fig. 1) [41]. This compound displayed a significant loss in SGLT2 activity (75-fold) as compared to compound **5a**. Link et al., reported that a similar modification of **2** to generate the carbon analog **4** produced a >20-fold loss in potency [42]. Together, these findings imply that the isosteric replacement of the oxygen glucosidyl link with a methylene greatly attenuated previously favorable ligand–protein interactions. Possibly, the greater conformational freedom of **4** and **5b** contributed to the reduction in SGLT2 affinity due to absence of the conformational constraints imposed by the *exo*-anomeric effect [43, 44].

Fortuitously, an alternative lead for *C*-glucoside-derived SGLT2 inhibitors surfaced upon characterization of a minor *C*-arylglucoside side-product that was generated during our SGLT2 program [45]. Of particular interest was the *meta* presentation of the glucosyl and benzyl appendages of **6** rather than the typical *ortho* presentation of *O*-glucoside-derived inhibitors. The activity (EC_{50} SGLT2 = 1,300 nM) and selectivity (>sixfold vs SGLT1) of **6** was unexpected especially since the SAR for all *O*-glucosides indicated that substitution of the aglycone with polar substituents markedly reduced SGLT2 activity. In hopes of achieving improved potency, **7a**, the counterpart of a potent *O*-glucoside SGLT2 inhibitor **5a**, was synthesized. The in vitro profile of **7a** was extremely encouraging: SGLT2 EC_{50} = 22 nM; >600-fold selectivity vs SGLT1. The importance of a *para* substituent on the distal ring became readily apparent upon comparison of **7a** to the parent **7b** or *meta* isomer **7c**, thereby underscoring the role of the substituent to properly orient the distal ring to achieve high affinity. A similar bias for *para* substitution of the distal ring had been observed for *O*-arylglucoside analogs of both **5b** and dihydrochalcones reported by Hongu et al. [46].

These findings prompted a systematic study of *meta*-*C*-arylglucosides with varying linkers to evaluate proper placement of the distal aryl ring. The assumption was that high affinity SGLT2 inhibitors require not only the distal aryl ring to bear a lipophilic substituent but also that the distal ring assumes an orientation such that the lipophilic substituent can occupy a favorable binding pocket. Incorporation of a small lipophilic substituent at C-4 of the central aryl ring further augmented SGLT2 potency with little apparent impact on selectivity versus SGLT1, thereby maximizing potency while minimizing the potential for gastrointestinal (GI) side effects mediated by inhibition of intestinal SGLT1. A number of compounds with preferred C-4 substitutions were synthesized and evaluated to further understand the consequences of introduction of a zero, one, two, or three methylene spacer and impact of distal ring substitution with *m*-methyl or *p*-methyl groups.

Table 1 summarizes the structure–activity consequences upon alteration of the spacer moiety between the *C*-glucoside proximal and distal rings. Variation of the spacer from one (**7**) to zero (**8**), two (**9**), or three methylenes (**10**) reduced affinity ~ threefold for the unsubstituted (R = H) and *m*-methyl substituted examples. In contrast, changing the methylene spacer of **7a** from one to zero (**8c**), two (**9c**), or to three methylenes (**10c**) reduced the binding affinity of the *p*-methyl substituted analogs 13-, 19-, and 29-fold, respectively. The unique advantage conferred by the single methylene of **7a** is further confirmed by the respective threefold and 25-fold reduction in affinity upon replacement with a sulfur (**12**) or oxygen (**11**) bridging atom. Significant inhibition of human SGLT1 was not observed for any of the *C*-arylglucosides tested. In particular, the demonstrably high level of selectivity of **7a** is expected to preclude gastrointestinal side effects.

Upon *i.v.* administration to rats and mice at 1 and 0.3 mg/kg, respectively, **7a** produced maximum glucosuric levels of 230 and 600 mg/dL.[1] In contrast, upon

[1] Fasted male Sprague–Dawley rats or Swiss–Webster mice were anesthetized with I.P. Ketamine: Xylazine (0.001 mL/g), an abdominal incision made, and their bladders cannulated with a 16 gauge catheter. Drug was administered intravenously, and urine was collected over 60 min in 10 min intervals. Glucose concentration in urine samples were determined by COBAS MIRA.

Table 1 *C*-Aryglucoside (**I**) SAR exploration of the aglycone spacing element (A) and distal ring substituent (R)[a]

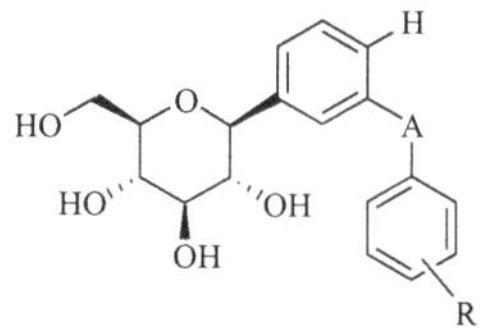

Compounds	A	R	*h*SGLT2 EC_{50} (nM)	Binding Select. vs. *h*SGLT1
7a	CH_2	4-Me	22	>600
7b	CH_2	3-Me	510	ND
7c	CH_2	H	190	>50
8a	bond	H	623	>13
8b	bond	3-Me	1,200	ND
8c	bond	4-Me	290	>30
9a	$(CH_2)_2$	H	710	ND
9b	$(CH_2)_2$	3-Me	970	ND
9c	$(CH_2)_2$	4-Me	430	>20
10a	$(CH_2)_3$	H	480	ND
10b	$(CH_2)_3$	3-Me	1,200	ND
10c	$(CH_2)_3$	4-Me	630	>13
11	O	4-Me	540	>15
12	S	4-Me	69	>100
1			35	10
4			8	350
1[b]			160	1
2[b]			50	4
13	CH2	4-Et	10	1,000

[a]The intracellular accumulation of the SGLT-selective glucose analog [^{14}C]-alpha-methyl glucopyranoside (AMG) by CHO cells expressing the human SGLT2 or the human SGLT1 transporter was quantified in vitro in the presence and absence of inhibitors, using the following conditions: Each inhibitor, dissolved in DMSO, was tested at eight concentrations in the presence of 137 mM NaCl and 10 μM [^{14}C] AMG, over a 120-minute incubation in protein-free buffer. Percent inhibition of transport activity was calculated based on a comparison of the activity of non-inhibited control cells treated with DMSO alone. The response curve was fitted to an empirical four-parameter model to determine the inhibitor concentration at half-maximal inhibition. The reported EC_{50} is the aggregate result of triplicate dose–response determinations.
[b]EC_{50} data from Oku et al. [25]

administration of *O*-glucoside **5** under the same conditions, efficacy in rats was reduced ~ 50-fold relative to that obtained in mice. *C*-Arylglucoside **7a** was found to be ~100-fold more stable in the presence of rat liver microsomes than the corresponding *O*-glucoside **5** [45]. We attribute the greater in vitro stability of **7a** and the diminished variability in glucosuric response across species to **7a** being impervious to glucosidase cleavage (unlike **5**). This profile of the *C*-glucoside **7** to that of **5** reveals greater selectivity versus SGLT1, as well as enhanced metabolic

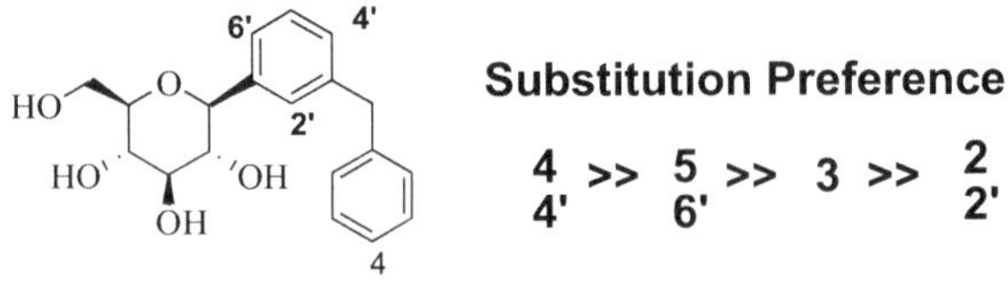

Fig. 2 Structure–activity relationship for diarylmethane *C*-glucoside SGLT2 inhibitors

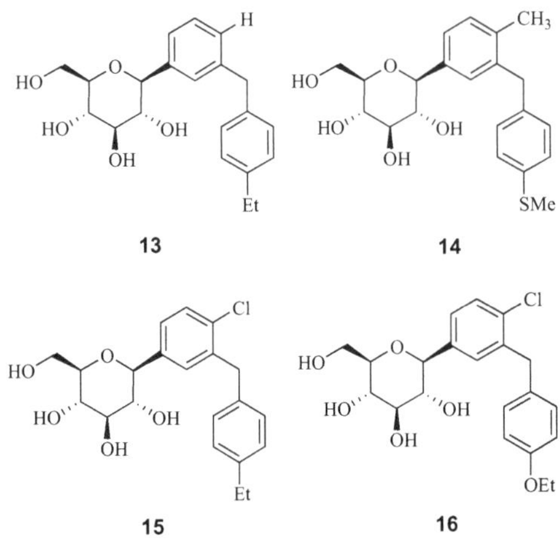

Fig. 3 Structures of advanced SGLT2 inhibitors

stability. *C*-Arylglucosides show enhanced glucosuric activity in rats compared to *O*-arylglucosides that we attribute, in part, to the metabolic stability of the arylglucosyl C–C bond.

SAR exploration revealed *meta*-substituted diarylmethanes to be superior SGLT2 ligands to their biphenyl and 1,2-diarylethane counterparts [41]. Only small, hydrophobic *para* (C-4) substituents such as ethyl, methoxy, thiomethyl, and difluoromethoxy enhanced SGLT2 inhibitory activity resulting in EC_{50} values of ~10 nM, whereas substitution at C-2 or C-3 was not beneficial. Compounds in this series exhibit >1,000-fold selectivity for SGLT2 than SGLT1. Further SAR exploration of the central aryl ring revealed that small lipophilic substituents at the C-4′ position of the central aryl ring further increased SGLT2 affinity such that EC_{50} values decreased to 1 nM. Although substitution at C-5′ or C-6′ modestly improved affinity, C-2′ substitution was deleterious (Fig. 2).

This SAR culminated in the discovery of several selective SGLT2 inhibitors such as compounds **13**, **14**, **15**, **16**, which exhibited properties warranting further progression as clinical candidates for the treatment of type 2 diabetes (Fig. 3).

The in vitro SGLT inhibitory potential (EC_{50}) of **16** and analogs were assessed by monitoring the inhibition of accumulation of radiolabeled α-methyl-D-glucopyranoside (AMG) by CHO cells stably expressing human or rat SGLT2

Table 2 *h*SGLT2 and *h*SGLT1 inhibitory activity for 1, 6, and 16[a]

No.	*h*SGLT2 EC_{50} (nM)	*h*SGLT1 EC_{50} (nM)	Selectivity vs. *h*SGLT1 (fold)
1	35	270	8
2	6	211	30
5a	9	8,000	90
6	500	8,000	16
16	1	1,200	1,200

[a]Assays were performed in protein-free buffer

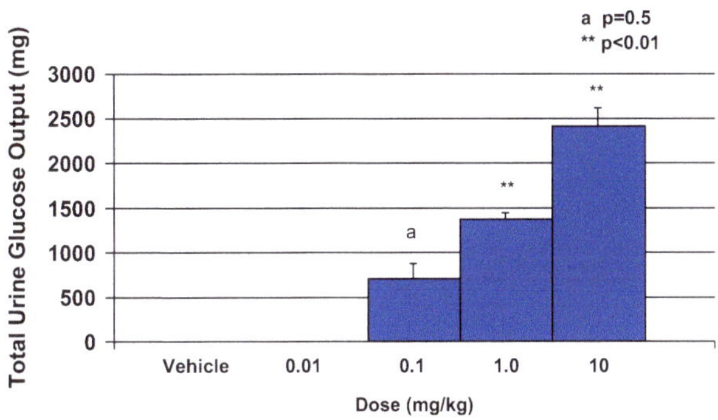

Fig. 4 Mean total urine glucose excretion over 24 h following a single oral dose of dapagliflozin 16 in normal SD rats. $n = 3$. Mean total urine glucose excretion was statistically significant vs. vehicle at the 0.1, 1, and 10 mg/kg dose groups

and SGLT1. As shown in Table 2, EC_{50} values of 1.0 nM for *h*SGLT2 and 1.6 μM for *h*SGLT1 were determined for **16** which corresponded to 1,200-fold selectivity for SGLT2 as compared with eightfold selectivity for phlorizin. The inhibitory potencies of **16** against rat and human SGLT2 were comparable, but the selectivity of **16** for rSGLT2 versus rSGLT1 decreased to 200-fold. At 20 μM, **16** was also found to weakly inhibit (8%) AMG uptake in human adipocyte mediated by GLUT1 and GLUT4 facilitative glucose transporter [47].

Statistically significant dose-dependent glucosuria occurred over a 24-hour period following oral administration of doses from 0.01 to 10 mg/kg of **16** to normal Sprague Dawley rats. This glucosuria was accompanied by dose-dependent increases in urine volume of 16–300% (Fig. 4). In this study, the loss of 700 and 2,400 mg of glucose per rat over 24 h following single oral dose of 0.1 and 10 mg/kg of **16**, respectively, corresponded to a 1,000- and 10,000-fold elevation in glucose disposal relative to vehicle controls. In a separate experiment with streptozotocin-induced (STZ) diabetic rats (starting blood glucose levels of 480–530 mg/dL), a single 0.1 mg/kg oral dose of **16** followed by food restriction produced a 55% reduction in the elevated blood glucose level of treated versus control rats at 5 h after dosing (Fig. 5). This level of efficacy surpassed that of **13** (34% reduction in hyperglycemic blood glucose levels) and matched the efficacy observed for other analogs **14** and **15** (59%, and 62%) at this dose [48, 49].

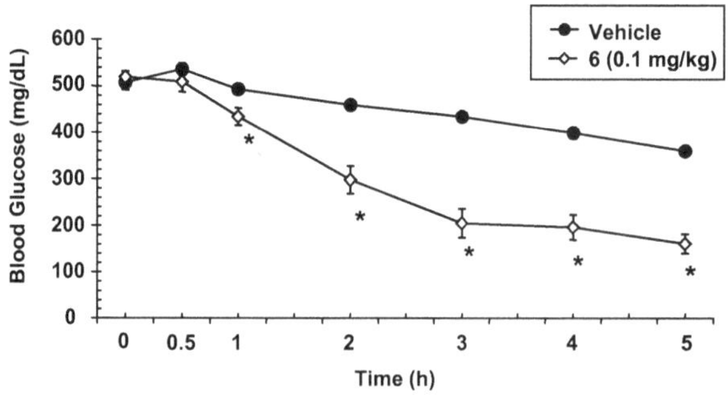

Fig. 5 Mean blood glucose values in streptozotocin-induced (STZ) diabetic SD rats following a single oral dose of 0.1 mg/kg dapagliflozin, 16. $n = 6$. *$p < 0.05$ vs. control group using a paired student *t*-test

Table 3 Pharmacokinetic profile of dapagliflozin **16** in rats

Dose (mg/kg)	1
C_{max} (PO dose, μg/mL)	0.6
T_{max} (PO dose, h)	1.7
$T_{1/2}$ (h)	4.6
F (%)	84
V_{ss} (L/kg)	1.6
Cl (mL/min/kg)	4.8

The above correlation of SGLT2 inhibition, glucosuria, and blood glucose lowering effects suggested that selective SGLT2 inhibition with compounds **13–16** held promise as a viable approach to treat type 2 diabetes.

Dapagliflozin **16** displayed a favorable ADME profile conducive to further development. At 10 μM in serum from rats and humans, the free fraction of **16** was 3 and 4%, respectively. Compound **16** is anticipated to be orally bioavailable in humans based on a high (>150 nm/s) caco-2 cell monolayer permeability value and 84% oral bioavailability in rats (Table 3). The steady-state volume of distribution value (1.6 L/kg) was greater than the total blood volume in rats, indicating that **16** distributed into the extravascular space. Low to intermediate in vitro metabolic rates were observed upon incubation of **16** with liver microsomes and hepatocytes from rats and humans. After oral administration of 1 mg/kg dose of **16** to rats, a C_{max} of 0.6 μg/mL was obtained at 1.7 h with a low systemic clearance rate of 4.8 mL/min/kg. The elimination half-life for **16** following intra-arterial administration was 4.6, 7.4, and 3.0 h in rats, dogs, and monkeys, respectively. These pharmacokinetic parameters for dapagliflozin in preclinical species revealed a compound with adequate oral exposure, clearance, and elimination half-life, consistent with the potential for single daily dosing in humans [49].

In summary, dapagliflozin, compound **16**, is a potent, selective SGLT2 inhibitor which stimulates glucosuria in normal rats. The promising significant reduction of blood glucose levels in diabetic STZ rats, combined with a favorable ADME profile, in vivo clinical data prompted nomination of dapagliflozin as new drug for the treatment of type 2 diabetes.

Clinical efficacy and safety of dapagliflozin were assessed in multiple Phase III placebo-controlled trials, and the results were recently reviewed [50]. Patients receiving dapagliflozin experienced improved glucose control as demonstrated by placebo-subtracted HbA1c reductions of >0.5%. Patients experienced additional benefits of weight loss (2–3%) and reductions in systolic blood pressure (4.4 mmHg) [50] that are attributed to caloric loss (glucosuria) and osmotic diuresis, respectively. Several studies have demonstrated the additional glucose-lowering effect of dapagliflozin on a background of insulin or oral glucose-lowering agents, indicating that the mechanism of action is complementary to other antidiabetic therapies [50].

2.2 Chemistry

To generate *C*-aryl glucosides, particularly with preferred C-4′ and C-4 substitutions, to support initial toxicological studies and Phase I clinical trials, a first generation synthesis was developed. This synthesis, based on the methodology developed by Kishi [51, 52] and Kraus [53], was used to gain rapid access to compounds **13**, **15**, and **16** starting with 2,3,4,5-tetra-*O*-benzyl-D-glucose **17** as outlined in Scheme 1. Lactone (**18**) [54–56] was prepared from compound **19** via TEMPO/bleach oxidation in >90% yield. On a large scale, purification of the syrupy product **20** was not practical; therefore, it was used directly in the next step without purification. Various *m*-bromo substituted aglycones [41] (**19**) employed in this route were lithiated and then added to 2,3,4,6-tetra-*O*-benzyl-gluconolactone (**18**). Reduction of the resultant lactols (**20**) with the sterically encumbered triisopropylsilane was preferred to generate predominantly β-linked glucosides [57] **21** that were deprotected by hydrogenolysis to give amorphous API (**7–11**, **13**, **16**) in which "A" is a methylene.

During process development of the first generation synthesis, we identified several disadvantages to using 2,3,4,6-tetra-*O*-benzyl-D-glucono-1,5-lactone **18** as a starting material. For instance, **18** was synthesized from commercially available, but expensive, 2,3,4,6-tetrabenzylglucose **17**. Product **18** was difficult to isolate and purify on scale due to its syrupy nature. The crude product also gradually decomposed over several months, thereby limiting its usefulness in a commercial process. The reduction of intermediate lactol **20** required sterically hindered silanes to give favorable β-selectivity, which were expensive and difficult to obtain on commercial scale. More critically, the synthesis of *C*-arylglycosides via 2,3,4,6-tetra-*O*-benzyl gluconolactone **18** required a hydrogenolytic deprotection in the final step of the synthesis. Due to the amorphous nature of the active pharmaceutical ingredient (API) (**7**, **8**, **9**, **10**, **11**, **13**, or **16**) chromatography was required to remove palladium and process impurities. Although this route enabled rapid delivery of API

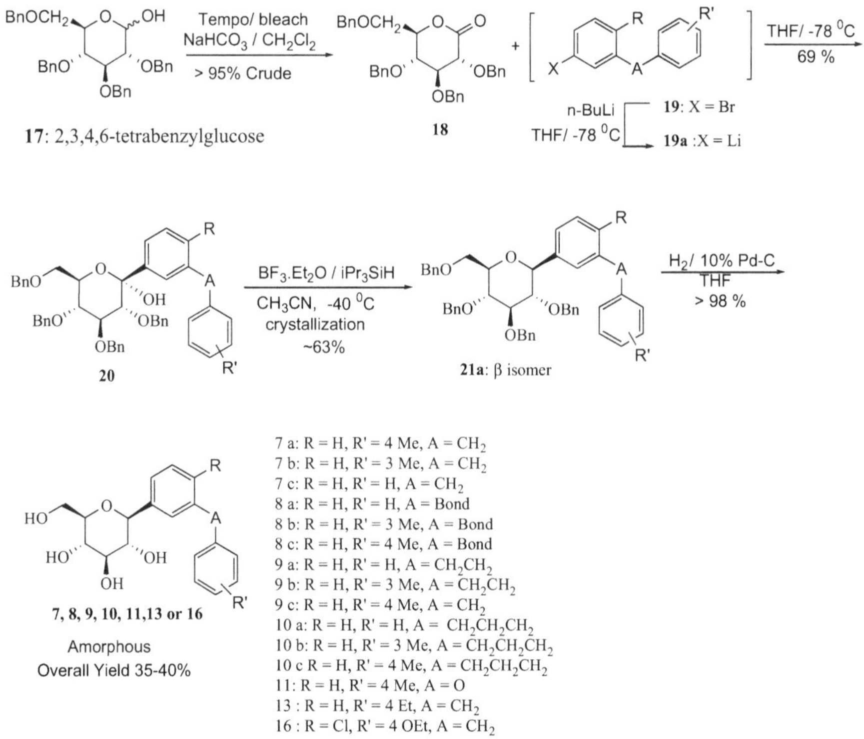

Scheme 1 General route employed in the synthesis of *C*-arylglucosides

to support early toxicology and preclinical studies, the inability to crystallize API was of paramount concern.

Despite extensive investigation, a neat crystalline form of the API has not been discovered, possibly due to the amphiphilic nature of the molecule. In addition, the amorphous, foamy API becomes a tacky or gummy solid on exposure to moisture. A syrupy, noncrystalline and hygroscopic form presented a formidable challenge to the development team. A lack of sufficient control over the isolation and purification of amorphous API, a high cost of goods, and unfavorable physical characteristics of several intermediates provided motivation for the development of a new process and a new final form for API, particularly to the nominated candidates **13**, **14**, **15**, and **16**, which required significant quantity for toxicology and clinical studies.

While some derivatives of glucose (such as **17**) can be expensive and as such not realistic starting materials, we did realize the advantages of employing materials from the naturally occurring chiral pool that eliminated the need to control four of the five stereocenters present in the API. This strategy suggested that we should retain the glycoside core of **17** or **18** as a starting material and explore protecting group modifications to circumvent the adverse physical characteristics of the intermediates in the original synthesis. In this manner, proper protecting group selection could offer cost advantages and improvements to the physical form of certain intermediates for ease of isolation. A final desired change would entail the discovery of a simple deprotection method to avoid the use of the hydrogenolysis that was employed in the first generation synthesis.

Avoidance of chromatographic step for the final purification of the amorphous API was critical. We discovered that per-acetylation of API, particularly for compounds **13**, **14**, **15**, and **16** yielded the corresponding tetraacetate **25** as a stable and pure crystalline solid that was amenable to purification by crystalization. The subsequent deprotection of this material to API (**13**, **14**, **15**, and **16**) was facile and the resulting purity was very high (>99.9 area percent purity). Although the advantages presented by this crystalline derivative were apparent, there was a concern that acetates would not be compatible with the organometallic chemistry used to generate aglycone **17**. As an alternative, we employed the transient protection of gluconolactone as the per-trimethylsilyl derivative [58–60], since the trimethylsilyl groups would resist lithioaromatics at low temperature but could also be cleaved during the work-up under mildly acidic conditions. Starting with D-gluconolactone **21**, silyl protection was achieved using TMSCl/N-methyl morpholine in THF to give **21a**. The work-up of this intermediate required a carefully controlled process as its oily nature and sensitivity to moisture precluded easy isolation. Therefore, the reaction mixture was diluted with toluene prior to aqueous washes; the per-silylated lactone **21a** was isolated as a solution. The selection of toluene as solvent was critical as it efficiently extracted the product and enabled azeotropic drying, required in the following step.

Bromo-aglycone **17** was transmetallated with *n*BuLi to form the lithiated species **17a** which was reacted at −78°C with the solution of **21a** prepared above to give lactol **22**. In this transformation, the toluene/THF solvent combination (3.5:1) proved ideal for yield and product purity; in THF solvent, a competing silyl transfer reaction dominates, forming trimethylsilyl-aglycone as a major by-product [61].

In all cases, lactol **22** was noncrystalline and problematic to isolate, therefore the toluene solution of **22** was quenched into methanol/methanesulfonic acid. This acidic quench served to: (1) neutralize any excess organometallic species, (2) promote in situ *trans*-ketalization with methanol, and (3) expedite concomitant removal of the TMS protecting groups to produce crude solutions of methyl glycosides **23**. Intermediates **23** were typically amorphous, foamy solids. Extensive investigation failed to produce crystalline forms of **23**. Thus, after an aqueous work-up, the solutions were dried via toluene azeotrope and further telescoped into the following step. Toluene solutions of **23** were typically isolated in >83% yield.

Based on our previous investigations [58–60], acetate groups were chosen to protect the hydroxyls in **23** since the reduction would provide advantageous crystalline per-acetylated API (**13**, **15** or **16**). Toluene solutions of **23** were exhaustively acetylated with Ac_2O/TEA, catalyzed by DMAP, to generate solutions of tetraacetate **24**. Following quench of the remaining anhydride by aqueous phosphoric acid and aqueous work-up, the partially concentrated dry toluene solutions were ready for the key transformation: reduction.

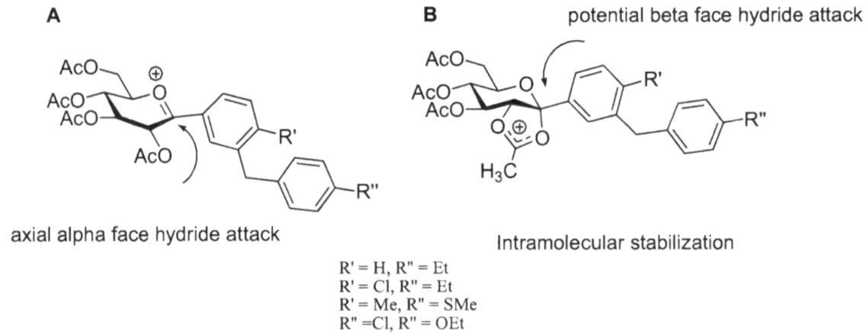

Fig. 6 Proposed stereochemistry of hydride attack

We screened a wide variety of anomeric reduction conditions, starting with acid-catalyzed silyl hydride process. We investigated both Lewis and Bronsted acids as potential alternatives to BF_3•Et_2O along with triethylsilane as an alternative to the tri-isopropyl derivative. While several alternative catalysts gave encouraging results, we decided to retain BF_3•Et_2O for a combination of cost, product purity, and process robustness. The less expensive and more readily available triethylsilane (Et_3SiH) replaced triisopropylsilane (iPr_3SiH) since the resulting isomer ratio (>20:1, β:α) was similar.

The optimized process entailed dilution of toluene solutions of **24** with acetonitrile followed by reduction with Et_3SiH, BF_3•Et_2O, and one equivalent of water. Water is critical for the reaction to progress to completion; and we believe that water and BF_3•OEt_2 generate a strong Bronsted acid, such as $H^+BF_3OH^-$, which accelerates [62, 63] the formation of an oxa-carbenium ion intermediate. These strongly acidic conditions are essential to compensate for the deactivating effect of the carbohydrate acetoxy-substituents. Based on the literature [64] and our own studies [58–60], we believe that predominant axial hydride attack from the α-face on the oxacarbenium ion intermediate "A" is preferred resulting in β-*C*-arylglucosides **25**[2] as the major products (>20:1 β:α). Intramolecular stabilization via the *C*-2-acetate group (B) which would block the *α*-face does not seem to result in β-hydride attack (Fig. 6).

The use of Et_3SiH on scale presented safety concerns. Unreacted silane can generate hydrogen gas in spent reaction streams, which would render the waste streams more hazardous to storage and transport. To mitigate this risk, 2,2-dimethoxypropane was charged to the mature reaction mixture to quench excess Et_3SiH. After complete disappearance of triethylsilane by GC analysis, aqueous work-up and crystallization of the product from ethanol produced a

[2] Crystallographic coordinates and unit cell data (excluding structure factors) for **25, 26, 27, 28, 29** and L-Proline forms **13**-P1, **13**-P2, **13**-P3, **13**-TP have been deposited with the Cambridge Crystallographic Data Centre as supplementary publication numbers CCDC 756030–756038. These data can be obtained free of charge via www.ccdc.cam.ac.uk/const/retrieving.html (or from the CCDC, 12 Union Road, Cambridge CB21EZ, UK; fax + 44 1223 336033; e-mail deposit@ccdc.cam.ac.uk).

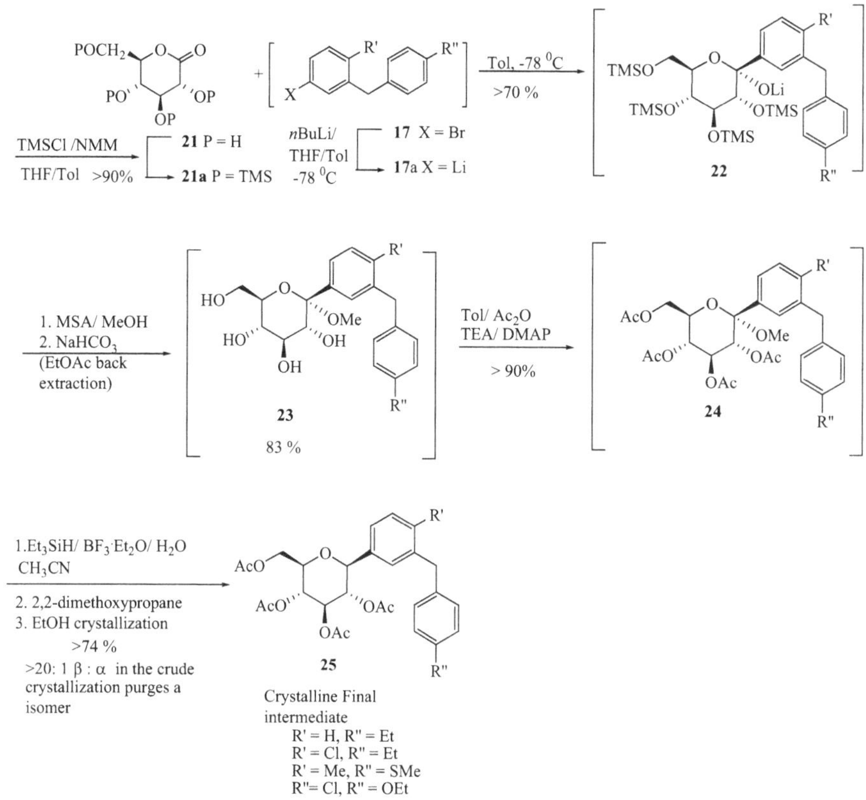

Scheme 2 Synthesis of tetraacetate **25**

~74% yield of the final intermediates **25** (>98.6 HPLC AP purity). Saponification of tetraacetates **25** (aq. NaOH/EtOH) produced API (**13**, **14**, **15**, and **16**) as amorphous powders (Scheme 2).

To solve the isolation and final crystal-form issues of the amorphous, amphiphatic drug substances **13**, **14**, **15**, and **16**, we first attempted traditional high-throughput crystallization screening studies, testing different solvents, solvent/antisolvent combinations and temperatures. These studies failed to produce a crystalline form. Subsequently, we focused on a co-crystallization approach [65] based on understanding of the structure of the drug substances. Although in principle virtually any pair of molecules may co-crystallize [66], we reasoned that another molecule containing structural elements that could favorably interact with both components of the drug substances should enhance the probability of crystallization. An amphiphilic co-crystallization component might interact with both the polar hydroxyls and the hydrophobic hydrocarbon regions of β-*C*-arylglucosides (**13**, **14**, **15**, and **16**). The ideal co-crystallization component would be a GRAS (Generally Regarded as Safe) list compound, mitigating safety concerns. We identified several natural amino acids as possible partners for

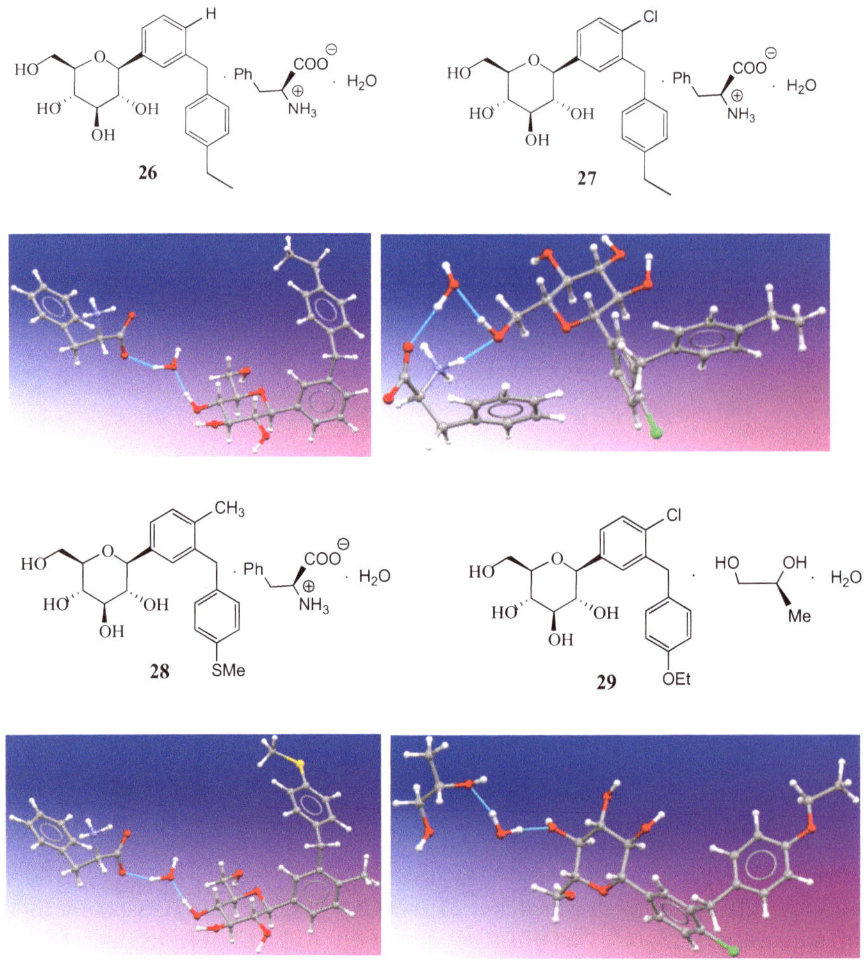

Fig. 7 X-ray structure of complexes **26**, **27**, **28**, and **29**

co-crystallization with **13**, **14**, **15**, and **16** but concentrated on L-Phenylalanine. These efforts provided crystal forms (complexes) **26**, **27**, and **28**, each composed of 1 mol of drug substance + 1 mol L-phenylalanine + 1 mol of water. In case of dapagliflozin, **16**, the L-Phenylalanine complex was isolated but its physical properties were deemed not suitable for development. An S-propylene Glycol (S-PG), **29**, complex was also identified. Crystals of **29** are fine needles that are stable under ambient conditions.

The structures of these hydrogen bonded complexes could only be solved from synchrotron X-ray diffraction data of the very fine, hair-like crystals (see footnote 2). The S-PG complex, comprised of 1 mol of drug + 1 mol S-PG + 1 mol of water, was found to be stable and suitable for development; in addition, it provided a significant weight advantage over the other complexes. Thus, the S-PG form was chosen for development (Fig. 7).

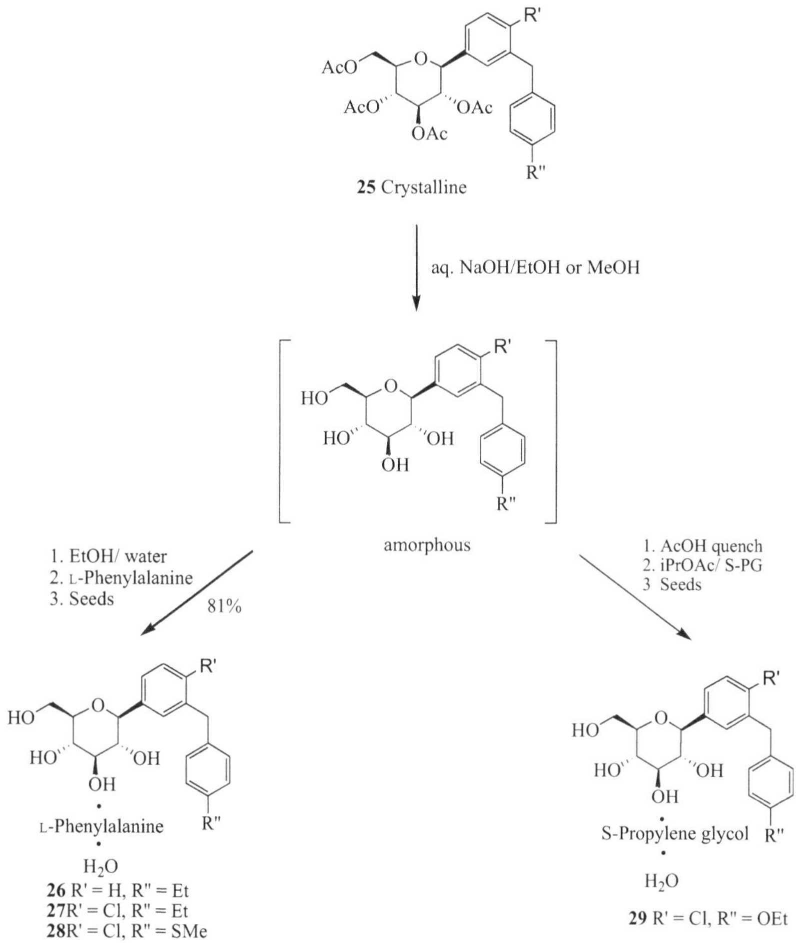

Scheme 3 Synthesis of co-crystalline complexes **26**, **27**, **28**, and **29**

Stable crystalline complexes with L-proline and D-phenylalanine were also prepared and their structures (see footnote 2) were determined via single X-ray analysis. The power of such "foreign" agents as co-crystallization "partners" in assembling crystal structures of pharmaceutical products that alone crystallize with difficulty (if at all) has been increasingly recognized in recent years (dedicated issue on "Pharmaceutical Co-crystal" please see: [67]).

The final step of the synthesis consists of deprotection of the acetyl protecting groups and formation of the crystalline complex. In contrast to the hydrogenation required to cleave the benzyl protecting groups of the final intermediate in the first generation synthesis (Scheme 1), the final step now required only aqueous base to remove the acetates. The saponification is rapid and irreversible and can be performed under a wide range of reagent concentration and temperature conditions.

There are no potential side reactions in this process which can impact the stereochemical integrity of the compound. Thus, in the final step the quality of the product was effectively preserved. Essentially no process changes were required to scale up the final intermediate **25**. Deacetylation using EtOH or MeOH and aq. NaOH (Scheme 3) followed by neutralization with aq. HCl to pH 7 generated a solution of compounds **13**, **14**, **15**, or **16** which was converted to the final complexes **26**, **27**, **28**, **or 29** respectively and crystallized as a monohydrate. After filtration and drying, the isolated product was obtained in 80–88% yield with 99.9 HPLC Area Percent purity.

In summary, a robust and practical synthesis of β-*1C*-arylglucosides was developed that includes a co-crystalization approach with L-Phenylalanine or S-Propylene glycol to circumvent isolation and purification challenges.

3 Summary, Conclusions, Outlook

In summary, β-*1C*-arylglucosides were identified as potent SGLT-inhibitors that are significantly more stable than their *O*-glycosides counterparts. Initial SAR studies of benzylaryl-*O*-glucosides led to discovery of *C*-arylglucosides that showed greater selectivity towards SGLT2 versus SGLT1 with increased metabolic stability. Further exploration of the in vivo SAR studies allowed us to identify superior SGLT2 inhibitors based on a β-*1C*-arylglucoside with a diarylmethane moiety and C-4′ and C-4 substitutions. Through this investigation, β-*1C*-arylglucoside dapagliflozin **16** was identified as a potent and selective *h*SGLT2 inhibitor with an EC_{50} for *h*SGLT2 of 1.0 nM and 1,600-fold selectivity over *h*SGLT. Initial toxicological studies, combined with a favorable ADME profile and vivo data, led to development of dapagliflozin **16** as a drug for the treatment of type 2 diabetes. Multiple placebo-controlled clinical trials demonstrated safety and efficacy of dapagliflozin. The results showed patients receiving dapagliflozin experienced improved glucose control, benefits of weight loss, and reductions in systolic blood pressure. The glucose-lowering effect of dapagliflozin occurs on a background of insulin or oral glucose-lowering agents, indicating that the mechanism of action is complementary to other antidiabetic therapies. Dapagliflozin was approved by the Australian TGA and the European Commission under the brand name Forxiga™ in a 10 mg daily dose (5 mg starting dose for severe hepatic impairment) as oral monotherapy for the treatment of diabetes. It was also approved by the US Food and Drug Administration (FDA) under the brand name Farxiga™, as a once-daily oral treatment to improve glycemic control in adults with type 2 diabetes. In addition, the European Commission granted Marketing Authorization for a fixed dose combination product of dapagliflozin and metformin hydrochloride, two anti-hyperglycemic products with complementary mechanisms of action to improve glycemic control, in a twice daily tablet (5 mg/850 mg and 5 mg/1,000 mg tablets) under the brand name Xigduo™. This is the first regulatory approval for a fixed dose combination of an SGLT2 inhibitor and metformin.

References

1. Skyler JS (2004) Diabetes mellitus: pathogenesis and treatment strategies. J Med Chem 47: 4113–4117
2. Porte D (2001) Clinical importance of insulin secretion and its interaction with insulin resistance in the treatment of type 2 diabetes mellitus and its complications. Diabetes Metab Res Rev 17:181–188
3. Kikkawa R (2000) Chronic complications in diabetes mellitus. Br J Nutr 84:S183–S185
4. Edelman SV (1998) Importance of glucose control. Med Clin N Am 82:665–687
5. Rotella DP (2004) Novel "second-generation" approaches for the control of type 2 diabetes. J Med Chem 47:4111–4112
6. Mueller G (2002) Concepts and options for current insulin research and future anti-diabetic therapy. Recent Res Dev Endocrin 3:199–218
7. UK Prospective Diabetes Study (UKPDS) Group (1998) Effect of intensive blood-glucose control with metformin on complications in overweight patients with type 2 diabetes (UKPDS 34). Lancet 352:854–865
8. The Diabetes Control and Complications Trial Research Group (1993) The effect of intensive treatment of diabetes on the development and progression of long-term complications in insulin-dependent diabetes mellitus. N Engl J Med 329:977–986
9. Deetjen P, von Baeyer H, Drexel H (1992) Renal glucose transport. In: Seldin DW, Giebisch G (eds) The kidney: physiology and pathophysiology, 2nd edn. Raven Press, New York, pp 2873–2888
10. Moe OW, Berry CA, Rector FC (2000) Renal transport of glucose, amino acids, sodium, chloride and water. In: Brenner BM, Rector FC (eds) The kidney, 5th edn. WB Saunders, Philadelphia, pp 375–415
11. Wright EM, Hirayama B, Hazama A, Loo DDF, Supplisson S, Turk E, Hager KM (1993) The sodium/glucose cotransporter (SGLT1). Soc Gen Physiol Ser 48:229–241
12. Wright EM, Turk E, Hager K, Lescale-Matys L, Hirayama B, Supplisson S, Loo DDF (1992) The sodium/glucose cotransporter (SGLT1). Acta Physiol Scand Suppl 146:201–207
13. Mackenzie B, Loo DDF, Panayotova-Heiermann M, Wright EM (1996) Biophysical characteristics of the Pig kidney Na^+/glucose cotransporter SGLT2 reveal a common mechanism for SGLT1 and SGLT2. J Biol Chem 271:32678–32683
14. Kanai Y, Lee WS, You G, Brown D, Hediger MA (1994) The human kidney low affinity Na^+/glucose cotransporter SGLT2. Delineation of the major renal reabsorptive mechanism for D-glucose. J Clin Invest 93:397–404
15. Scheepers A, Joost H-G, Schuermann A (2004) The glucose transporter families SGLT and GLUT: molecular basis of normal and aberrant function. J Parent Enter Nutr 28:364–371
16. Wright EM (2001) Renal Na^+-glucose cotransporters. Am J Physiol Renal Physiol 280: F10–F18
17. Wallner EI, Wada J, Tramonti G, Lin S, Kanwar YS (2001) Status of glucose transporters in the mammalian kidney and renal development. Ren Fail 23:301–310
18. You G, Lee WS, Barros EJG, Kanai Y, Huo TL, Khawaja S, Wells RG, Nigam SK, Hediger MA (1995) Molecular characteristics of Na^+-coupled glucose transporters in adult and embryonic Rat kidney. J Biol Chem 270:29365–29371
19. Hediger MA, Rhoads DB (1994) Molecular physiology of sodium-glucose cotransporters. Physiol Rev 74:993–1026
20. Wells RG, Pajor AM, Kanai Y, Turk E, Wright EM, Hediger MA (1992) Cloning of a human kidney *c*DNA with similarity to the sodium-glucose cotransporter. Am J Physiol 263: F459–F465
21. Kasahara M, Maeda M, Hayashi S, Mori Y, Abe T (2001) A missense mutation in the Na^+/glucose cotransporter gene SGLT1 in a patient with congenital glucose-galactose malabsorption: normal trafficking but inactivation of the mutant protein. Biochim Biophys Acta 1536:141–147

22. Turk E, Zabel B, Mundlos S, Dyer J, Wright EM (1991) Glucose/galactose malabsorption caused by a defect in the Na^+/glucose cotransporter. Nature 350:354–356
23. van den Heuvel LP, Assink K, Willemsen M, Monnens L (2002) Autosomal recessive renal glucosuria attributable to a mutation in the sodium glucose cotransporter (SGLT2). Hum Genet 111:544–547
24. Santer R, Kinner M, Schneppenheim R, Hillebrand G, Kemper M, Ehrich J, Swift P, Skovby F, Schaub J (2000) The molecular basis of renal glucosuria: mutations in the gene for a renal glucose transporter (SGLT2). J Inherit Metab Dis 23(Suppl 1):178
25. Oku A, Ueta K, Arakawa K, Ishihara T, Nawano M, Kuronuma Y, Matsumoto M, Saito A, Tsujihara K, Anai M, Asano T, Kanai Y, Endou H (1999) T-1095, an inhibitor of renal Na^+-glucose cotransporters, may provide a novel approach to treating diabetes. Diabetes 48: 1794–1800
26. Ehrenkranz JRL, Lewis NG, Kahn CR, Roth J (2005) Phlorizin: a review. Diabetes Metab Res Rev 21:31–38
27. Jonas J-C, Sharma A, Hasenkamp W, Ilkova H, Patanè G, Laybutt R, Bonner-Weir S, Weir G (1999) Chronic hyperglycemia triggers loss of pancreatic β-cell differentiation in an animal model of diabetes. J Biol Chem 274:14112–14121
28. Dimitrakoudis D, Vranic M, Klip A (1992) Effects of hyperglycemia on glucose transporters of the muscle: use of the renal glucose reabsorption inhibitor phlorizin to control glycemia. J Am Soc Nephrol 3:1078–1091
29. Rossetti L, Smith D, Shulman GI, Papachristou D, DeFronzo RA (1987) Correction of hyperglycemia with phlorizin normalizes tissue sensitivity to insulin in diabetic rats. J Clin Invest 79:1510–1515
30. Doggrell SA, Castaner J (2001) T-1095 antidiabetic sodium-glucose cotransporter inhibitor. Drugs Future 26:750–753
31. Arakawa K, Ishihara T, Oku A, Nawano M, Ueta K, Kitamura K, Matsumoto M, Saito A (2001) Improved diabetic syndrome in C57BL/KsJ-*db/db* mice by oral administration of the Na^+-glucose cotransporter inhibitor T-1095. Br J Pharmacol 132:578–586
32. Oku A, Ueta K, Arakawa K, Kano-Ishihara T, Matsumoto M, Adachi T, Yasuda K, Tsuda K, Saito A (2000) Antihyperglycemic effect of T-1095 *via* inhibition of renal Na^+-glucose cotransporters in streptozotocin-induced diabetic rats. Biol Pharm Bull 23:1434–1437
33. Adachi T, Yasuda K, Okamoto Y, Shihara N, Oku A, Ueta K, Kitamura K, Saito A, Iwakura T, Yamada Y, Yano H, Seino Y, Tsuda K (2000) T-1095, a renal na^+-glucose transporter inhibitor, improves hyperglycemia in streptozotocin-induced diabetic rats. Metab Clin Exp 49:990–995
34. Tsujihara K, Hongu M, Saito K, Kawanishi H, Kuriyama K, Matsumoto M, Oku A, Ueta K, Tsuda M, Saito A (1999) Na^+-glucose cotransporter (SGLT) inhibitors as antidiabetic agents. 4. Synthesis and pharmacological properties of 4′-dehydroxyphlorizin derivatives substituted on the B ring. J Med Chem 42:5311–5324
35. Oku A, Ueta K, Nawano M, Arakawa K, Kano-Ishihara T, Matsumoto M, Saito A, Tsujihara K, Anai M, Asano T (2000) Antidiabetic effect of T-1095, an inhibitor of Na^+-glucose cotransporter, in neonatally streptozotocin-treated rats. Eur J Pharmacol 391:183–192
36. Nawano M, Oku A, Ueta K, Umebayashi I, Ishirahara T, Arakawa K, Saito A, Anai M, Kikuchi M, Asano T (2000) Hyperglycemia contributes insulin resistance in hepatic and adipose tissue but not skeletal muscle of ZDF rats. Am J Physiol Endocrinol Metab 278: E535–E543
37. Kikuchi N, Fujikura H, Tazawa S, Yamato T, Isaji M (2004) Preparation of pyrazole glycoside compounds as SGLT inhibitors. PCT Int. Appl. WO2004113359; (2004) Chem Abstr 142: 94061
38. Fushimi N, Yonekubo S, Muranaka H, Shiohara H, Teranishi H, Shimizu K, Ito F, Isaji M (2004) Preparation of glucopyranoside compounds having fused heterocycle as SGLT inhibitors. PCT Int. Appl. WO2004087727; (2004) Chem Abstr 141:332411

39. Fujikura H, Nishimura T, Katsuno K, Isaji M (2004) Preparation of D-glucose derivatives as human SGLT2 inhibitors. PCT Int. Appl. WO2004058790; (2004) Chem Abstr 141:123854
40. Fushimi N, Ito F, Isaji M (2003) Preparation of glucopyranosyloxybenzylbenzene derivatives as inhibitors of human SGLT2 (sodium-dependent glucose-transporter 2), medicinal composition containing the same, medicinal use thereof, and intermediate for production thereof. PCT Int. Appl. WO2003011880; (2003) Chem Abstr 138:153771
41. Ellsworth BA, Meng W, Patel M, Girotra RN, Wu G, Sher PM, Hagan DL, Obermeier MT, Humphreys WG, Robertson JG, Wang A, Han S, Waldron TL, Morgan NN, Whaley JM, Washburn WN (2008) Aglycone exploration of *C*-arylglucoside inhibitors of renal sodium-dependent glucose transporter SGLT2. Bioorg Med Chem Lett 18:4770–4773
42. Link JT, Sorensen BK (2000) A method for preparing *C*-glycosides related to phlorizin. Tetrahedron Lett 41:9213–9217
43. Kishi Y (1993) Preferred solution conformation of marine natural product palytoxin and of *C*-glycosides and their parent glycosides. Pure Appl Chem 65:771–778
44. Wang Y, Goekjian PG, Ryckman DM, Miller WH, Babirad SA, Kishi Y (1992) Preferred conformation of *C*-glycosides 9. conformational analysis of 1,4-linked carbon disaccharides. J Org Chem 57:482–489
45. Meng W, Ellsworth BA, Nirschl AA, McCann PJ, Patel M, Girotra RN, Wu G, Sher PM, Morrison EP, Biller SA, Zahler R, Robertson JG, Wang A, Han S, Wetterau JR, Janovitz EB, Flint OP, Whaley JM, Washburn WN (2008) Discovery of dapagliflozin: a potent, selective renalsodium-dependent glucose cotransporter 2 (SGLT2) inhibitor for the treatment of type 2diabetes. J Med Chem 51:1145–1149
46. Hongu M, Tanaka T, Funami N, Saito K, Arakawa K, Matsumoto M, Tsujihara K (1998) Na + glucose cotransporter inhibitors as a antidiabetic agents. II. Synthesis and structure-activity relationships of 4′-dehydroxyphlorizin derevatives. Chem Pharm Bull 46:22–33
47. Wood IS, Trayhurn P (2003) Glucose transporters (GLUT and SGLT): expanded families of sugar transport proteins. Br J Nutr 89:3–9
48. Washburn WN, Meng W (2006) *C*-aryl glucoside SGLT2 inhibitors and method for the treatment of diabetes and related diseases. US. Pat. Appl. Publ. US 20060063722 A1 20060323
49. Obermeier MT, Yao M, Khanna A, Koplowitz B, Zhu M, Li W, Komoroski B, Kasichayanula S, Discenza L, Washburn W, Meng W, Ellsworth BA, Whaley JM, Humphreys WG (2009) In vitro characterization and pharmacokinetics of dapagliflozin (BMS-512148), a potent sodium-glucose cotransporter type II (SGLT2) inhibitor, in animals and humans. Drug Metab Dispos 38:405–414
50. Washburn WN, Poucher SM (2013) Differentiating sodium-glucose co-transporter-2 inhibitors in development for the treatment of type 2 diabetes mellitus. Expert Opin Investig Drugs 22: 463–486
51. Lewis MD, Cha K, Kishi Y (1982) Highly stereoselective approaches to α- and β-*C*-glycopyranosides. J Am Chem Soc 104:4976–4978
52. Babirad SA, Wang Y, Kishi Y (1987) Synthesis of *C*-disaccharides. J Org Chem 52: 1370–1372
53. Kraus GA, Molina MT (1988) A direct synthesis of *C*-glycosyl compound. J Org Chem 53:752–753
54. Rajanikanth B, Seshadri R (1989) A facile synthesis of nojirimycin. Tetrahedron Lett 30: 755–758
55. Benhaddou R, Czernecki S, Farid W, Ville G, Xie J, Zegar A (1994) Tetra-n-propylammonium tetra-oxoruthenate(VII): a reagent of choice for the oxidation of diversely protected glycopyranoses and glycofuranoses to lactones. Carbohydr Res 260:243–250
56. Olah GA, Wang Q, Prakash GKS (1992) Synthetic methods and reactions 177. Ionic hydrogenation with triethylsilane-trifluoroacetic acid-ammonium fluoride or triethylsilane-pyridinium poly(hydrogen fluoride). Synlett 647–650

57. Ellsworth BA, Doyle AG, Patel M, Caceres-Cortes J, Meng W, Deshpande PP, Pullockaran A, Washburn WN (2003) *C*-Arylglucoside synthesis: triisopropylsilane as a selective reagent for the reduction of an anomeric *C*-phenyl ketal. Tetrahedron Asymmetry 14:3243–3247
58. Deshpande PP, Ellsworth BA, Buono FG, Pullockaran A, Singh J, Kissick TP, Huang M-H, Lobinger H, Denzel T, Mueller RH (2007) Remarkable β-1-*C*-arylglucosides: stereoselective reduction of acetyl-protected methyl 1-*C*-aryl-glucosides without acetoxy-group participation. J Org Chem 72(25):9746–9749
59. Horton D, Priebe W (1981) Synthetic routes to higher-carbon sugars. Reaction of lactones with 2-lithio-1,3-dithiane. Carbohydr Res 94(1):27–41
60. Deshpande PP, Singh J, Pullockaran A, Kissick T, Ellsworth BA, Gougoutas JZ, Dimarco J, Fakes M, Reyes M, Lai C, Lobinger H, Denzel T, Ermann P, Crispino G, Randazzo M, Gao Z, Randazzo R, Lindrud M, Rosso V, Buono F, Doubleday WW, Leung S, Richberg P, Hughes D, Washburn WN, Meng W, Volk KJ, Mueller RH (2012) A practical stereoselective synthesis and novel cocrystallization of an amphiphatic SGLT2 inhibitor. Org Process Res Dev 16:577–585
61. Barrett AGM, Pena M, Willardsen AJ (1996) Total synthesis and structural elucidation of the antifungal agent Papulacandin D. J Org Chem 61:1082–1100
62. Akiyama T, Takaya J, Kagoshima H (1999) Chemoselctive activation of aldimines in preference to aldehydes by the combination of BF3.OEt2 and water: novel catalyst for the Mannich-type reation. Chem Lett 28:947–948
63. Ribe S, Wipf P (2001) Water-accelerated organic tranformations. Chem Commun 299–307
64. Deslongchamps P (1993) Intramolecular strategies and stereoelectronic effects. Glycosides hydrolysis revisited. Pure Appl Chem 65:1161–1178
65. Gougoutas JZ (2002) Amino acid complexes of *C*-aryl glucosides for treatment of diabetes PCT Int. Appl, p 80, WO 2002083066
66. Mutig T, Kemnitz E, Troyanov SI (2010) Synthesis and molecular structures of heptafluoropropylated fullerenes: $C_{70}(n\text{-}C_3F_7)8$, $C_{70}(n\text{-}C_3F_7)_6O$, and $C_{70}(C_3F_7)$. J Flour Chem 131: 861–866
67. Rodríguez-Hornedo N (2007) Molecular design of pharmaceutical materials. Mol Pharm 4:299–434

Top Med Chem (2014) 12: 95–122
DOI: 10.1007/7355_2014_49

Published online: 27 April 2014

Design, Synthesis, and Applications of Galectin Modulators in Human Health

Alison Mackinnon, Wei-Sheng Chen, Hakon Leffler, Noorjahan Panjwani, Hans Schambye, Tariq Sethi, and Ulf J. Nilsson

Abstract Over the last decade, the family of galectin proteins has been identified as key regulators of important biological processes. They bind β-D-galactopyranoside residues in glycoconjugates, and by presenting multiple binding sites, within one galectin or by forming dimers or multimers, they can cross-link glycoproteins and form galectin-glycoprotein lattices. Such lattices formed on the cell surface or in vesicles have been shown to control, for example, surface residence time and signaling by receptors. Hence, compounds modulating galectin binding to their glycoprotein ligands are of potential clinical interest. This chapter describes the design and development of disubstituted thiodigalactoside derivatives that form optimal interactions with the galectin-3 binding site resulting in double-digit nanomolar affinities. Studies are discussed in which such galectin-3-modulating

A. Mackinnon
MRC Centre for Inflammation Research, The Queen's Medical Research Institute, University of Edinburgh, Edinburgh, UK

W.-S. Chen
Sackler School of Graduate Biomedical Sciences, Tufts University, Boston, MA, USA

H. Leffler
Section MIG (Microbiology, Immunology, Glycobiology), Department of Laboratory Medicine, Lund University, 223 62 Lund, Sweden

N. Panjwani
Sackler School of Graduate Biomedical Sciences, Tufts University, Boston, MA, USA

Department of Ophthalmology, New England Eye Center, Tufts University, Boston, MA, USA

H. Schambye
Galecto Biotech AB, Cobis, Ole Maaloes vej 3, Copenhagen, N DK-2200, Denmark

T. Sethi
Department of Respiratory Medicine and Allergy, Kings College, Hill Campus, London, UK

U.J. Nilsson (✉)
Department of Chemistry, Centre for Analysis and Synthesis, Lund University, POB 124, SE-22100, Lund, Sweden
e-mail: ulf.nilsson@chem.lu.se

compounds have been important in elucidating galectin-3 mechanisms, including galectin-3 trafficking, cancer, inflammation, fibrosis, and angiogenesis. Medically relevant models using the galectin-3 modulators in characterizing macrophage alternative activation and chronic inflammation, myofibroblast activation and fibrosis, and ocular angiogenesis are discussed in more detail. In summary, the high galectin-3 affinity and definitive effects in relevant models of the disubstituted thiodigalactosides identify them as promising as lead compounds for drug development, albeit leaving a challenge in terms of optimizing PK/ADME properties.

Keywords Angiogenesis, Cancer, CD98, Fibrosis, Galectin, Macrophage, Modulator, Myofibroblast, Small molecule, TGF-β, VEGF

Contents

Abbreviations

ADME	Absorption-distribution-metabolism-excretion
ApoE	Apolipoprotein E
bFGF	Basic fibroblast growth factor
CD31	Cluster of differentiation 31 protein
CD8	Cluster of differentiation 8 protein
CD98	Cluster of differentiation 98 protein
CRD	Carbohydrate recognition domain
DMSO	Dimethylsulfoxide
EGF	Epidermal growth factor
EGF-R	Epidermal growth factor receptor
EMT	Epithelial–mesenchymal transition
FDA	Food and Drug Administration
galectin-3C	Galectin-3 C-terminal domain
galectin-8N	Galectin-8 N-terminal domain
galectin-9N	Galectin-9 N-terminal domain
Grb2	Growth factor receptor-bound protein 2
I.T.	Intratracheal

IL4	Interleukin 4
IPF	Idiopathic pulmonary fibrosis
IRS-1	Insulin receptor substrate-1
JAK	Janus kinase
K_d	Dissociation constant
LacNAc	*N*-acetyllactosamine
Mgat5	β1,6-*N*-acetylglucosaminyl transferase 5
MHC-1	Major histocompatibility complex 1
MMFFs	Merck molecular force field s
PBS	Phosphate buffer saline
PDB	Protein data bank
PI3K	Phosphatidylinositide 3-kinase
PK	Pharmacokinetic
siRNA	Small interfering ribonucleic acid
Smad2	Mothers against decapentaplegic homolog 2
Smad3	Mothers against decapentaplegic homolog 3
STAT6	Signal transducer and activator of transcription 6
TCR	T-cell receptor
TGF-β	Transforming growth factor β
TGF-βRII	Transforming growth factor β receptor 2
Th2	Type 2 helper T cell
VEGF	Vascular epithelial growth factor
VEGF-R	Vascular epithelial growth factor receptor
VEGF-RII	Vascular epithelial growth factor receptor 2
WT	Wild type

1 Introduction

Modulators in human health acting via chemical means have largely focused on receptor messenger concentrations and downstream signal pathway enzyme activities. Disease related to too high or too low messenger levels has been counteracted by the use of antagonists and agonists, respectively, of receptor activation. Virtually no chemical modulators have targeted the concentration, localization, or availability of active receptors, although such targeting in principle could be viewed as an equally valid strategy. The biological mechanisms behind the control of concentration, localization, and availability of receptors and other glycoproteins on cell surfaces and intracellularly have during the last decade been the subject of a number of ingenious studies and a coherent picture related to protein glycosylation is emerging [1–4]. A key aspect in this picture is lectins that by interacting with glycoprotein glycans influence glycoprotein localization and/or interactions with their binding partners, thus influencing glycoprotein biological activities. The family of galectin proteins appears to play a central role in such mechanisms.

Galectins are small soluble proteins defined by a carbohydrate recognition domain (CRD) with affinity for β-galactose and a conserved amino acid sequence motif [5, 6]. One galectin type consists of only one ~130 aa CRD with little other sequence, which may form dimers over a certain concentration; they are also named prototype and include galectin-1, -2, -5, -7, -10, -11, -13, -14, and -15 in mammals. Another galectin type has two different CRDs in the same peptide chain joined by a linker peptide sequence; they are also referred to as tandem-repeat galectins and include galectin-4, -6, -8, -9, and -12 in mammals. Galectin-3 has one CRD and an unusual ~100 aa N-terminal domain containing in glycine, proline, and tyrosine rich repeats. It is the only galectin of this chimeric type in vertebrates, but other chimeric galectins occur in invertebrates. The N-terminal domain of galectin-3 confers efficient oligomerization of the protein upon ligand encounter and is required for many of its biological effects. Galectin-3 is often depicted as a pentamer in recent reviews, but the evidence for this is weak [7] and alternative models of oligomerization have been proposed [8]. Dimerization of prototype galectins, oligomerization of galectin-3, and the dual CRD of tandem-repeat galectins allow for cross-linking of glycoproteins carrying galectin-binding glycans and formation of membrane-associated lattices [1–4]. Lattice formation between glycoproteins and galectins has been observed [9] and is suggested to influence cell surface residence time and thus activity of receptors, interactions between different glycoproteins, and organization of glycoproteins in vesicles (lysosomes) targeting extracellular locations. Altogether, galectins are thus different from other molecules modulating glycoproteins (e.g., a cytokine specifically activating one membrane-bound receptor) in that they influence cellular signaling rather than inducing it and that a galectin can modulate several different glycoproteins as long as the glycoprotein carries a common glycan structural motif. Hence, such modulation depends on the particular galectin expressed and enzymatic glycan-producing activities in a certain situation. Selected important and medically relevant examples of galectin-3 lattices roles include organizing non-raft glycoprotein cargo in vesicles targeting the apical membrane in epithelial cells, retention of EGF-R, TGF-β-R, CD98, TCR, and VEGF-R at the cell surface modulating downstream effects, and prevention of TCR co-localizing with CD8 required for MHC-1 binding and immunological synapse formation (Table 1). Other galectins are also interesting targets for medical therapy, but they will not be covered in detail here. At the cellular level also other galectins are likely to have the role in directing intracellular glycoproteins and forming membrane lattices as described for galectin-3, but different based on their different fine specificity. At the tissue and organismal level, galectin-1, the most studied, has mainly drawn attention because of its immunosuppressive [20] and pro-angiogenic activities [21]. The immunosuppressive effect is proposed to be due to induction of apoptosis in certain activated T cells and also to effects on dendritic cells. Inhibition of the carbohydrate recognition site of galectin-1 has been proposed to have anticancer effects by preventing tumor immune evasion [22]. Galectin-1 itself may act as an immunosuppressive agent in autoimmune disease [20]. The pro-angiogenic effect may also be inhibited by the antiangiogenic peptide anginex, which interacts with galectin-1, although the exact mechanism remains unknown [23–25]. Some studies suggest that biological effects of some plant polysaccharides are due to interaction with

Table 1 Selected examples of galectin-3 lattice formation with glycoproteins and proposed effects

Glycoprotein	Cell or cell function	Physiological effect
Non-raft glycoproteins [10, 11]	Epithelial cells	–
EGF-R [12]	Cell growth	Tumor growth
CD98 [13]	M2 macrophages	Excessive fibrosis in chronic inflammation
TGF-β [14]	Myofibroblasts	Excessive fibrosis in chronic inflammation
TCR [15–17]	Cytotoxic tumor-infiltrating T lymphocytes	Immunosuppression in cancer
VEGF-R [18, 19]	Angiogenesis	Excessive neovascularization in eye

another site, different from the canonical carbohydrate recognitions site, on galectin-1 [26]. Galectin-9 also has immunosuppressive effects and most attention has been given to its interaction with the ligand Tim-3 to promote regulatory T cells [27], although this interaction has been called into question [28]. The other galectins have been studied much less, but it is likely that also they are potentially interesting therapeutic targets by different or similar mechanisms as found for galectin-1, -3, and -9, specified by the different carbohydrate-binding fine specificity and tissue expressions in each case [29]. Altogether, these observations point to galectins being relevant targets for modulation in clinical situations.

This chapter summarizes our chemistry effort over the last decade towards development of small-molecule galectin-antagonizing modulators that target specific subsites, as well as recent examples of effects of such modulators in medically relevant systems of cancer [30, 31], fibrosis [13, 14], angiogenesis [18, 19], and inflammation [13, 32, 33]. The antagonist development described below is largely focused on galectin-3 since a number of medically relevant functions have been suggested for this galectin and since structural data was published [34] early for this galectin.

2 Galectin CRD Structure and Ligand Subsites

The galectin CRD is of a β-sandwich type with two antiparallel β-sheets and an extended groove or valley along one of the β-sheets forms the carbohydrate binding site [35] which is long enough to hold about a tetrasaccharide. For orientation, it may be described in terms of the subsites A–E in order to facilitate analysis and discussions on ligand binding [6, 36–38] (Fig. 1). The subsite C is the galectin-defining β-D-galactoside-binding site and is built largely from the conserved amino acid sequence motif. Subsites A–B extend from the β-D-galactopyranoside HO_3, these two sites are structurally more varied than subsite C, and this structural variation in subsites A–B among the galectins confers different selectivities towards natural complex glycoconjugate ligands. Subsite D harbors a second

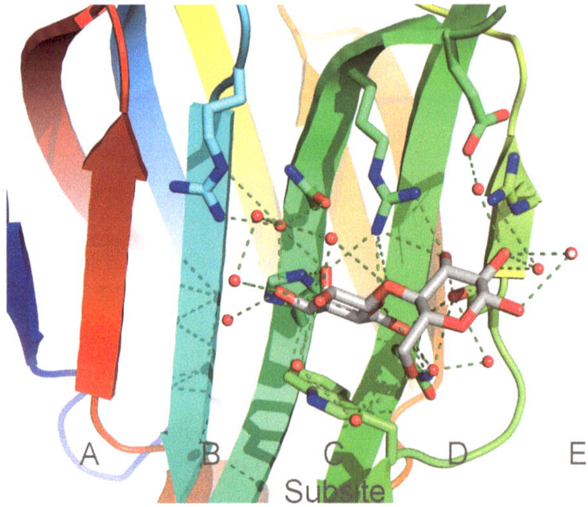

Fig. 1 Visualization of a high-resolution X-ray structure of galectin-3 CRD in complex with lactose ([39], PDB id 3ZSJ). The figure was generated with PyMOL Molecular Graphics System (Version 1.5.0.4 Schrödinger, LLC)

pyranose moiety that consistently provides one secondary hydroxyl for hydrogen to the CRD. This subsite D pyranose is most commonly glucose or *N*-acetylglucosamine as in lactose and LacNAc, but can also be other pyranoses. Subsite E typically does not form critical interactions with endogenous saccharides but may be important for interactions with aglycons or synthetic structures.

Galectin-3C does easily and reproducibly crystallize in complex with natural and synthetic ligands, and a recent study reported ultra-high-resolution structures of apo galectin-3C and its complex with glycerol and lactose of 1.08, 0.9, and 0.86 Å resolution, respectively (Fig. 1) [39]. The near atomic ultrahigh resolution also revealed a number of water sites in the apo structure. These water sites occupied positions that in the glycerol and lactose complexes were occupied with ligand oxygens or, in a few positions, carbons. Furthermore, dynamic simulations did identify water sites that agreed with the crystal structure data [39]. Hence, the galectin-3C binding appears to be pre-organized to recognize a sugar-like framework of atoms.

An overall reflection is that the β-sheet spanning the comparatively shallow CRD-binding valley comprising of subsites A–E does not resemble a so-called drugable binding site, which are typically smaller and deeper with significant hydrophobic character providing efficient affinity-enhancing interactions with a larger part of a ligand surface. Hence, in this respect, development of galectin-modulating small synthetic molecules can be regarded as a major challenge.

3 Galectin Modulators

Various strategies for modulating or inhibiting galectin activities have been developed and investigated involving natural saccharides and fragments thereof (e.g., LacNAc and lactose), modified polysaccharides [40–43], glycodendrimers [44–46], antibodies, siRNA, agonistic peptides [25, 47–49], antagonistic peptides [50–57], dominant negative fragments (e.g., the C-terminal CRD of galectin-3) [58, 59], and synthetic small molecule antagonists. The latter has typically been investigations with synthetically modified carbohydrate derivatives by others [60–69] and by us targeting subsites A–B, C, D, or E, or several subsites simultaneously [70–92]. This has been reviewed elsewhere [93–96]. Glycosylation with glycosyltransferase inhibitors also serves as a possible route to indirectly enhance or diminish galectin binding to glycoproteins.

Natural saccharides, polysaccharides, and fragments thereof can have the advantage of being more readily available from natural sources, but suffer from the drawbacks of large size, high polarity, and hydrolytic lability. Glycodendrimers can confer high affinity but mostly also have the drawback of large size and high polarity. Antibodies are attractive due to having high affinity and selectivity for a specific galectin, but suffer from disadvantages associated with protein structures, such as size, production challenges, limitation to extracellular activity, and short in vivo half-life. Furthermore, for example, galectin-3 antibodies typically do not interfere with carbohydrate ligand binding. Dominant negative fragments suffer from the same protein-associated disadvantages. From a pharmaceutical development view, small molecules are particularly attractive due to more straightforward route towards optimal pharmacodynamics/pharmacokinetic profile by means of synthetic modifications. Research on small molecule galectin modulators is exemplified below based on the discoveries and developments of structural motifs interacting with each of the subsites A–B and D–E while retaining the core subsite C-binding β-D-galactopyranose structure intact.

3.1 Small Molecule Galectin-3 Modulators Targeting Subsites A–D

The first galectin-3 X-ray structure [34] made it evident that the groove extending from O3′ of bound LacNAc disaccharide forming subsites A–B (Fig. 1) was an attractive target for accommodating affinity-enhancing ligand structural modifications. Fortunately, the core subsite C-bound galactose HO_3 does not form any critical direct hydrogen bonds to galectin-3, which promised that chemical manipulation of galactose O3 or C3 would be well tolerated by galectin-3. Indeed, already the first studies of C3′-modified LacNAc derivative compounds with significantly enhanced affinity were identified [84, 85] (Fig. 2a–c). While compounds carrying a variety of anionic moieties at the LacNAc C3′ were intended to ion pair with galectin-3 Arg144, the best compounds were identified among a panel of randomly selected LacNAc C3′-benzamides. A C3′ 2,3,5,6-tetrafluoro-4-methoxy-benzamide [86] derivatization proved to be particularly beneficial with almost two orders of magnitude affinity enhancement over the parent LacNAc disaccharide structure (Fig. 2c) [84]. A larger collection of second-generation LacNAc C3′-benzamide

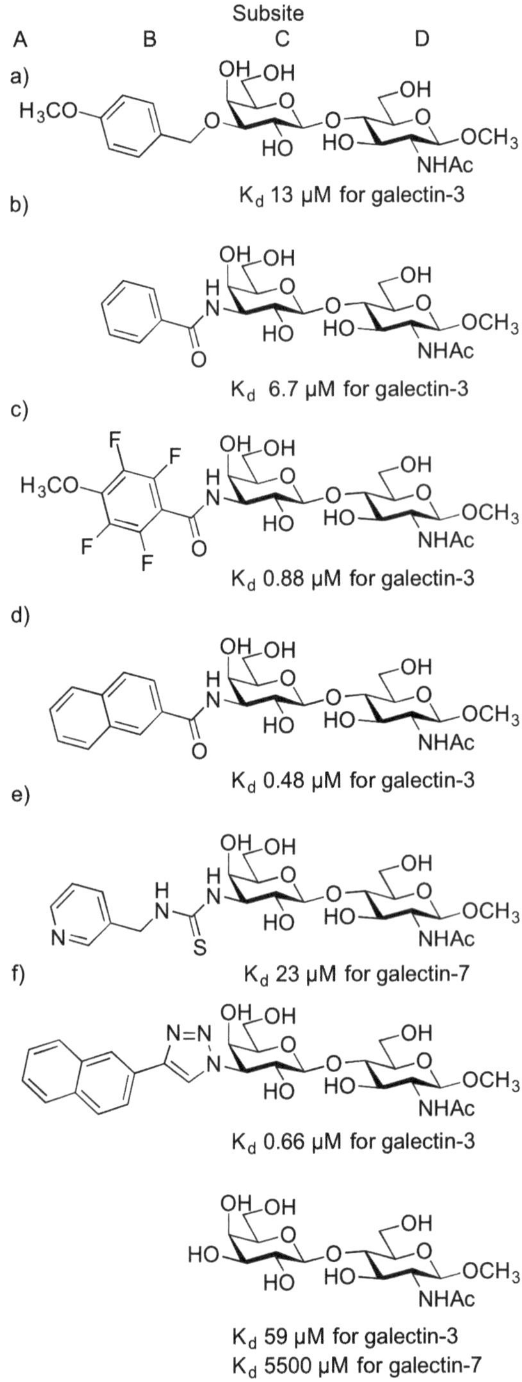

Fig. 2 Selected examples of LacNAc and galactose-derived structure binding subsites A–D: 3′-ethers [85], 3′-amides [84, 86], 3′-thioureas [83], and 3′-triazoles [81, 82]. LacNAc is shown for reference

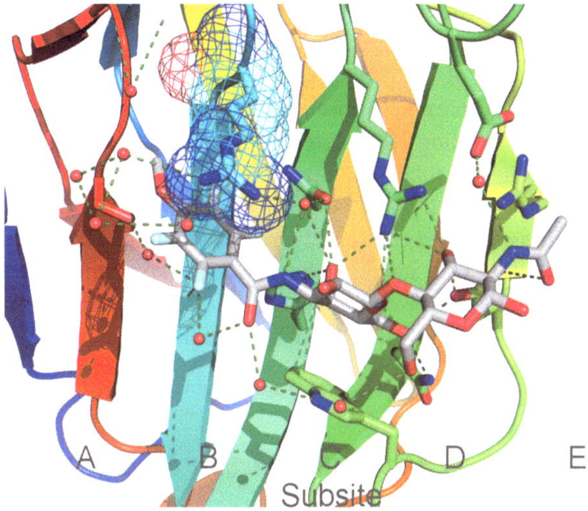

Fig. 3 Visualization of an X-ray structure the galectin-3 CRD complexing a 3′-benzamido-LacNAc derivative binding subsites A–D ([84], PDB id 1KJR). The Arg144 side chain stacking face-to-face onto the left-hand 3,5-dimethoxybenzamide is illustrated in mesh. The figure was generated with PyMOL Molecular Graphics System (Version 1.5.0.4 Schrödinger, LLC)

derivatives further improved the affinity for galectin-3, which is exemplified with a 2-naphthamido derivative in Fig. 2d. The X-ray structure of the subsite A–B binding 3′-(2,3,5,6-tetrafluoro-4-methoxybenzamido)-LacNAc derivative (Fig. 2c), in complex with galectin-3 [84], revealed a surprise in that Arg144 had undergone a conformational change moving the guanidinium functionality >2.5 Å to create a new pocket accommodating the 2,3,5,6-tetrafluoro-4-methoxy-benzamide moiety (Fig. 3). This structural data demonstrated that LacNAc 3′-derivatizations indeed target and bind subsites A–B in galectin-3 and that it enhances affinity significantly. The affinity enhancement was hypothesized to mainly originate from a high surface complementary between the 2,3,5,6-tetrafluoro-4-methoxy-benzamide moiety and the newly created pocket, specific fluorine-amide carbonyl interactions [97, 98], highly favorable desolvation of a heavily fluorinated structure, and to a arginine-π interaction. A later X-ray investigation of a complex between the corresponding unsubstituted 3′-benzamido-LacNAc derivative and galectin-3 revealed a structure in which the Arg144 side chain remained unmoved and the 3′-benzamido instead resided face-to-face stacking onto Arg144 on the surface of the protein [99]. Hence, the structure and electronic properties of the 3′-benzamido moiety apparently influence the subsite B Arg144 conformation and thus the complex structure and this has to be taken into consideration when designing novel galectin-3-modulating compounds.

Following the early discoveries of 3′-benzamido-LacNac derivatives showing sub-μM affinity, other 3′-derivatizations proved to enhance affinity for galectins. 4-Carbamido- and 4-aryl-1,2,3-triazol-1-yl derivatives [81, 82] were about as efficient binders of galectin-3 as the 3′-benzamides described above, while aryl-thiourea derivatives displayed enhanced affinity towards galectin-7 and galectin-9N [83].

Recently in-depth studies of thermodynamics and structure of galectin-3-ligand binding have revealed intriguing and to some extent counterintuitive features. An NMR-relaxation study of ^{13}C- and ^{15}N-labelled galectin-3 CRD (galectin-3C) provided order parameters for backbone amide, arginine and tryptophan NH bonds, as well as methyl CH bonds [99]. The order parameters in turn allowed for estimation of the protein conformational entropy of apo galectin-3C and of three ligand complexes, lactose and two 3C′-benzamido-LacNAc derivatives (Fig. 2b, c). Ligand binding did increase galectin-3C conformational entropy to an extent equaling the total enthalpy of binding. This may suggest that proteins that bind polar and well-solvated carbohydrates have evolved to have a free energy of binding to a larger extent being driven by protein conformational entropy increase [99–101]. This suggests that further studies on how carbohydrate binding by proteins influence protein dynamics is of importance and may provide key information on how to efficiently design lectin antagonists.

3.2 Small Molecule Galectin Modulators Targeting Subsite C–E

A key feature of galectin-ligand binding is that the β-face of the core β-D-galactopyranoside residue stacks onto a conserved Trp residue. The opposite side of the binding site, encompassing the α-face of the core β-D-galactopyranoside, is lined up with 2–3 basic amino acid residues. In case of galectin-3, the three basic amino acids are Arg144, Arg162, and Arg186. The Arg144 was involved in affinity-enhancing stacking with the aromatic moiety of the 3′-benzamido-LacNAc compounds described above. The highly conserved Arg186 in galectin-3, corresponding to Arg74, Arg186, Arg75, and Arg87 in galectin-1, -3, -7, and -9N, respectively, lies beneath the *N*-acetyl of bound LacNAc, which lead to the hypothesis that an Arg186-arene interaction could be created with aromatic 2-*O*-esters, i.e., compounds binding subsites C–E (Fig. 4). Computer modeling supported this hypothesis (Fig. 5) and synthetic aromatic 2-*O*-esters did show almost two orders of magnitude higher galectin-3 affinity than the parent lactoside. Thus, aromatic substituents at 3′- or 2-position of lactosides or LacNAc both do form greatly affinity-enhancing interactions with conserved arginine side chains in subsites A–B and E in the galectin-3 binding site. These observations point to the possibility of obtaining even further improved galectin-3 binding ligands by designing novel molecules that present properly positioned aromatic rings interacting with both Arg144 and Arg186.

3.3 Small Molecule Galectin Modulators Targeting Subsites A–E

Thiodigalactoside (di-β-D-galactopyranosyl sulfane) can be regarded as a lactose or LacNAc mimic. Indeed, in experiments where thiodigalactoside has been

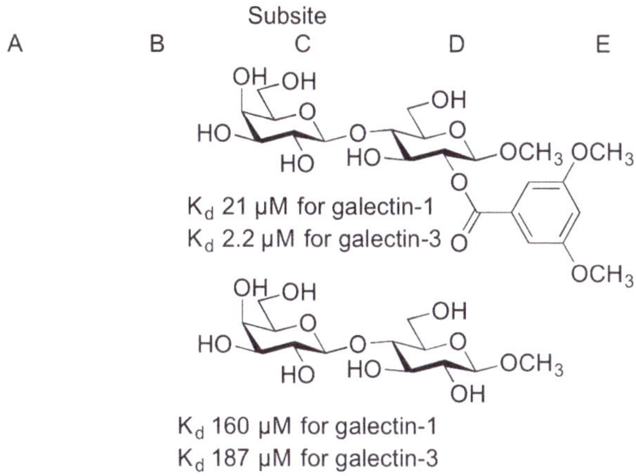

Fig. 4 Example of a lactoside ester binding subsite C–E [72]. The galactoside binds subsite C and the glucoside subsite D, while the benzoate aromatic ring stacks face-to-face onto Arg186 in subsite E of galectin-3

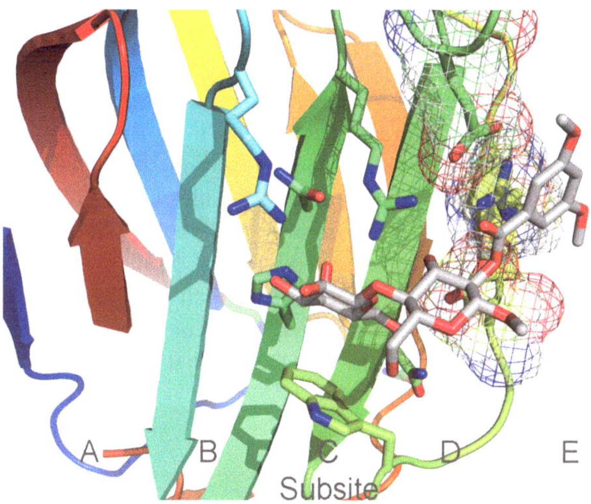

Fig. 5 A model of the galectin-3 CRD complexing a lactoside ester binding subsite C–E. A 3,5-dimethoxybenzoate stacking face-to-face onto a π-surface, created by in-plane Glu-Arg-Glu ion pairs, is illustrated with mesh. The model was built from an X-ray structure of galectin-3 CRD in complex with LacNAc (PDB id 1KJL [84]). A π-surface, created by in-plane Glu-Arg-Glu ion pairs, stacking face-to-face with one 3,5-dimethoxybenzamide, is illustrated in mesh. Modeling was performed with the MMFFs force field with water implemented in MacroModel (version 9.1, Schrödinger, LLC, New York). The figure was generated with PyMOL Molecular Graphics System (Version 1.5.0.4 Schrödinger, LLC)

co-crystallized with both galectin-1 [31] and galectin-3 [91], the thiodigalactoside occupies subsites C–D. One galactopyranose expectedly binds the core galactose subsite C and the second galactopyranose binds subsite D with its 2-OH forming a hydrogen bond analogously to 3-OH of lactosides or LacNAc. A 3,3′-disubstituted thiodigalactoside bound to the galectin-3 subsites C–D would thus place its two substituents in ideal positions to stack with Arg144 and Arg186 to create double arginine–arene interactions. This hypothesis was verified by synthesizing 3,3′-benzoyl-[75] and 3,3′-bis-benzamido-thiodigalactosides [73, 74], which indeed had strong affinities for galectin-3 down to sub-μM and double-digit nanomolar, respectively (Fig. 6a, b). The 1,2,3-triazol-1-yl substituents, earlier proven to be as efficient affinity enhancers as 3-benzamides towards galectin-3, were also implemented on the thiodigalactoside core [14, 32, 81, 82] (Fig. 6c, d), which resulted in even further improved affinity for galectin-3, but at the same time also with less selectivity as galectin-1 was inhibited with equal affinity.

With nanomolar affinities for galectin-3, the 3,3′-bis-benzamido- and 3,3′-bis-1,2,3-triazol-1-yl thiodigalactosides constitute powerful research tools for elucidating galectin biology mechanisms as well as being promising lead structures for development of galectin-modulating drugs.

3.4 Small Molecule Galectin Modulators Targeting Alternative Subsite Combinations

While the nanomolar galectin-3 modulator binding subsites A–E have been used to investigate galectin-3 functions and mechanisms in biological systems (vide infra), compounds targeting subsites C–D and A–D with noncarbohydrate subsite D binders [87, 88] C-C′ and C-C′-D with galactosides [77, 79] or talosides [70, 76, 80, 91] presenting nonsugar groups in subsite C′ (perpendicular to subsite C) and A–C with β-mannosides [89] have been developed. None of these compounds have hitherto been used in biological systems due to their weaker affinities for galectins. However, it is probable that further development of such compounds may lead to smaller molecules with acceptable PK/PD space that will appear in future galectin biology investigations (Fig. 7).

4 Galectin-3 Modulator Biological Effects

A large number of studies using galectin-modulating natural saccharide fragments and polysaccharides in biological in vitro and in vivo assays and model have been instrumental in contributing to the current understanding of galectin functions and mechanisms. The small synthetic subsite A–E binding molecules described above have been used to a lesser extent, but have, however, been employed to demonstrate carbohydrate-dependent galectin functions in highly clinically relevant systems, e.g., endocytosis, tumor migration, tumor cell apoptosis, antitumor immunity,

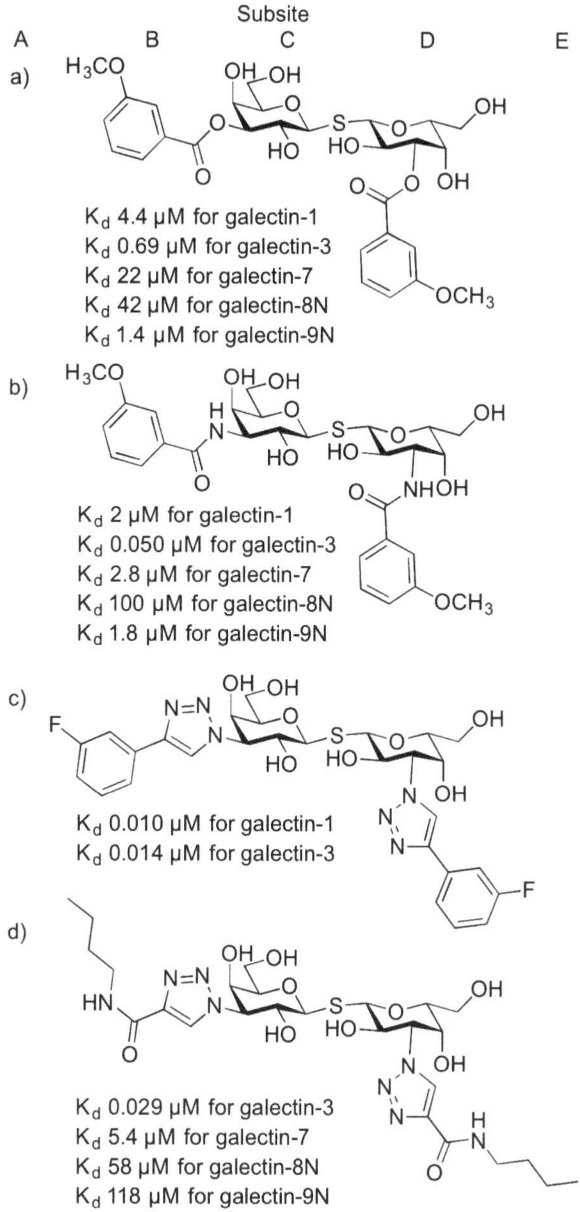

Fig. 6 Example of 3,3′-disubstituted thiodigalactosides, esters [75], amides [13, 73, 74] and triazoles [14, 32, 33], and binding subsites A–E. The two galactoside moieties bind subsites C–D, while the esters, amides, and triazole groups bind subsites A and E via stacking with Arg144 and Arg186, respectively

experimental hepatitis, pancreatic islet cell apoptosis, macrophage differentiation, myofibroblast activation, and angiogenesis (Table 2). The latter three are examples with a clear clinical need due to lack of satisfactory treatments of fibrotic diseases and ocular angiogenesis, and they are discussed more in detail below.

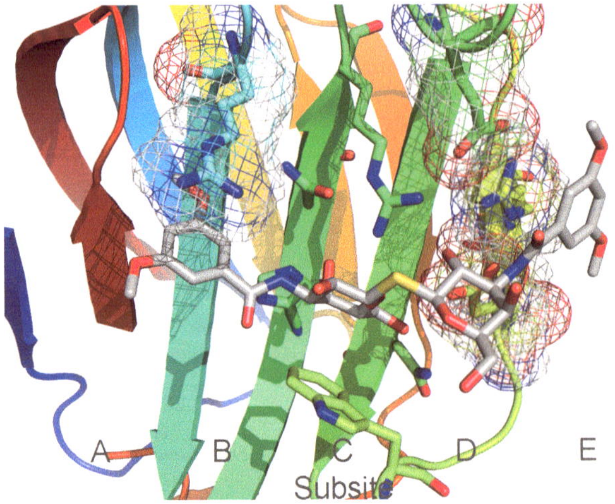

Fig. 7 A model of the galectin-3 CRD complexing a 3,3′-dibenzamido-thiodigalactoside binding subsites A–E. The model was built from an X-ray structure of galectin-3 CRD in complex with a 3′-benzamido-LacNAc derivative (PDB id 1KLR [84]). A π-surface, created by in-plane Glu-Arg-Glu ion pairs, stacking face-to-face with the right-hand 3,5-dimethoxybenzamide and Arg144 stacking face-to-face onto the left-hand 3,5-dimethoxybenzamide, is illustrated in mesh. Modeling was performed with the MMFFs force field with water implemented in MacroModel (version 9.1, Schrödinger, LLC, New York). The figure was generated with PyMOL Molecular Graphics System (Version 1.5.0.4 Schrödinger, LLC)

4.1 Galectin-3 Modulators in Chronic Inflammation and Fibrosis

Fibrotic diseases represent a significant clinical need that to a large extent still today is unmet. Cellular and molecular mechanisms are complex and involve a number of different cells and signaling pathways [104]. Fibrosis can be considered to be dysregulated wounding response to chronic injury characterized by myofibroblast activation and extracellular matrix deposition leading to scarring and organ dysfunction (reviewed in Wynn and Ramalingam [104]). This general response underlies the pathophysiology of many common diseases, including cardiovascular disease, cancer, and chronic diseases of the liver, lung, and kidney. There are very few therapeutic options for the treatment of fibrosis and there are currently no FDA-approved anti-fibrotic therapies. Transplantation remains the only option for end-stage disease so the characterization of novel anti-fibrotic therapies is urgently required.

In recent years, it has become clear that galectin-3 has profound effects on tissue fibrosis in a variety of organs [14, 105–107]. Using mice deficient in galectin-3, it was observed that there was reduced collagen accumulation in experimental models of fibrosis of the liver, kidney, lung, and blood vessels. In WT animals and in human fibrotic disease, there is an increase in galectin-3 expression within the areas of scarring (Fig. 8).

Table 2 Small molecule galectin modulators in experiments demonstrating roles of galectin-3 carbohydrate binding

Galectin-3 inhibitor (see Fig. 4)	Cell or animal organ	Effect
3,3′-Ditriazolyl thiodigalactoside [102]	M2 macrophages	Galectin-3 endocytosis attenuated
3,3′-Benzoyl thiodigalactoside [75]	Lung and prostate tumor cell lines	Tumor cell motility attenuated
3,3′-Ditriazolyl thiodigalactoside [30]	Thyroid tumor cell line	Enhanced sensitivity to proapoptotic drug
Thiodigalactoside [31]	Mouse breast cancer model	Stimulation of antitumor immunity
3,3′-Ditriazolyl thiodigalactoside [32]	Mouse liver	Attenuation of conA-induced hepatitis
3,3′-Ditriazolyl thiodigalactoside [33]	Mouse pancreas	Attenuation of pancreatic islet cell apoptosis
3,3′-Benzamido thiodigalactoside [13]	M2 macrophages	Attenuation of macrophage M2 polarization
3,3′-Ditriazolyl thiodigalactoside [14]	Mouse lung	Attenuation of bleomycin-induced lung fibrosis
3,3′-Ditriazolyl thiodigalactoside [103]	Mouse eye	Reduced neovascularization in eye

Although the myofibroblasts themselves upregulate galectin-3 expression [106], the source of galectin-3 in diseased organs may be principally derived from tissue resident and recruited immune cells principally macrophages [13, 105]. Macrophages are key cells which orchestrate the evolution and resolution of fibrosis in liver [108, 109], kidney [105], and lung [110], and galectin-3 has been shown to induce macrophage differentiation towards a pro-fibrotic phenotype leading to the promotion of scar formation [13] (Fig. 9). Moreover the key roles of macrophage differentiation in kidney [105] and of bone marrow-derived, lung, and peritoneal macrophages [13] and myofibroblast differentiation in hepatic [106], lung [14, 91], and cardiac fibrotic development are linked to enhanced galectin-3 expression. This has raised the question of whether the role of galectin-3 in the macrophages/myofibroblast axis during fibrotic development is carbohydrate dependent and may be amenable to modulation by galectin-3 CRD-binding molecules.

4.1.1 Modulating Macrophage M2 Differentiation

Classical macrophage differentiation (M1-polarization) and function are associated with acute inflammatory processes resulting in eradication of, e.g., pathogens and tumor cells. The wound-healing phase on the other hand is characterized by alternative macrophage M2 polarization, which eventually leads to tissue remodeling and fibrosis. M2 polarization is stimulated by the Th2 cytokines IL-4 and IL-13 inducing their respective receptor dimerization, which in turn activates JAK tyrosine kinases resulting in STAT6 and insulin receptor substrate-1 (IRS-1)

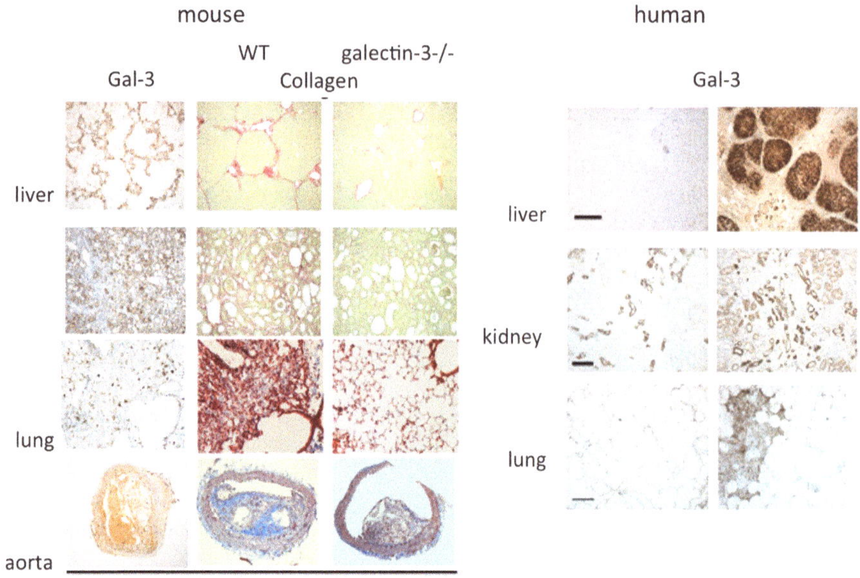

Fig. 8 Galectin-3 expression in WT and collagen deposition in WT and galectin-3-/-fibrotic mouse tissue (*left*). Galectin-3 expression in normal (*left*) and fibrotic (*right*) human liver (cirrhosis), kidney (glomerulonephritis), and lung (usual interstitial pneumonia (UIP))

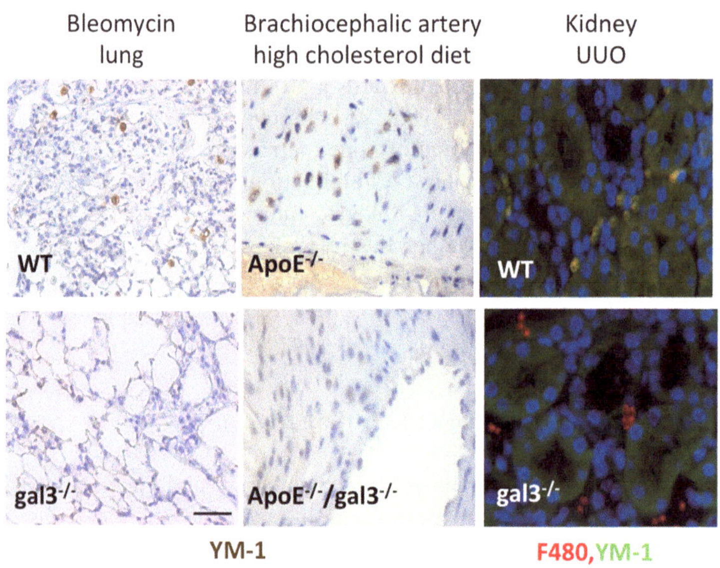

Fig. 9 Alternatively activated macrophages (YM-1 positive) are recruited to the injured areas of lung (bleomycin), artery (ApoE−/−), and kidney (unilateral ureteric obstruction) in WT but not galectin-3−/− mice

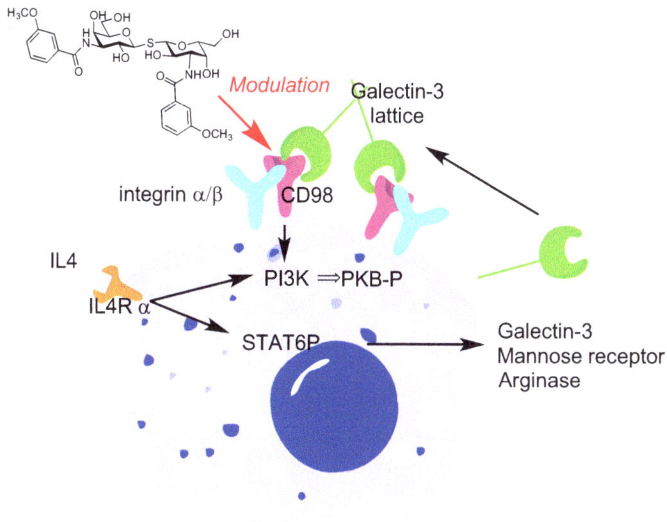

Fig. 10 Proposed mechanism for a carbohydrate-dependent galectin-3 feedback loop in macrophage M2 polarization and its modulation by a small molecule binding galectin-3 subsites A–E with nanomolar affinity [13]. IL4- and IL13-induced activation of macrophages leads to PI3K activation and STAT6 phosphorylation and nuclear translocation, which in turn induce galectin-3, mannose receptor, and arginase expression. Extracellular galectin-3 cross-links CD98, which enhances CD98 binding to integrins and thus even further PI3K activation

family phosphorylation. Nuclear translocation of phosphorylated STAT6 initiates transcription and phosphorylated IRS-1 activates PI3K and Grb2. Galectin-3 and arginase are key genes upregulated during M2 polarization. Galectin-3 is known to bind cell surface CD98, an inflammation-related transmembrane glycoprotein that is highly expressed in macrophages [111]. CD98 in turn associates with β1 integrins leading to further enhanced PI3K activation. Activation of arginase leads to increased arginine catabolism and polyamine and proline production (required for collagen production by myofibroblasts, vide infra). Furthermore, macrophages are a significant source of TGF-β, a central regulator of myofibroblast activation and fibrosis (vide infra). On a cellular level, it was observed that M2 polarization of bone-marrow-derived macrophages in vitro and resident lung and recruited peritoneal macrophages in vivo was reduced in galectin-3-deficient mice [13]. These observations were also corroborated by the discovery that a subsite A–E nanomolar-binding galectin-3 modulator (bis-(3-deoxy-3-(3-methoxybenzamido)-β-D-galactopyranose) sulfane, Fig. 6) blocked M2 polarization (Fig. 10). Altogether, these data suggest that macrophage M2 polarization is enhanced by a carbohydrate-dependent galectin-3 feedback mechanism, which points to galectin-3 being a relevant target within the galectin-3/CD98/PI3K pathway for therapeutic strategies based on small molecules modulating galectin-3 carbohydrate binding for treating chronic inflammation, fibrosis, and possibly cancer.

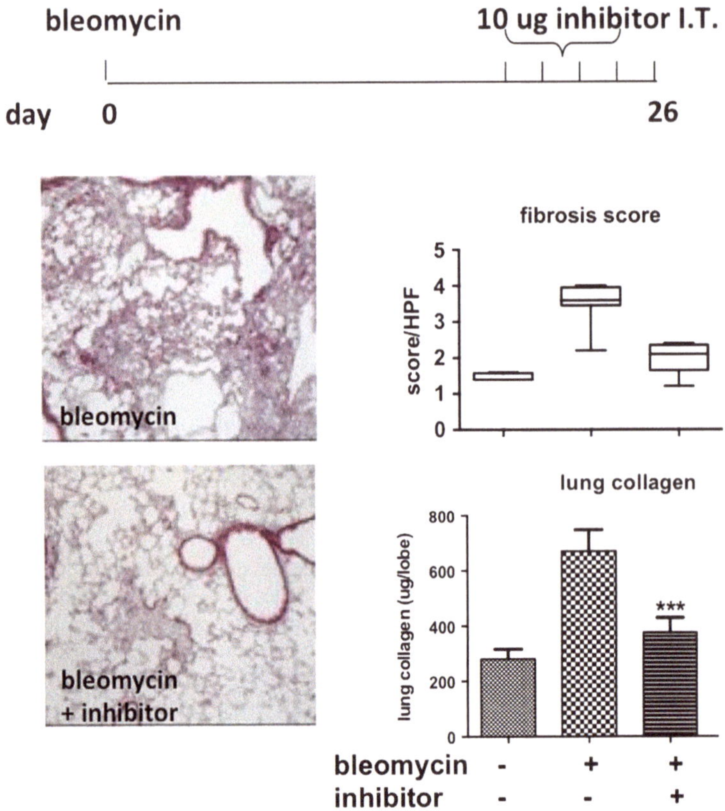

Fig. 11 Subsite A–E nanomolar-binding galectin-3 modulator (bis-(3-deoxy-3-(4-(3-fluorophenyl)-1,2,3-triazol-1-yl)-β-D-galactopyranose) sulfane, Fig. 6) was administered I.T during the fibrotic phase of bleomycin-induced murine lung fibrosis. Mice treated with inhibitor showed decreased fibrosis score and lung collagen content [14]

4.1.2 Modulating Myofibroblast Activation and Fibrosis

Galectin-3 expression is consistently enhanced in fibrotic tissue (vide supra) and lung macrophages are known as a major source in pulmonary inflammation and fibrosis. Mice deficient in galectin-3 were demonstrated to display a marked decrease in pulmonary fibrosis in two models, fibrosis induced by bleomycin or by the pro-fibrotic cytokine TGFβ [14]. This result was also mirrored in an experiment where TGFβRII cell surface residence and pulmonary fibrosis were greatly decreased by treatment with a subsite A–E nanomolar-binding galectin-3 modulator (bis-(3-deoxy-3-(4-(3-fluorophenyl)-1,2,3-triazol-1-yl)-β-D-galactopyranose) sulfane, Fig. 6) during the fibrotic phase of bleomycin-induced injury (Fig. 11).

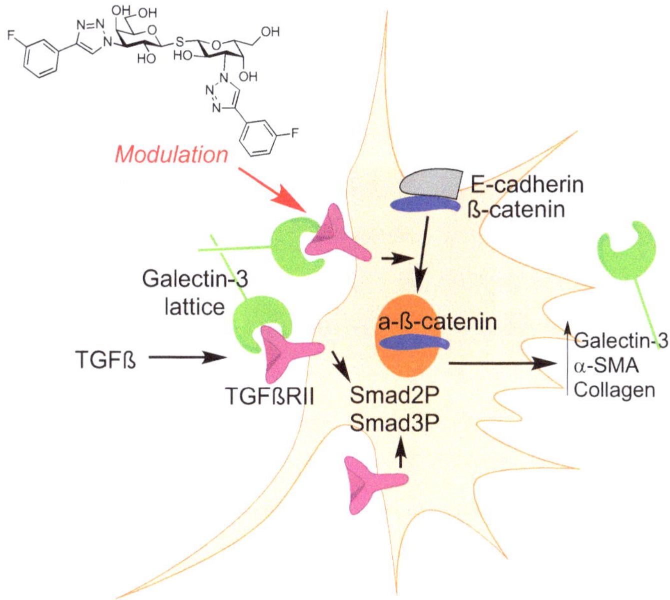

Fig. 12 Proposed mechanism for a carbohydrate-dependent galectin-3-mediated enhancement of TGFβ signaling and its modulation by a small molecule binding galectin-3 subsites A–E with nanomolar affinity [14]

While TGFβ is known to induce epithelial to mesenchymal transition (EMT), myofibroblast activation, and collagen production via downstream signaling involving Smad2 and Smad3 phosphorylation, galectin-3 deletion or inhibition had no effect on Smad phosphorylation. However, β-catenin phosphorylation and β-catenin nuclear translocation were reduced suggesting a link between TGFβRII galectin-3 binding and the noncanonical Wnt/β-catenin pathway, which is commonly aberrantly activated in idiopathic pulmonary fibrosis (IPF). It was hypothesized that TGFβRII-galectin-3 interactions and lattice formation enhance TGFβ-retention at the cell surface promoting β-catenin activation (Fig. 12). To sum up, galectin-3 manifests itself as a unique anti-fibrosis target as it is rate limiting for fibrotic activities by two key cell types in the development of fibrosis, M2-polarized macrophages, and myofibroblasts. This is of great clinical importance with respect to IPF, as this chronic condition is today lacking efficient methods for treatment and rapidly decreases lung function and thus ultimately leads to respiratory failure and death.

4.2 *Galectin-3 Modulators in Angiogenesis*

Formation of new blood vessels, angiogenesis, is a process needed for a variety of normal organism function as well as for progression of a large number of pathological states, including ocular diseases. Angiogenesis is initiated by epithelial cells

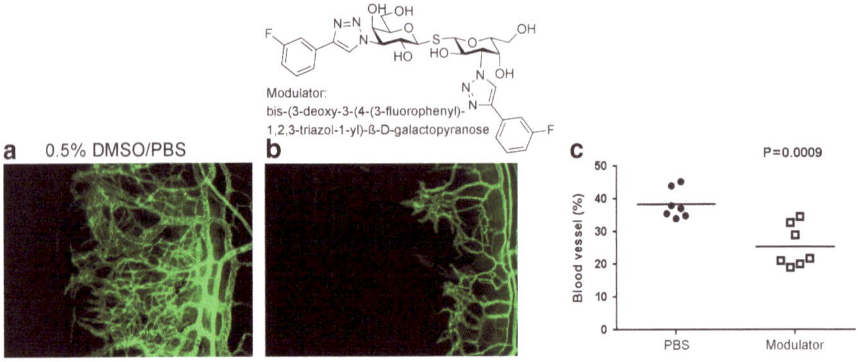

Fig. 13 Treatment with the galecin-3-modulator bis-(3-deoxy-3-(4-(3-fluorophenyl)-1,2,3-triazol-1-yl)-β-D-galactopyranose markedly reduces corneal neovascularization. Neovascularization was induced in mouse corneas by silver nitrate cautery. The modulator (325 ng in 10 μL) or vehicle (10 μL PBS containing 0.5% DMSO) was administered by subconjunctival injections every other day. In addition, a 10 μL of vehicle alone or 50 μM modulator eye drop was applied once every day. After 5 days, mice were sacrificed, and flat mounts of corneas were stained with anti-CD31 to visualize blood vessels (*green*). (**a**) Administration of vehicle; (**b**) administration of modulator. Representative segments from corneal flat mounts stained with anti-CD31 (A and B). The density of blood vessels (**c**) covering the whole cornea was quantified by ImageJ. The neovascular response after cautery is significantly diminished when the modulator is administered (Magnification: **a** and **b**, ×100) [103]

responding to cytokines, e.g., bFGF and VEGF, which result in increased migration, proliferation, and finally blood vessel formation. Galectin-3 was recently demonstrated by siRNA knockdown experiments and in galectin-3 knockout mice to be a regulator of bFGF and VEGF signaling [19]. VEGFRII cell surface residence time and thus neovascularization were later shown to be related to galectin-3, and carbohydrate binding with the VEGFRII glycoprotein, presumably by forming lattices as an interaction between galectin-3 and VEGFRII at the plasma membrane, could be detected [18]. Furthermore, these observations were supported by experiments showing reduced neovascularization in galectin-3 and in Mgat5 knockout mice. As Mgat5 induces branching of *N*-glycan polylactosamine chains on glycoproteins and thus increases the prevalence of galectin-binding epitopes, the latter in vivo result suggests carbohydrate dependence. Indeed, a subsite A–E nanomolar-binding galectin-3 modulator bis-(3-deoxy-3-(4-(3-fluorophenyl)-1,2,3-triazol-1-yl)-β-D-galactopyranose) sulfane (Fig. 6) diminished VEGF-induced endothelial cell sprouting and migration in a dose-dependent manner (Fig. 13), which further strongly supports the view that galectin-3-VEGFRII interactions are carbohydrate dependent [103]. This was also corroborated in vivo in an experiment where subconjunctival administration of the same galectin-3 modulator reduced corneal neovascularization in a mouse model of silver nitrate-induced cautery [103]. All these data lead to the intriguing suggestion that galectin-3 modulators may be of value in treating not only angiogenesis-related ocular conditions but also other diseases in which VEGFRII signaling and angiogenesis is rate limiting.

5 Summary, Conclusions, Outlook

Systematic investigations with synthetically modified galactopyranose derivatives designed to specifically interact with galectin-3 CRD subsites have proven that it is possible to develop galectin modulators with high (single-digit nanomolar) affinities and high selectivities and that such compounds indeed can antagonize galectin functions in relevant in vitro and in vivo models. As galectin-3 appears rate limiting in a number of pathological processes, these discoveries hold great promise for future galectin-3-targeting pharmaceuticals. However, although several examples of galectin-3 modulators with in vivo efficacy are known, most have used models with local administration as ADME/PK properties of galectin-3 modulators are typically far from optimal due to their hydroxylated and polar nature. Hence, with an increased focus on ADME/PK optimization, the field can expect to witness the development of future galectin-3 modulators with good systemic exposure and efficacy in a broader range of pathological models.

References

1. Boscher C, Dennis JW, Nabi IR (2011) Glycosylation, galectins and cellular signaling. Curr Opin Cell Biol 23:383–392
2. Delacour D, Koch A, Jacob R (2009) The role of galectins in protein trafficking. Traffic 10:1405–1413
3. Dennis JW, Brewer CF (2013) Density-dependent lectin-glycan interactions as a paradigm for conditional regulation by posttranslational modifications. Mol Cell Proteomics 12:913–920
4. Dennis JW, Lau KS, Demetriou M, Nabi IR (2009) Adaptive regulation at the cell surface by N-glycosylation. Traffic 10:1569–1578
5. Houzelstein D, Goncalves IR, Fadden AJ, Sidhu SS, Cooper DN, Drickamer K, Leffler H, Poirier F (2004) Phylogenetic analysis of the vertebrate galectin family. Mol Biol Evol 21:1177–1187
6. Leffler H, Carlsson S, Hedlund M, Qian Y, Poirier F (2004) Introduction to galectins. Glycoconj J 19:433–440
7. Ahmad N, Gabius H-J, André S, Kaltner H, Sabesan S, Roy R, Liu B, Macaluso F, Brewer CF (2004) Galectin-3 precipitates as a pentamer with synthetic multivalent carbohydrates and forms heterogeneous cross-linked complexes. J Biol Chem 279:10841–10847
8. Lepur A, Salomonsson E, Nilsson UJ, Leffler H (2012) Ligand induced galectin-3 self-association. J Biol Chem 287:21751–21756
9. Nieminen J, Kuno A, Hirabayashi J, Sato S (2007) Visualization of galectin-3 oligomerization on the surface of neutrophils and endothelial cells using fluorescence resonance energy transfer. J Biol Chem 282:1374–1383
10. Delacour D, Greb C, Koch A, Salomonsson E, Leffler H, Le Bivic A, Jacob R (2007) Apical sorting by galectin-3-dependent glycoprotein clustering. Traffic 8:379–388
11. Schneider D, Greb C, Koch A, Straube T, Elli A, Delacour D, Jacob R (2010) Trafficking of galectin-3 through endosomal organelles of polarized and non-polarized cells. Eur J Cell Biol 89:788–798

12. Partridge EA, Le Roy C, Di Guglielmo GM, Pawling J, Cheung P, Granovsky M, Nabi IR, Wrana JL, Dennis JW (2004) Regulation of cytokine receptors by Golgi N-glycan processing and endocytosis. Science 306:120–124
13. MacKinnon AC, Farnworth SL, Hodkinson PS, Henderson NC, Atkinson KM, Leffler H, Nilsson UJ, Haslett C, Forbes SJ, Sethi T (2008) Regulation of alternative macrophage activation by galectin-3. J Immunol 180:2650–2658
14. Mackinnon AC, Gibbons MA, Farnworth SL, Leffler H, Nilsson UJ, Delaine T, Simpson AJ, Forbes SJ, Hirani N, Gauldie J, Sethi T (2012) Regulation of TGF-β1 driven lung fibrosis by galectin-3. Am J Resp Crit Care Med 185:537–546
15. Demetriou M, Granovsky M, Quaggin S, Dennis JW (2001) Negative regulation of T-cell activation and autoimmunity by Mgat5 N-glycosylation. Nature 409:733–739
16. Demotte N, Stroobant V, Courtoy PJ, Van Der Smissen P, Colau D, Luescher IF, Hivroz C, Nicaise J, Squifflet J-L, Mourad M, Godelaine D, Boon T, van der Bruggen P (2008) Restoring the association of the T cell receptor with CD8 reverses anergy in human tumor-infiltrating lymphocytes. Immunity 28:414–424
17. Demotte N, Wieërs G, Van Der Smissen P, Moser M (2010) A galectin-3 ligand corrects the impaired function of human CD4 and CD8 tumor-infiltrating lymphocytes and favors tumor rejection in mice. Cancer Res 70:7476–7488
18. Markowska AI, Jefferies KC, Panjwani N (2011) Galectin-3 protein modulates cell surface expression and activation of vascular endothelial growth factor receptor 2 in human endothelial cells. J Biol Chem 286:29913–29921
19. Markowska AI, Liu F-T, Panjwani N (2010) Galectin-3 is an important mediator of VEGF- and bFGF-mediated angiogenic response. J Exp Med 207:1981–1993
20. Rabinovich GA, Toscano MA (2009) Turning 'sweet' on immunity: galectin–glycan interactions in immune tolerance and inflammation. Nat Rev Immunol 9:338–352
21. Thijssen VL, Barkan B, Shoji H, Aries IM, Mathieu V, Deltour L, Hackeng TM, Kiss R, Kloog Y, Poirier F, Griffioen AW (2010) Tumor cells secrete galectin-1 to enhance endothelial cell activity. Cancer Res 70:6216–6224
22. Lefranc F, Mathieu V, Kiss R (2011) Galectin-1 as an oncotarget in gliomas and melanomas. Oncotarget 2:892–893
23. Dings RPM, Miller MC, Nesmelova I, Astorgues-Xerri L, Kumar N, Serova M, Chen X, Raymond E, Hoye TR, Mayo KH (2012) Antitumor agent calixarene 0118 targets human galectin-1 as an allosteric inhibitor of carbohydrate binding. J Med Chem 55:5121–5129
24. Salomonsson E, Thijssen VL, Griffioen AW, Nilsson UJ, Leffler H (2011) The anti-angiogenic peptide anginex greatly enhances galectin-1 binding affinity for glycoproteins. J Biol Chem 286:13801–13804
25. Thijssen VLJL, Postel R, Brandwijk RJMGE, Dings RPM, Nesmelova I, Satijn S, Verhofstad N, Nakabeppu Y, Baum LG, Bakkers J, Mayo KH, Poirier F, Griffioen AW (2006) Galectin-1 is essential in tumor angiogenesis and is a target for antiangiogenesis therapy. Proc Natl Acad Sci 103:15975–15980
26. Miller MC, Klyosov A, Mayo KH (2009) The alpha-galactomannan Davanat binds galectin-1 at a site different from the conventional galectin carbohydrate binding domain. Glycobiology 19:1034–1045
27. Zhu C, Anderson AC, Schubart A, Xiong H, Imitola J, Khoury SJ, Zheng XX, Strom TB, Kuchroo VK (2005) The Tim-3 ligand galectin-9 negatively regulates T helper type 1 immunity. Nat Immunol 6:1245–1252
28. Leitner J, Rieger A, Pickl WF, Zlabinger G (2013) TIM-3 does not act as a receptor for galectin-9. PLoS Pathog 9:e1003253. doi:10.1371/journal.ppat.1003253
29. Carlsson S, Carlsson MC, Leffler H (2007) Intracellular sorting of galectin-8 based on carbohydrate fine specificity. Glycobiology 17:906–912
30. Lin C-I, Whang EE, Donner DB, Jiang X, Price BD, Carothers AM, Delaine T, Leffler H, Nilsson UJ, Nose V, Moore FD, Ruan DT (2009) Galectin-3 targeted therapy with a small

molecule inhibitor activates apoptosis and enhances both chemosensitivity and radiosensitivity in papillary thyroid cancer. Mol Cancer Res 7:1655–1662
31. Stannard KA, Collins PM, Ito K, Sullivan EM, Scott SA, Gabutero E, Darren Grice I, Low P, Nilsson UJ, Leffler H, Blanchard H, Ralph SJ (2010) Galectin inhibitory disaccharides promote tumour immunity in a breast cancer model. Cancer Lett 299:95–110
32. Volarevic V, Milovanovic M, Ljujic B, Pejnovic N, Arsenijevic N, Nilsson U, Leffler H, Lukic ML (2012) Galectin-3 deficiency prevents concanavalin A- induced hepatitis in mice. Hepatology 55:1954–1964
33. Saksida T, Nikolic I, Vujicic M, Nilsson UJ, Leffler H, Lukic ML, Stojanovic I, Stosic-Grujicic S (2013) Galectin-3 deficiency protects pancreatic islet cells from cytokine-triggered apoptosis in vitro. J Cell Phys 228:1568–1576
34. Seetharaman J, Kanigsberg A, Slaaby R, Leffler H, Barondes SH, Rini JM (1998) X-ray crystal structure of the human galectin-3 carbohydrate recognition domain at 2.1-A resolution. J BiolChem 273:13047–13052
35. Yang R-Y, Rabinovich GA, Liu F-T (2008) Galectins: structure, function and therapeutic potential. Exp Rev Mol Med 10:1–24
36. Hirabayashi J, Hashidate T, Arata Y, Nishi N, Nakamura T, Hirashima M, Urashima T, Oka T, Futai M, Muller WEG, Yagi F, K-i K (2002) Oligosaccharide specificity of galectins: a search by frontal affinity chromatography. Biochem Biophys Acta 1572:232–254
37. Knibbs R, Agrwal N, Wang JL, Goldstein IJ (1993) Carbohydrate-binding protein 35. II. Analysis of the interaction of the recombinant polypeptide with saccharides. J Biol Chem 268:14940–14947
38. Salomonsson E, Carlsson M, Osla V, Hendus-Altenburger R, Kahl Knutson B, Oberg C, Sundin A, Nilsson R, Nordberg-Karlsson E, Nilsson U, Karlsson A, Rini JM, Leffler H (2010) Mutational tuning of galectin-3 specificity and biological function. J Biol Chem 285:35079–35091
39. Saraboji K, Håkansson M, Genheden S, Diehl C, Qvist J, Weininger U, Nilsson U, Leffler H, Ryde U, Akke M, Logan D (2011) The carbohydrate-binding site in galectin-3 is pre-organized to recognize a sugar-like framework of oxygens: implications for drug design. Biochemistry 51:296–306
40. Platt D, Raz A (1992) Modulation of the lung colonization of B16-F1 melanoma cells by citrus pectin. J Natl Cancer Inst 84:438–442
41. Nangia-Makker P, Hogan V, Honjo Y, Baccarini S, Tait L, Bresalier R, Raz A (2002) Inhibition of human cancer cell growth and metastasis in nude mice by oral intake of modified citrus pectin. J Natl Cancer Inst 94:1854–1862
42. Courts FL (2013) Profiling of modified citrus pectin oligosaccharide transport across Caco-2 cell monolayers. Biochem Pharm 1–39
43. Morris VJ, Belshaw NJ, Waldron KW, Maxwell EG (2013) The bioactivity of modified pectin fragments. Bioact Carbohydr Diet Fibre 1:21–37
44. André S, Kaltner H, Furuike T, Nishimura S-I, Gabius H-J (2004) Persubstituted cyclodextrin-based glycoclusters as inhibitors of protein-carbohydrate recognition using purified plant and mammalian lectins and wild-type and lectin-gene-transfected tumor cells as targets. Bioconj Chem 15:87–98
45. André S, Specker D, Bovin NV, Lensch M, Kaltner H, Gabius H-J, Wittmann V (2009) Carbamate-linked lactose: design of clusters and evidence for selectivity to block binding of human lectins to (neo)glycoproteins with increasing degree of branching and to tumor cells. Bioconj Chem 20:1716–1728
46. André S, Lahmann M, Gabius H-J, Oscarson S (2010) Glycocluster design for improved avidity and selectivity in blocking human lectin/plant toxin binding to glycoproteins and cells. Mol Pharm 7:2270–2279
47. Dings R, Laar E, Webber J, Zhang Y, Griffin R, Waters SJ, MacDonald JR, Mayo KH (2008) Ovarian tumor growth regression using a combination of vascular targeting agents anginex or topomimetic 0118 and the chemotherapeutic irofulven. Cancer Lett 265:270–280

48. Dings R, van der Schaft D, Hargittai B, Haseman J, Griffioen AW, Mayo KH (2003) Anti-tumor activity of the novel angiogenesis inhibitor anginex. Cancer Lett 194:55–66
49. Dings RPM, Van Laar ES, Loren M, Webber J, Zhang Y, Waters SJ, Macdonald JR, Mayo KH (2010) Inhibiting tumor growth by targeting tumor vasculature with galectin-1 antagonist anginex conjugated to the cytotoxic acylfulvene, 6-hydroxylpropylacylfulvene. Bioconjug Chem 21:20–27
50. Andre S, Arnusch C, Kuwabara I, Russwurm R, Kaltner H, Gabius H-J, Pieters R (2005) Identification of peptide ligands for malignancy- and growth-regulating galectins using random phage-display and designed combinatorial peptide libraries. Bioorg Med Chem 13:563–573
51. Andre S, Maljaars C, Halkes K, Gabius H-J, Kamerling J (2007) Discovery of galectin ligands in fully randomized combinatorial one-bead-one-compound (glyco)peptide libraries. Bioorg Med Chem Lett 17:793–798
52. Arnush CJ, André S, Valentini P, Lensch M, Russwurm R, Siebert H-C, Fischer MJE, Gabius H-J, Pieters RJ (2004) Interference of the galactose-dependent binding of lectins by novel pentapeptide ligands. Bioorg Med Chem 14:1437–1440
53. Maljaars CEP, André S, Halkes KM, Gabius H-J, Kamerling JP (2008) Assessing the inhibitory potency of galectin ligands identified from combinatorial (glyco) peptide libraries using surface plasmon resonance spectroscopy. Anal Biochem 378:190–196
54. Zou J, Glinsky VV, Landon LA, Matthews L, Deutscher SL (2005) Peptides specific to the galectin-3 carbohydrate recognition domain inhibit metastasis-associated cancer cell adhesion. Carcinogenesis 26:309–318
55. Guha P, Kaptan E, Bandyopadhyaya G, Kaczanowska S, Davila E, Thompson K, Martin SS, Kalvakolanu DV, Vasta GR, Ahmed H (2013) Cod glycopeptide with picomolar affinity to galectin-3 suppresses T-cell apoptosis and prostate cancer metastasis. Proc Natl Acad Sci USA 110:5052–5057
56. Moise A, Andr S, Eggers F (2011) Towards bioinspired galectin mimetics: identification of ligand-contacting peptides by proteolytic-excision mass spectrometry. J AM Chem Soc 133:14844–14847
57. Newton-Northup JR, Dickerson MT, Ma L, Besch-Williford CL, Deutscher SL (2012) Inhibition of metastatic tumor formation in vivo by a bacteriophage display-derived galectin-3 targeting peptide. Clin Exp Metastasis 30:119–132
58. John CM, Leffler H, Kahl-Knutsson B, Svensson I, Jarvis GA (2003) Truncated galectin-3 inhibits tumor growth and metastasis in orthotopic nude mouse model of human breast cancer. Clin Cancer Res 9:2374–2383
59. Mirandola L, Yu Y, Chui K, Jenkins M, Cobos E, John CM, Chiriva-Internati M (2011) Galectin-3C inhibits tumor growth and increases the anticancer activity of bortezomib in a murine model of human multiple myeloma. PloS one 6:e21811
60. Fort S, Kim HS, Hindsgaul O (2006) Screening for galectin-3 inhibitors from synthetic lacto-N-biose libraries using microscale affinity chromatography coupled to mass spectrometry. J Org Chem 71:7146–7154
61. Giguère D, André S, Bonin M-A, Bellefleur M-A, Provencal A, Cloutier P, Pucci B, Roy R, Gabius H-J (2011) Inhibitory potential of chemical substitutions at bioinspired sites of β-D-galactopyranose on neoglycoprotein/cell surface binding of two classes of medically relevant lectins. Bioorg Med Chem 19:3280–3287
62. Giguere D, Bonin M, Cloutier P, Patnam R, St-pierre C, Sato S, Roy R (2008) Synthesis of stable and selective inhibitors of human galectins-1 and -3. Bioorg Med Chem 16:7811–7823
63. Giguere D, Patnam R, Bellefleur M-A, St-Pierre C, Sato S, Roy R (2006) Carbohydrate triazoles and isoxazoles as inhibitors of galectins-1 and -3. Chem Commun 2379–2381
64. Giguere D, Sato S, St-pierre C, Sirois S, Roy R (2006) Aryl O- and S-galactosides and lactosides as specific inhibitors of human galectins-1 and -3: Role of electrostatic potential at O-3. Bioorg Med Chem Lett 16:1668–1672

65. Glinsky GV, Price JE, Glinsky VV, Mossine VV, Kiriakova G, Metcalf JB (1996) Inhibition of human breast cancer metastasis in nude mice by synthetic glycoamines. Cancer Res 56:5319–5324
66. Glinsky VV, Kiriakova G, Glinskii OV, Mossine VV, Mawhinney TP, Turk JR, Glinskii AB, Huxley VH, Price JE, Glinsky GV (2009) Synthetic galectin-3 inhibitor increases metastatic cancer cell sensitivity to taxol-induced apoptosis in vitro and in vivo. Neoplasia 11:901–909
67. Ingrassia L, Nshimyumukiza P, Dewelle J, Lefranc F, Wlodarczak L, Thomas S, Dielie G, Chiron C, Zedde C, Tisnes P, van Soest R, Braekman JJ, Darro F, Kiss R (2006) A lactosylated steroid contributes in vivo therapeutic benefits in experimental models of mouse lymphoma and human glioblastoma. J Med Chem 49:1800–1807
68. Rabinovich GA, Cumashi A, Bianco GA, Ciavardelli D, Iurisci I, D'Egidio M, Piccolo E, Tinari N, Nifantiev N, Iacobelli S (2006) Synthetic lactulose amines: novel class of anticancer agents that induce tumor-cell apoptosis and inhibit galectin-mediated homotypic cell aggregation and endothelial cell morphogenesis. Glycobiology 16:210–220
69. St-Pierre C, Ouellet M, Giguere D, Ohtake R, Roy R, Sato S, Tremblay MJ (2011) Galectin-1-specific inhibitors as a new class of compounds to treat HIV-1 infection. Antimicrob Agents Chemother 56:154–162
70. Collins PM, Oberg CT, Leffler H, Nilsson UJ, Blanchard H (2012) Talosides in complex with galectin-1 and galectin-3. Chem Biol Drug Design 79:339–346
71. Cumpstey I, Carlsson S, Leffler H, Nilsson UJ (2005) Synthesis of a phenyl thio-β-D-galactopyranoside library from 1,5-difluoro-2,4-dinitrobenzene: discovery of efficient and selective monosaccharide inhibitors of galectin-7. Org Biomol Chem 3:1922–1932
72. Cumpstey I, Salomonsson E, Sundin A, Leffler H, Nilsson UJ (2007) Studies of arginine–arene interactions through synthesis and evaluation of a series of galectin-binding aromatic lactose esters. ChemBioChem 8:1389–1398
73. Cumpstey I, Salomonsson E, Sundin A, Leffler H, Nilsson UJ (2008) Double affinity amplification of galectin-ligand interactions through arginine-arene interactions: synthetic, thermodynamic, and computational studies with aromatic diamido-thiodigalactosides. Chem Eur J 14:4233–4245
74. Cumpstey I, Sundin A, Leffler H, Nilsson UJ (2005) C_2-Symmetrical thiodigalactoside bis-benzamido derivatives as high-affinity inhibitors of galectin-3: efficient lectin inhibition through double arginine–arene interactions. Angew Chem Int Ed 44:5110–5112
75. Delaine T, Cumpstey I, Ingrassia L, Le Mercier M, Okechukwu P, Leffler H, Kiss R, Nilsson UJ (2008) Galectin-inhibitory thiodigalactoside ester derivatives have anti-migratory effects in cultured lung and prostate cancer cells. J Med Chem 51:8109–8114
76. Öberg CT, Blanchard H, Leffler H, Nilsson UJ (2008) Protein subtype-targeting through ligand epimerization: talose-selectivity of galectin-4 and galectin-8. Bioorg Med Chem Lett 18:3691–3694
77. Öberg CT, Leffler H, Nilsson UJ (2008) Arginine binding motifs: design and synthesis of galactose-derived arginine tweezers as galectin inhibitors. J Med Chem 51:2297–2301
78. Öberg CT, Noresson A-L, Leffler H, Nilsson UJ (2014) Structural and thermodynamic studies of arene-anion based arginine tweezers. In preparation
79. Öberg CT, Noresson A-L, Leffler H, Nilsson UJ (2011) Arene–anion based arginine-binding motif on a galactose scaffold: structure–activity relationships of interactions with arginine-rich galectins. Chem Eur J 17:8139–8144
80. Öberg CT, Noresson A-L, Leffler H, Nilsson UJ (2011) Synthesis of 3-amido-3-deoxy-β-D-talopyranosides: all-cis-substituted pyranosides as lectin inhibitors. Tetrahedron 67:9164–9172
81. Salameh BA, Cumpstey I, Sundin A, Leffler H, Nilsson UJ (2010) 1*H*-1,2,3-Triazol-1-yl thiodigalactoside derivatives as high affinity galectin-3 inhibitors. Bioorg Med Chem 18:5367–5378

82. Salameh BA, Leffler H, Nilsson UJ (2005) 3-(1,2,3-Triazol-1-yl)-1-thio-galactosides as small, efficient, and hydrolytically stable inhibitors of galectin-3. Bioorg Med Chem Lett 15:3344–3346
83. Salameh BA, Sundin A, Leffler H, Nilsson UJ (2006) Thioureido N-acetyllactosamine derivatives as potent galectin-7 and 9N inhibitors. Bioorg Med Chem 14:1215–1220
84. Sörme P, Arnoux P, Kahl-Knutsson B, Leffler H, Rini JM, Nilsson UJ (2005) Structural and thermodynamic studies on cation-Π interactions in lectin-ligand complexes: high-affinity galectin-3 inhibitors through fine-tuning of an arginine–arene interaction. J Am Chem Soc 127:1737–1743
85. Sörme P, Kahl-Knutsson B, Wellmar U, Magnusson B-G, Leffler H, Nilsson UJ (2003) Design and synthesis of galectin inhibitors. Meth Enz 363:157–169
86. Sörme P, Qian Y, Nyholm P-G, Leffler H, Nilsson UJ (2002) Low micromolar inhibitors of galectin-3 based on 3′-derivatization of N-acetyllactosamine. ChemBioChem 3:183–189
87. Tejler J, Leffler H, Nilsson UJ (2005) Synthesis of O-galactosyl aldoximes as potent LacNAc-mimetic galectin-3 inhibitors. Bioorg Med Chem Lett 15:2343–2345
88. Tejler J, Salameh B, Leffler H, Nilsson UJ (2009) Fragment-based development of triazole-substituted O-galactosyl aldoximes with fragment-induced affinity and selectivity for galectin-3. Org Biomol Chem 7:3982–3990
89. Tejler J, Skogman F, Leffler H, Nilsson UJ (2007) Synthesis of galactose-mimicking1H-(1,2,3-triazol-1-yl)-mannosides as selective galectin-3 and 9N inhibitors. Carbohydr Res 342:1869–1875
90. Tejler J, Tullberg J, Frejd T, Leffler H, Nilsson UJ (2006) Synthesis of multivalent lactose derivatives by 1,3-dipolar cycloadditions: selective galectin-1 inhibition. Carbohydr Res 34:1353–1362
91. Bum-Erdene K, Gagarinov IA, Collins PM, Winger M, Pearson AG, Wilson JC, Leffler H, Nilsson UJ, Grice ID, Blanchard H (2013) Investigation into the feasibility of thioditaloside as a novel scaffold for galectin-3-specific inhibitors. ChemBioChem 14:1331–1342
92. Van Hattum H, Branderhorst HM, Moret EE, Nilsson UJ, Leffler H, Pieters RJ (2013) Tuning the preference of thiodigalactoside- and lactosamine-based ligands to galectin-3 over galectin-1. J Med Chem 56:1350–1354
93. Camby I, Le Mercier M, Mathieu V, Ingrassia L, Lefranc F, Kiss R (2008) Galectin-1 as potential therapeutic target for cancer progression. Drug Future 33:1057–1069
94. Ingrassia L, Camby I, Lefranc F, Mathieu V, Nshimyumukiza P, Darro F, Kiss R (2006) Anti-galectin compounds as potential anti-cancer drugs. Curr Med Chem 13:3513–3527
95. Pieters RJ (2006) Inhibition and detection of galectins. ChemBioChem 7:721–728
96. Téllez-Sanz R, García-Fuentes L, Vargas-Berenguel A (2013) Human galectin-3 selective and high affinity inhibitors. Present state and future perspectives. Curr Med Chem 20:2979–2990
97. Müller K, Faeh C, Diederich F (2007) Fluorine in pharmaceuticals: looking beyond intuition. Science 317:1881–1886
98. Paulini R, Müller K, Diederich F (2005) Orthogonal multipolar interactions in structural chemistry and biology. Angew Chem (Int Engl) 44:1788–1805
99. Diehl C, Engström O, Delaine T, Håkansson M, Genheden S, Modig K, Leffler H, Ryde U, Nilsson UJ, Akke M (2010) Protein flexibility and conformational entropy in ligand design targeting the carbohydrate recognition domain of galectin-3. J Am Chem Soc 132:14577–14589
100. MacRaild C, Daranas A, Bronowska A, Homans S (2007) Global changes in local protein dynamics reduce the entropic cost of carbohydrate binding in the arabinose-binding protein. J Mol Biol 368:822–832
101. Nesmelova IV, Ermakova E, Daragan VA, Pang M, Menéndez M, Lagartera L, Solís D, Baum LG, Mayo KH (2010) Lactose binding to galectin-1 modulates structural dynamics, increases conformational entropy, and occurs with apparent negative cooperativity. J Mol Biol 397:1209–1230

102. Lepur A, Carlsson MC, Novak R, Dumic J, Nilsson UJ, Leffler H (2012) Galectin-3 endocytosis by carbohydrate independent and dependent pathways in different macrophage like cell types. Biochim Biophys Acta 1820:804–818
103. Chen W-S, Leffler H, Nilsson UJ, Panjwani N (2013) TDX, a galectin-1 and galectin-3-specific inhibitor, mitigates VEGF-A-induced angiogenesis. FASEB J 27:828.1
104. Wynn TA, Ramalingam TR (2012) Mechanisms of fibrosis: therapeutic translation for fibrotic disease. Nat Med 18:1028–1040
105. Henderson NC, Mackinnon AC, Farnworth SL, Kipari T, Haslett C, Iredale JP, Liu FT, Hughes J, Sethi T (2008) Galectin-3 expression and secretion links macrophages to the promotion of renal fibrosis. Am J Pathol 172:288–298
106. Henderson NC, Mackinnon AC, Farnworth SL, Poirier F, Russo FP, Iredale JP, Haslett C, Simpson KJ, Sethi T (2006) Galectin-3 regulates myofibroblast activation and hepatic fibrosis. Proc Natl Acad Sci USA 103:5060–5065
107. Mackinnon A, Forbes S (2013) Bone marrow contributions to fibrosis. Biochim Biophys Acta Mol Basis Dis 1832:955–961
108. Duffield JS, Forbes SJ, Constandinou CM, Clay S, Partolina M, Vuthoori S, Wu S, Lang R, Iredale JP (2005) Selective depletion of macrophages reveals distinct, opposing roles during liver injury and repair. J Clin Invest 115:56–65
109. Ramachandran P, Pellicoro A, Vernon MA, Boulter L, Aucott RL, Ali A, Hartland SN, Snowdon VK, Cappon A, Gordon-Walker TT, Williams MJ, Dunbar DR, Manning JR, van Rooijen N, Fallowfield JA, Forbes SJ, Iredale JP (2012) Differential Ly-6C expression identifies the recruited macrophage phenotype, which orchestrates the regression of murine liver fibrosis. Proc Natl Acad Sci 109:E3186–E3195
110. Gibbons MA, MacKinnon AC, Ramachandran P, Dhaliwal K, Duffin R, Phythian-Adams AT, van Rooijen N, Haslett C, Howie SE, Simpson AJ, Hirani N, Gauldie J, Iredale JP, Sethi T, Forbes SJ (2011) Ly6Chi monocytes direct alternatively activated profibrotic macrophage regulation of lung fibrosis. Am J Resp Crit Care Med 184:569–581
111. Henderson NC, Collis EA, MacKinnon AC, Simpson KJ, Haslett C, Zent R, Ginsberg M, Sethi T (2004) CD98hc (SLC3A2) interaction with beta 1 integrins is required for transformation. J Biol Chem 279:54731–54741

Top Med Chem (2014) 12: 123–168
DOI: 10.1007/7355_2014_52

Published online: 10 August 2014

Discovery and Application of FimH Antagonists

Sébastien G. Gouin, Goedele Roos, and Julie Bouckaert

Abstract Just like bacteria need to be mobile to seek for nutrients, bacteria need to adhere to biotic and abiotic surfaces to enable their progression. Most bacteria regulate the expression of a multitude of fimbrial adhesins that display varying specificities and architectures. FimH at the tip of type 1 fimbriae is one of the first recognized lectins on *Escherichia coli*. FimH evokes through its binding symptomatic and chronic *E. coli* infections in the urinary tract, in the intestine, and beyond. The mannose specificity of type 1 fimbriae has been the lead to the discovery of the FimH adhesin more than 32 years ago and presents today a role model as the template for anti-adhesive drug design. Curiously, the specificity of the FimH lectin had been defined very early on toward a Manα1,3Manβ1,4GlcNAc trisaccharide isolated from the urine of mannosidase-deficient patients. Indeed, a much larger dependence of bacterial adhesion can be attributed to structural differences in the mannosidic receptors than based on amino acid variance in FimH. The crystal structure of FimH in complex with oligomannoside-3 presented a breakthrough that enhanced the rational design of mannose-based anti-adhesives against FimH. In this overview, we will provide insights gained from a plethora of FimH antagonists. Crystal structures of FimH in complex with anti-adhesives and applications in vivo in mouse models for metabolic diseases reveal unexpected features and alternative routes for improved molecules.

S.G. Gouin
Chimie Et Interdisciplinarité, Synthèse, Analyse, Modélisation (CEISAM), LUNAM Université, UMR 6230 du CNRS, 2, rue de la Houssinière, BP 92208, 44322 Nantes Cedex 3, France

G. Roos
VUB department of General Chemistry and VIB department of Structural Biology, Vrije Universiteit Brussel, VIB, Building G, Pleinlaan 2, 1050 Brussels, Belgium

J. Bouckaert (✉)
Unité de Glycobiologie Structurale et Fonctionnelle, Université Lille 1, UMR 8576 du CNRS, Avenue Mendeleiev building C9, 59655 Villeneuve d'Ascq, France
e-mail: julie.bouckaert@univ-lille1.fr

Keywords Anti-adhesives, Bacterial adhesion, *E. coli*, Enterobacteriaceae, FimH adhesin, Mannosides, Structure-based drug design, Type 1 fimbriae

Contents

Abbreviations

AIEC	Adherent-invasive *E. coli*
ATP	Adenosine triphosphate
bpMan	Biphenyl α-D-mannose
BM	Butyl α-D-mannose
cAMP	Cyclic adenosine monophosphate
CDK2	Cyclin dependent kinase 2
CEACAM6	Carcinoembryonic antigen-related cell adhesion molecule 6
CRP	cAMP receptor protein
DC-SIGN	Dendritic cell-specific intercellular adhesion molecule-3-grabbing nonintegrin
DFT	Density functional theory
DP	Degree of polymerization
E. coli	*Escherichia coli*
ER	Endoplasmic reticulum
ESI-MS	Electrospray ionization mass spectrometry
GPI	Glycosylphosphatidylinositol
HDAC6	Histone deacetylase 6
HM	Heptyl α-D-mannose
HSA	Human serum albumin

IBC	Intracellular bacterial community
ITC	Isothermal titration calorimetry
LB	Luria Bertani
M cells	Microfold cells, these are cells found in the follicle-associated epithelium of the Peyer's patch
MD	Molecular dynamics
MM	Methyl α-D-mannose
MO	Molecular orbital
NMR	Nuclear magnetic resonance
PDB	Protein data bank
RR	Recurrency rate
SAXS	Small-angle X-ray solution scattering
SPR	Surface plasmon resonance
SQM	Semi-empirical quantum mechanical
TazMan	Thiazolyl α-D-mannose
TNF-α	Tumor necrosis factor alpha
UTI	Urinary tract infection

1 Introduction

Escherichia coli is the most prevalent causative agent of bladder infection or *cystitis*, in more than 75% of the cases. Other bacterial agents include *Klebsiella*, *Proteus*, *Enterococcus*, and *Pseudomonas*. *E. coli* belongs to the family of *enterobacteriaceae* and only opportunistically enters the urinary tract: at times, bacteria invade it from the intestinal tract and cause urinary tract infections (UTIs). Most *E. coli* express a few hundreds of about 1 μm-long rod-shaped proteinaceous organelles on their cell surface to adhere in a multivalent fashion to the superficial bladder cell linings. On surface, *E. coli* forms colonies, and when under stress it forms communities in a protective biofilm matrix. The fimbriae carry at the edge of a flexible tip fibrillum, an adhesin that is a lectin with high-affinity for mannose, FimH. A successful settlement of an *E. coli* seed in the urinary tract enholds binding of FimH to high-mannosylated glycoprotein receptors, such as uroplakin Ia [1], and α3β1 integrins [2] by the binding of the type-1 fimbrial FimH adhesin. FimH binding provokes large conformational changes in the glycoprotein receptor, translating into signaling in the cytoplasm of the infected epithelial cell [3] and uptake of *E. coli*. In the epithelial cells, *E. coli* grows dynamically in intracellular bacterial communities (IBCs) in a biofilm-like matrix [4–6], which sometimes bulges out of broken epithelial cells [7]. Type 1 fimbriae are thus necessary for bacterial adherence and invasion [8–10], but were also found essential for biofilm formation and for maturation of the IBCs [11]. Maturation of the biofilms allows the bacteria to flux out of the cells and restart their pathogenic life cycle by adhering to neighboring cells. Without fimbriae, the bacteria have a much reduced fitness in the infected host [12].

Usual antibiotics cannot reach the bacteria inside the cells and the IBCs. The recurrence and persistence of urinary tract infections is due to this internalization of *E. coli* in the superficial epithelial cells [5]. Because the superficial cell layer is damaged and sloughed off, *E. coli* may also penetrate in the underlying cell layers and resides there quiescently for a long time, to only reappear and reseed the infection upon underlying cell growth and differentiation [13–15]. Remarkably, the latter process is itself activated by the binding of FimH, but not by chemical substances that can remove bacteria-loaded superficial cells such as protamine sulfate [13], and is the onset to tissue remodeling [16]. Severe early inflammatory responses to the bladder cell invasion by the uropathogenic bacterium are the prelude to recurrent and persistent urinary tract infections, because they speed up tissue remodeling [16, 17]. Patients suffering from diabetes mellitus are more prone to contract a urinary tract infections caused by the type-1 fimbriated *E. coli*, and the recurrency rate of these infections is higher (RR = 1.40) compared to women without diabetes mellitus, despite longer treatments with a more potent initial antibiotic such as norfloxacin [18]. It appears that the increased prevalence of urinary tract infections in diabetic women is not the result of a difference in the bacteria, but due to an increased adherence of *E. coli* expressing type-1 fimbriae to uroepithelial cells of women with diabetes mellitus [19]. Not only is the adherence of type 1-piliated *E. coli* higher to diabetic than to control epithelial cells, but it was also found that there is a correlation between the degree of bacterial adherence and the levels of glycosylated hemoglobin in the blood of the diabetic patients. This parallel observation prompted the authors to suggest that the receptors of FimH in diabetic uroepithelial cells have a different glycosylation profile than in those of healthy controls. Therapeutics against bacterial infections linked to metabolic changes may become a fast developing field in targeted anti-adhesive drug design [20].

2 Discoveries

2.1 *The Mannose-Binding Lectin on the Surface of* E. coli

The very first report on the isolation of a mannose-specific lectin from *E. coli* and its role in the adherence of the bacteria to epithelial cells was submitted more than 32 years ago by Nathan Sharon [21]. The purified lectin, with a molecular weight of about 36,500 Da, was located on the *E. coli* surface but its amino acid composition was markedly different from the type 1 pili protein or from the K99 (nowadays called F5) pili proteins reported earlier. In this same seminal paper, inhibition of yeast agglutination by *E. coli*, but not *Salmonella typhimurium*, by *p*-nitrophenyl α-D-mannoside was 60-fold more powerful than by yeast mannan (minimal inhibitory concentration of 0.25 mg/mL) or methyl α-D-mannoside [22]. This was indicative of a hydrophobic site near the mannose binding site of the *E. coli* but not the *S. typhimurium* type-1 fimbrial adhesin [23]. The much increased inhibitory potential of aglycons *p*-nitro-*o*-chlorophenyl and 4-methyl umbelliferyl over methyl was further confirmed among a panel of substituents of the phenyl group [24]. The location of the hydrophobic site in the *E. coli* adhesin could be determined as juxtaposed to the binding pocket for the reducing α-anomeric D-mannoside by distinguishing the lower inhibitory activity on yeast agglutination by *E. coli* O25 of 4-methylumbelliferyl α-mannobioside (twofold increase relative to the methyl aglycon) over 4-methylumbelliferyl α-mannose that was 600-fold more potent. By means of mannose derivatives or oligosaccharides isolated from urine of patients with diseases of glycoprotein catabolism [25], the carbohydrate binding site of FimH was suggested to be an extended pocket with optimal fit for a trisaccharide, Manα1,3Manβ1,4GlcNAc, and with a hydrophobic region in or close to the combining site [26]. The hydrophobic site located next to the α-D-mannose-binding pocket became the basis for the development of generalized anti-adhesives (Fig. 1).

2.2 *FimH Recognizes Short* N-*Linked High-Mannose Glycans on Protein Receptors*

The gene responsible for the mannose-specific adhesion protein of type-1 fimbriae was identified [28]. Its gene product, FimH, appeared antigenically conserved among *E. coli*, *Klebsiella pneumoniae*, *Serratia marcescens*, and *Citrobacter freundii* species, that each belongs to the Enterobaceriaceae family [29]. In congruence with these findings, FimH variants from uropathogenic, faecal, and enterohaemorrhagic isolates expressed the same specificities and affinities for high-mannose glycan structures [30, 31]. The only exception to this rule FimH adhesins from O157 strains with a mutation, Asn135Lys, which annihilates all mannose binding [30]. Affinity measurements demonstrated a strong preference of FimH toward oligomannosides exposing Manα1,3Man ($K_d = 300$ nM) at their nonreducing end [30, 32]. Binding is further enhanced by the β1,4-linkage to a chitobiose by a 100-fold over that of α-D-mannose. Epitope mapping on high-mannose glycan receptors revealed the highest affinity of the FimH lectin domain for oligomannose-3 and oligomannose-5 [30]. Both these oligomannosides expose Manα1,3Manβ1,4GlcNAc at the nonreducing end of the D1 branch and have an increased affinity for FimH over mannotriose and mannopentaose that lack the chitobiose unit (Fig. 1). The increased affinity was partially due to the β-anomeric linkage of the mannotriose to the chitobiose GlcNAcβ1,4GlcNAc, because Manα1,3ManβOMe had an elevated affinity ($K_d = 112$ nM) over the anomeric mixture of Manα1,3Man ~ OH ($K_d = 281$ nM). FimH thus prefers the β-anomeric configuration on the core mannose. However, the presence of the β-linkage alone is not sufficient to explain the significant increase in affinity between mannotriose ($K_d = 204$ nM) and mannopentaose ($K_d = 127$ nM) on one hand, both lack the

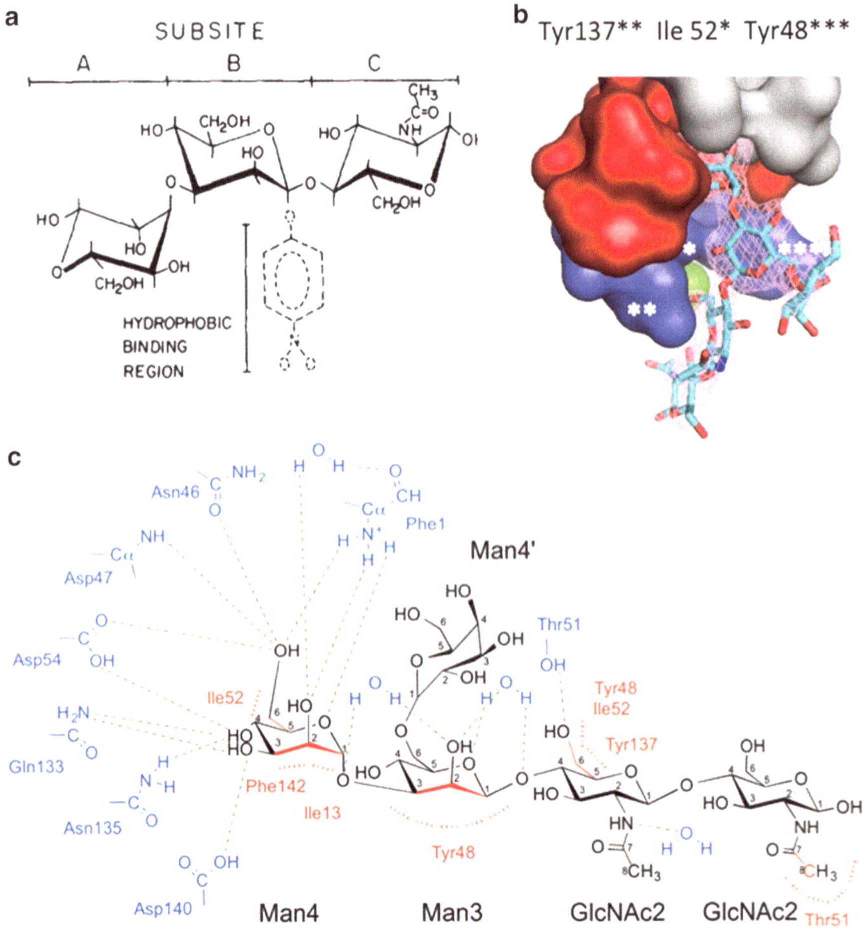

Fig. 1 (**a**) Nurit Firon and Nathan Sharon predicted the existence and the position of a hydrophobic subsite on the *E. coli* mannose-binding lectin and its importance in adhesion to α1,3-terminating short high-mannose glycan receptors (figure from [24]). Predicted subsite A corresponds to Man4, B to Man3, and C to GlcNAc2 of Fig. 1c. (**b**) The crystal structure of FimH in complex with the oligomannoside-5 *N*-linked glycan showed stacking of the central core mannose between two tyrosines, Tyr48 and Tyr137, over Ile52 and with its nonreducing end *N*-acetylglucosamine at Thr51 (*bright green*). [27]. (**c**) Interaction network of oligomannoside-3 with FimH, where polar ligand residues are shown in *blue* with their hydrogen bonds in *green*, and aliphatic or aromatic ligand residues are shown in *red* with hydrophobic stacking and van der Waals interactions in *red* [27]

chitobiose core; on the other hand, oligomannose-3 and oligomannose-5 ($K_d = 20$ nM) contain the chitobiose unit. Instead, the GlcNAcβ1,4GlcNAc chitobiose sequence, bridging the mannosides to the asparagine side chain in the Asn-X-Ser/Thr motif of glycoprotein receptors, is a crucial contributor to the interaction with FimH, in agreement with previous results.

2.3 Serendipitous Discovery of Butyl α-D-Mannose as FimH Ligand Was the Onset of New Drug Design Ventures

The FimH lectin domain was first expressed and purified to co-crystallize it with differently linked mannosides; nevertheless all the crystals obtained turned out to have the same space group and unit cell content. In two independent crystallography laboratories, the crystals of the FimH lectin domain were found to contain a clear electron density in the binding pocket of FimH. The electron density corresponded to mannose with a tail with the size of a butyl on the α-anomeric oxygen, butyl α-D-mannose [33]. Although no such sugar had been advertently added with the objective of crystallization, butyl α-D-mannose could be well-fitted in the electron density of the crystal and refined (Fig. 2b). We hypothesized that a mannosidic ligand had been picked up from the Luria–Bertani (LB) medium used to grow bacteria for the expression of recombinant FimH. We analyzed the sample using electrospray ionization mass spectrometry (ESI-MS), and as a control, butyl α-D-mannose was synthesized and analyzed in the same manner to reassure its size and the atomic nature of the anomeric linkage (Fig. 2a). When ESI-MS was run on FimH lectin expressed in bacteria grown in minimal medium, no peak corresponding to butyl α-D-mannose could be detected, indicating that the ligand

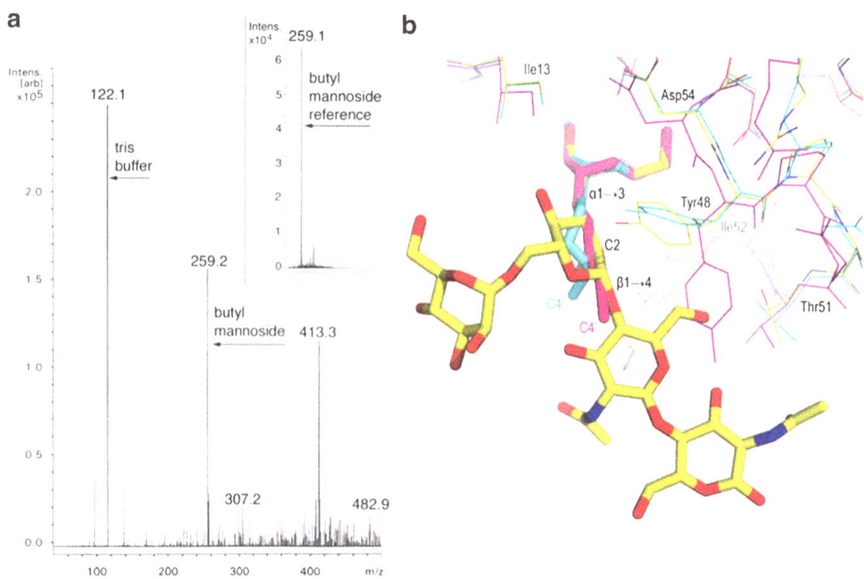

Fig. 2 The finding of butyl α-D-mannoside as the first anti-adhesive for pathogenic *E. coli* based on a simple alkyl-derivatized α-D-mannoside was the onset for further investigations. (**a**) Mass spectrometric analysis of the recombinant FimH lectin domain produced by *E. coli* grown in Luria Bertani broth. The inset shows the mass spectrum of the synthesized *O*-linked butyl α-D-mannose (*m/z* of 259 Da). (**b**) Overlay of the two crystal structures of butyl α-D-mannose–FimH complexes (PDB entries 1tr7 (*cyan*) and 1uwf (*magenta*)) with FimH–oligomannoside-3 (PDB entry 2vco)

was not present and that it indeed originated from the LB medium, although its biological origin is not understood. From then on, FimH lectins to be used for interaction studies were extracted from *E. coli* grown in minimal medium, which lacks the yeast mannans and remnants of GPI-anchored glycoproteins.

Butyl α-D-mannose conserves the crucial interactions of FimH with the α1,3 glycosidic linkage and the first β1,4 linkage to the chitobiose of oligomannose-3 [33]. First, the strong stacking of the central mannose with the aromatic ring of Tyr48 is congruent with the high affinity for synthetic inhibitors in which this central mannose is substituted by an aromatic group (Fig. 1a) [34]. Second, the FimH ligand butyl α-D-mannose, serendipitously co-crystalized with FimH, was perfectly mimicking the β1,4 linkage of the mannotriose to the first *N*-acetylglucosamine residue (Fig. 2b) [33].

The X-ray crystal structure, at 2.1 Å resolution, of the lectin domain of the FimH adhesin is complex with its most specific and minimal substructure of high-mannose *N*-glycans, oligomannoside-3 ($K_d = 18$ nM) [27], also illustrated that the chitobiose anchor to Asn-glycosylated receptors is important for specificity. This returns the story back to the specificity of the trisaccharide, Manα(1,3)Manβ(1,4) GlcNAc, identified as best FimH ligand by Nurit Firon and Nathan Sharon [23]. The potential of ligand-based design of antagonists of urinary tract infections is thus regulated by structural mimicry of the natural carbohydrate epitope. Extension of the butyl chain to longer alkyl revealed the potency of *n*-heptyl as the most suitable aglycon on α-D-mannoside, with a high affinity defined by different experimental methods ($K_d = 7.3 \pm 1.8$ nM using isothermal titration calorimetry (ITC) [35]; $K_d = 5$ nM using surface plasmon resonance (SPR)). The FimH ligand *n*-heptyl α-D-mannoside, which we call a first generation anti-adhesive targeting FimH, replaced methyl α-D-mannose as the new reference molecule for comparison of affinities and IC_{50} with second generation FimH antagonists (Fig. 3) [36–39].

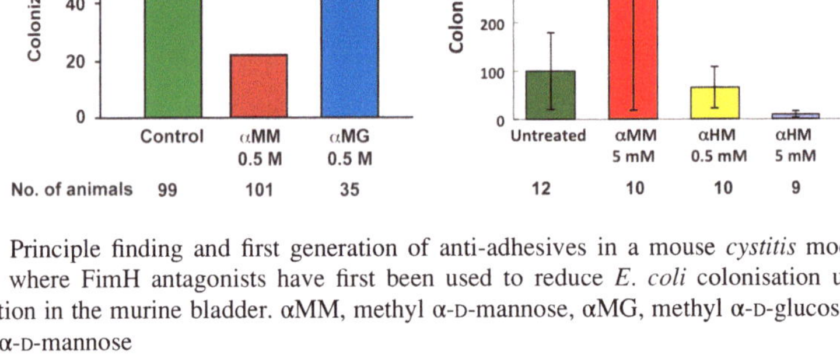

Fig. 3 Principle finding and first generation of anti-adhesives in a mouse *cystitis* model. The reports where FimH antagonists have first been used to reduce *E. coli* colonisation upon co-instillation in the murine bladder. αMM, methyl α-D-mannose, αMG, methyl α-D-glucose, αHM, heptyl α-D-mannose

3 Factors That Enhance Bacterial Adhesion

3.1 Glycoprotein Receptors of the FimH Adhesin

FimH is the type 1 fimbrial adhesin and invasin of *E. coli* and specifically recognizing high-mannosylated glycoprotein receptors. FimH can equally recognize the single high-mannose glycan on uroplakin Ia [40], on the urinary defense protein uromodulin or Tamm-Horsfall glycoprotein [41] and intriguingly enough also on the α3β1 integrin despite its presence on basolateral rather than lumen-exposed apical membranes [2]. The luminal side of the large superficial urothelial cells is abundantly seeded with uroplakins. When uropathogenic *E. coli* enters the bladder, the initial encounter of the bacterium with the urothelium is made with these uroplakins. The uroplakin Ia receptor is in ring-shaped complexes with three other uroplakins on the bladder epithelium and carries a single high-mannosylated glycosylation site. Uroplakin Ia from embryonic tissue carries a mixture of oligomannosides-6, -7, -8, and -9 [42]. Binding to the bladder epithelium is proposed to be mediated by binding to the uroplakin Ia's single *N*-linked glycan of the high-mannose type [42]. The same virulence factors of *E. coli*, the type-1 fimbriae with the FimH adhesin, recognize the CEACAM6 glycoprotein on enterocytes, which was found to be upregulated with the overgrowth (dysbiosis) of adherent-invasive *E. coli* (AIEC) [43], which in inflammatory bowel diseases developing into ulcerative colitis or Crohn's disease [40, 44]. Glycoprotein 2 exposed on M cells is another intestinal receptor for FimH in lymphoid structures, called the Peyer's patches [45]. Bacteria can transcytose across the M cells upon binding. Transcytosis is a prerequisite in mucosal immunization to evoke an antigen-specific immune response.

The *N*-glycans that are preferentially recognized by the FimH adhesin are paucimannosidic. The FimH adhesin has an unusually high affinity for truncated high-mannose glycans [20]. Truncated oligomannosides can be found in the urine of patients [25] and in cells [46] with mannosidase deficiencies due to their inability to achieve high-mannose structures. The specificity of FimH for high-mannose type branched *N*-glycans as they are present in glycoproteins is fine-tuned toward the shorter structures: the affinity increases from $K_d = 900$ nM for the largest-size oligomannoside-9 down to $K_d = 20$ nM for oligomannose-3 and oligomannose-5. This is not only true for clinical *E. coli* isolates from the urinary tract but also for enterohaemorraghic and faecal *E. coli*, excluding some enterohaemorrhagic O157: H7 [30]. FimH lectin variants isolated from uropathogenic or intestinal *E. coli* strains have similar affinity profiles regardless of their origin and display the highest affinity toward oligomannosides-3 and -5. The recent crystal structure of FimH in complex with oligomannoside-3 disclosed the reason why this is so: the tetrasaccharide Manα1,3Manβ1,4GlcNAcβ1,4GlcNAcβ of oligomannoside-3 forms a close and extended interaction with amino acid residues that are conserved among *E. coli* strains (Fig. 1c) [9, 33].

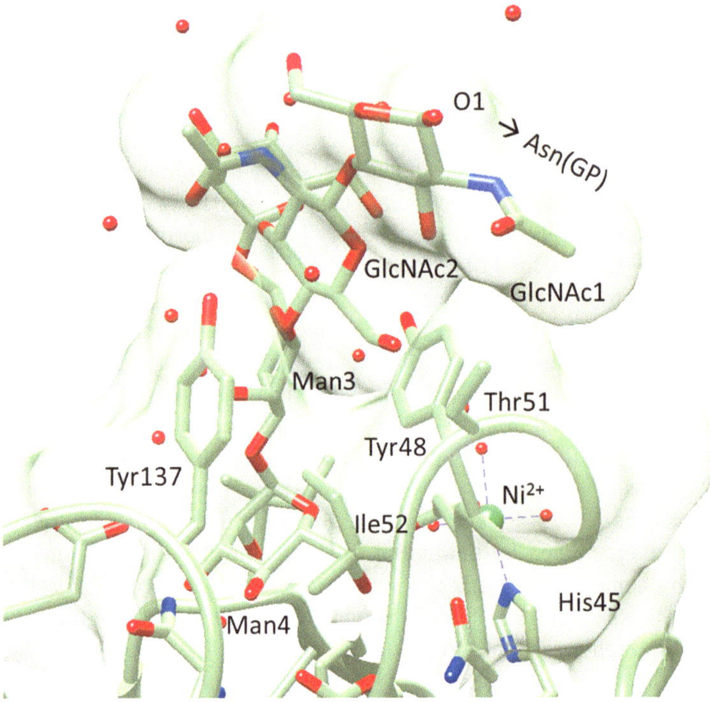

Fig. 4 Paucimannosylated glycoproteins (GP) are receptors for FimH and serve as a template for FimH-targetted drug design

The physiological receptors of FimH are paucimannosidic membrane glycoproteins (Fig. 4). FimH binds to high-mannosidic glycan receptors and this binding leads to signaling in the cytoplasm of the infected epithelial cell and invasion of the bacteria, with the induction of apoptosis [3]. However, high-mannose type *N*-glycans generally are not part of glycoproteins embedded in the membrane of eukaryotic cells. Therefore, the glycocalyx of a mammalian cell is not really compatible with FimH binding. The current paradigm being pursued is that the *N*-glycosylation profile of the membrane proteins and of the specific receptors for FimH shifts to more of the high-mannose type in diseased cells. It was recently recognized that apoptotic cells possess an altered surface glycopattern and release distinct types of blebs derived from the endoplasmic reticulum or the plasma membrane [47]. Both types of blebs exhibit desialylated glycotopes, resulting from surface exposure of immature endoplasmic reticulum (ER)-derived high-mannose glycoproteins, or from surface-borne sialidase activity resulting in exposure of galactosyl glycotopes, respectively [48]. Apoptotic cells or ER-derived blebs possessing high-mannosylated glycans can thus be the site of bacterial entry into epithelial cells [49]. Endocytic adaptors moreover facilitate the active uptake of FimH from the luminal cell surfaces seeded with high concentrations of glycoprotein receptors [50]. Possibly at later stages in the pathogenesis cycle or dependent on tissue remodeling, FimH is also interactive with the cytoskeleton via an HDAC6-modulated microtubule-dependent pathway [51]. The FimH adhesin thus potentially interferes with several different glycoproteins playing diverse roles in cellular processes, in favor of *E. coli* progression, but that all share one common receptor module: the mannose monosaccharide that served as the basis of FimH-targetted anti-adhesive drug design 1.

3.2 FimH Conformational Changes Under Shear Force

From early on, it was found that *E. coli* tropism depends strongly on the contents of the fimbrial shaft [31] and was unrelated to the specificity of FimH in hybrid fimbriae [52]. Factors other than the FimH adhesin were responsible for *E. coli* strain-dependent differential binding mediated by type 1 fimbriae. For example, *E. coli* expressing the CI#4 *fimH* variant from uropathogenic *E. coli* adheres tightly to yeast mannan, A498 human kidney cells, and J82 human bladder cells, whereas bacteria expressing *fimH* from the fecal *E. coli* F-18 strain show poor adhesion to all these three substrates. Allelic variation in *fimH* has been long held responsible for altered mannoside binding and bacterial adhesion profiles [53]. However, allelic variation in *fimH* is very limited among *E. coli* isolates with an intestinal or extraintestinal origin such as the urinary tract [9, 54]. Moreover, when expressed as a soluble lectin domain, in the absence of the other fimbrial components [30], or as a translational fusion protein with maltose-binding protein [52], FimH lectins failed to demonstrate any strain-dependent tropism and soluble FimH binds equally to all cells, mannosylated proteins, and mannosides [30, 31].

A breakthrough came when it was found that fecal, commensal F-18 *E. coli* do not agglutinate guinea pig red blood cells under static conditions but only co-aggregate when rocked [55]. The dependence of this strain on shear was further examined using parallel flow chambers. The studies revealed that type-1 piliated *E. coli* bacteria show a stick-and-roll adhesion pattern with increasing shear rates: they roll slowly along a mannose-coated surface at low shear and convert into stationary binding when shear is enhanced [56]. In general, it is expected that shear force would weaken the lectin–mannose interaction, because the interacting molecules are pulled apart. These interactions are called slip bonds, as the ligands slip out of the binding pocket at higher forces. In contrast to this type of binding, FimH at the fimbrial tip increases the lifetime of its noncovalent bonds with mannose when under shear force. Such bond formation is called catch bond. The stick-and-roll adhesion of type-1 fimbriated *E. coli* to mannose-coated surfaces can be explained by the transition of a low- to a high-affinity conformational state of FimH with allosteric regulation of the mannose-binding pocket [57]. The adhesion of type-1 piliated *E. coli* to mannose-coated surfaces is longest lived at intermediate forces of 60–100 pN, while binding is weakened or abolished below 40 pN and above 140 pN, respectively [55, 58].

Importantly, FimH does not show a shear-force dependent behavior when exposed to surfaces coated with mannotriose. Mannotriose contains the Manα1,3Man sequence at a nonreducing end recognized specifically by FimH and explaining its tenfold higher affinity (K_d ~ 300 nM) than mannose (K_d ~ 2.3 μM) [30]. Higher affinity ligands thus manage to induce the conformational change in FimH and induce the fit in the mannose-binding pocket independently from shear force [59]. However, the absence of these nonsubstituted mannotriose receptors, combined with the common presence of more weakly binding, nonreducing end α-D-mannose [9] and Manα1,2Man (K_d ~ 1,400 nM [30]) on high-mannose glycans [42] of healthy epithelial cells, explains the evolutionary adaptation of *E. coli* toward shear-force dependent bacterial adhesion. Another evolutionary benefit gained from shear-force enhanced adhesion of *E. coli* to surfaces is protection of *E. coli* against removal by oligosaccharides, antibodies, and natural defense proteins such Tamm-Horsfall glycoprotein. These natural inhibitors cannot apply shear force because of their soluble form [60].

In selectins, an interdomain communication regulates the strength of the adhesion. Congruent with this idea, bacteria expressing type 1 pili with either allelic variations or engineered mutations that were predicted to destabilize the linker between lectin and pilin domain of FimH and to facilitate its extension were shown to exhibit decreased shear-dependent adhesion. Steered molecular dynamics simulations predicted that application of tensile force along the long axis of FimH, as might result from shear forces, disrupts the interdomain contacts in the FimH adhesin, resulting in the extension of its flexible linker containing the Gly159–Gly160 sequence and pulling the FimH lectin domain in a high-affinity conformation [61]. These studies indicated that the short, stubby type-1 pilus tip fibrillum is in its low-affinity conformation when not in contact with potential receptors [62]. Meanwhile, the crystal structures of mannose [9] and butyl α-D-mannoside bound to FimH [33] showed that mannose binds snugly into a highly specific mannose-binding pocket at the tip of the FimH lectin domain (Fig. 5c). The pocket appeared near-optimal for mannose binding with almost all of the hydrogen-bonding potential satisfied, as indicated by the unusually high monosaccharide affinity of FimH for mannose (K_d = 2.3 μM). It was thus difficult to envision how conformational changes could additionally enhance mannose binding.

A second large breakthrough was made, again by the group of Wendy Thomas, with the crystal structure of the whole type-1 fimbrial tip (Fig. 5a, b, PDB entry code 3jwn). The crystal structure of the full fimbrial tip showed FimH in its resting, condensed conformation and low-affinity state (the fimbrial tip cannot be subject to shear forces in solution or in a crystal structure) [58, 63], with Ile13, Asp140, and Phe142 too far away from a potential mannosidic ligand to make interactions (Fig. 5b). This same study confirmed the high-affinity state of the isolated FimH lectin domain. Comparison of FimH in its low-affinity state with its extended, high-affinity conformation revealed FimH lectin–pilin interdomain contacts that break under shear force, paired with conformational rearrangements that propagate up to the binding pocket to allosterically improve affinity for the ligand (Fig. 5).

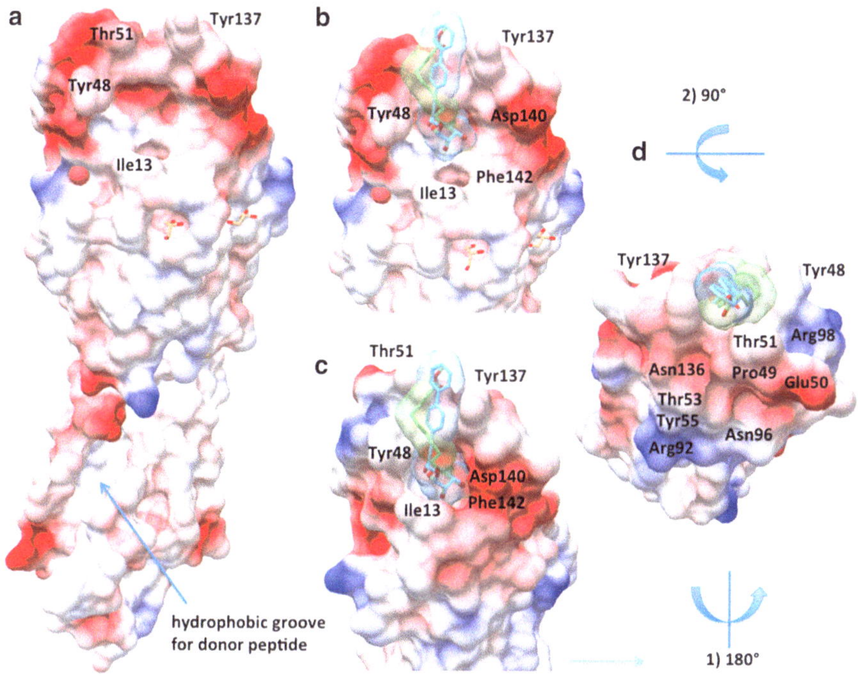

Fig. 5 Surfaces of FimH lectin domain crystal structures in the low-affinity, compressed state (**a** and **b**, PDB entry 3jwn) and its high-affinity, extended form (**c** and **d**, PDB entry 4av5, ligands from 4v5 (*blue*) and 4avi (*green*). The surfaces are colored for electrostatic potential calculated according to Coulomb's law, with red representing negative charges and blue representing positive charges. (**a**) The whole, compressed FimH conformation as present in the type-1 fimbrial tip (PDB entry code 3jwn), with pilin domain (*down*) and lectin domain up. Tyr48, Thr51, Tyr137, and Ile13 residues are indicated. Two glycerol molecules are shown as sticks. (**b**) Two ligands in their position in the open tyrosine gate superimposed on the low-affinity conformation of the lectin domain of FimH (3jwn): methylester octyl α-D-mannose (stick, *green* for carbon, *red* for oxygen, from 4avi) and the propynyl biphenyl α-D-mannose (*light blue* for carbon, entry 4av5) in the open gate configuration. Ile13 is in a retracted lip of the binding pocket and not available for the important van der Waals interactions with the C1–C2 bond of mannose. Also residues Phe142 and Asp140 are too far away from a potential ligand to make van der Waals interactions with the C2–C3 side of the mannose and a hydrogen bond to O_3, respectively (Fig. 7). (**c**) The Coulombic potential surface becomes less negatively charged when the lectin domain catches on the mannosidic ligand. The lectin domain extends and thins out. FimH ligand residues Ile13, Phe142, and Asp140 have leveraged toward the ligand to enable interaction. Shown are two ligands in their respective crystal structures (from 4avi at 2.4 Å resolution and from 4av5 at 1.4 Å resolution) with an open tyrosine gate [35]. (**d**) Surface with the foregoing ligands protruding from the mannose-binding pocket and peeking over the top of the lectin domain. FimH residues potentially interacting with the protein part of glycoprotein receptors are labeled

This new knowledge may relate to the previous findings that the affinity and specificity of bacterial adhesion are strongly influenced by the type-1 fimbrial shaft [31].

FimH lectins from *S. typhimurium*, *K. pneumoniae*, and *E. coli*, expressed in fusion with maltose-binding protein, bind indifferently to a broad range of mannose-presenting compounds, thus not explaining the distinct cell adhesion and agglutination profiles of these three bacterial species [52]. Intriguingly, the adhesion profile of bacteria expressing hybrid type 1 fimbriae (e.g., *K. pneumoniae* FimH at the tip of *E. coli* type 1 fimbriae or vice versa) resembles the adhesion profile of the bacteria from which the fimbrial rod was derived [31]. Electron microscopic pictures have revealed that the helical rods of type 1 fimbriae and P pili can extend significantly as a result of either mechanical shear during preparation or from exposure to 50% glycerol [64]. The extension is due to the breaking of the quaternary interactions between nonadjacent pilin subunits within the right-handed helical rod, leading to its uncoiling. The uncoiling of the helical structure of P and type 1 fimbriae to a linear conformation is fully reversible and the fimbrial rod can work as a spring under shear forces [65, 66]. The spring-like qualities of the rods have been suggested to be fine-tuned to support the formation of long-lived catch bonds that promote tight adhesion under conditions of shear stress. Type-1 fimbrial mechanical properties are expected to contribute considerably to adapting bacterial adhesion [67], in agreement with significant interstrain sequence variation in the FimA pilin, the major component of type 1 fimbriae.

3.3 Phase Variation of Fimbrial Expression Regulates Multiplicity of Bacterial Adhesins

Bacteria coordinate fimbrial gene expression at the single cell level to prevent coexpression of different types of fimbriae [68]. Phase variation allows individual bacteria to turn on and off the expression of specific virulence factors [69, 70]. Phase variation and coordinated transcriptional regulation are presumably employed by the pathogenic bacteria to adapt to sequentially changing environments during infection, colonization, and/or invasion. The proportion of bacteria in a population expressing a particular fimbrial adhesin can be influenced by a number of environmental factors such as temperature, pH, osmolarity, and the presence of specific ligands in the surrounding medium. Regulation of fimbriation is necessary because a high level of fimbriation on bacteria triggers inflammatory host responses, putting the entire bacterial colony at risk of being eliminated. Lowering the proportion of fimbriated bacteria decreases the inflammatory response to levels where it may even be beneficial for the bacterial population.

Phase variation of type-1 fimbrial expression in *E. coli* is controlled by the site-specific recombination of a 314-base pair invertible element [71], also called the *fim* switch or *fimS*. The *fim* switch is a transcriptional control element of type-1 fimbrial expression that contains the promoter for the *fimA* gene encoding the major subunit of type 1 fimbriae. The cAMP receptor protein, or CRP, is a metabolic sensor that regulates type-1 fimbrial expression [72]. cAMP is a second messenger derived from adenosine triphosphate (ATP) and used for intracellular signal transduction in many organisms. When CRP is bound to cAMP (cyclic adenosine monophosphate), it changes into a conformation that can bind tightly to DNA promotor regions, thus repressing the switching of nonfimbriated *E. coli* to the type-1 fimbriated phase. Type-1 fimbrial repression by CRP is alleviated by lowering the intracellular concentrations of cAMP. The presence of glucose in the growth medium, paired with a higher ATP production, or the entering of the bacteria into stationary growth, paired with a lower glucose consumption, each decreases cellular cAMP concentrations and thus de-repress type-1 fimbrial expression. Type-1 fimbrial expression as a response to high glucose and growth stop is consistent with the adaptation of *E. coli*, which evades extracellular stresses by entering into the host epithelial cells and forming intracellular bacterial communities (IBCs) therein [6]. Type 1 fimbriae have been shown to be crucial for the organization and maturation of the protective biofilm-like matrices of IBCs and, driven by intracellular stresses, to prepare *E. coli*'s way out back out of the cell [11]. The absence of type 1 fimbriae seriously compromises in vivo fitness of *E. coli*. The presence of the CRP-cAMP protein in pathogenic *E. coli* that can sense glucose contents to regulate type-1 fimbrial expression and bacterial adhesion is thus an important parameter to control infections in patients with metabolic diseases such as type-2 diabetes [73].

The complexity and importance of the regulation of bacterial fimbriation implicate that certain allelic variations in *fimH* could have resulted due to evolutionary constraints steered by fimbrial biogenesis rather than on affinity and specificity of the FimH adhesin. The FimH adhesin is indeed also the initiator of type 1 pilus biogenesis. FimH activates and opens the outer membrane usher of *E. coli* to reach upon transfer the bacterial surface as the first of the fimbrial proteins [74–76]. Examples of allelic variation that may affect fimbrial biogenesis are a single Gly73Glu amino acid difference between two FimH lectins, from the F-18 commensal *E. coli* strain and from the uropathogenic *E. coli* strain CI#4. When FimH is bound in the fimbrial export channel of the C-terminal domain 1 of the usher (PDB entry 3rfz), this variant residue is juxtaposed to the side chains of Thr706 and Arg709. Other mutations that can affect fimbrial biogenesis are the natural variant Ala/Val27, the double mutation engineered to inhibit FimH domain linker extension, Gln32Ala/Ser124Ala, and the point mutation Val156Pro, engineered to facilitate FimH domain linker extension [55], which are all located on the surface of FimH in the chaperon binding cleft. In conclusion, the ability to regulate the expression of type-1 fimbrial genes has important functional significances [77] and has been positively selected for during the evolution of pathogenic *E. coli* [78]. An altered efficiency in fimbrial assembly introduced by allelic variation in lectin and pilin sequences could result in different multitudes or ultrastructures of type 1 pili and offers another way to adapt adhesion to the particular needs of a bacterial strain.

3.4 Density of Glycan Receptors on Host Surfaces Mediates Multivalent Adhesion

FimH-mediated bacterial adhesion comprises a multivalent interaction, with a number of type 1 fimbriae on *E. coli* interacting with a number of receptors molecules presented on a cell surface. The recognition by bacterial lectins of glycosylation patterns on host cells is strongly enhanced by higher densities of those glycans on surfaces [54, 79]. This enhancement is especially remarkable for adhesins with a low affinity for their receptors such as F17-G [54]. The expression of glycosylated receptors on cells, such as glucose transporters, and also of the type of glycans and the degree of GlcNAc-initiated branching (by GnT-IV) of the glycans, depends strongly on the hexosamine flux and regulates cell growth and differentiation. It has been shown that the hexosamine flux upregulates the presence of cell surface glycoproteins with a low number of glycosylation sites in a linear-concentration dependent manner, whereas those receptors with a higher valency of glycosylation (both in number of sites and in degree of branching) display a hyperbolic response curve [80]. Only cell surface glycoproteins that are sufficiently dense glycosylated can get involved into the cross-linked lattices with galectin-3. A low glycan density on the glycoprotein receptors weakens their cell surface retention by means of the cross-linked lattices with galectin and induces their endocytosis for further glycan biogenesis and growth. Cross-linking with galectin-3 mediates signaling for the transcription of glycosyl transferases to maintain glycan biosynthesis, thereby suppressing cell proliferation and constitutive endocytosis [80]. Endocytosis is one of the mechanisms used by bacterial adhesins to enter host cells [81].

Multivalent binding is a mechanism employed by bacteria to enhance adhesion. This multivalency can potentially be used to create more potent inhibitors of bacterial adhesion. Small-molecule adhesin inhibitors can be competitive with the binding of multiple fimbriae simultaneously to epithelial surfaces only when their affinity is high enough to displace the bacteria. Multivalent anti-adhesives have the extra advantage over monovalent inhibitors in that they can target multiple fimbrial adhesins simultaneously and are a better mimic of bacteria adhesion. However, the design of multivalent inhibitors for fimbrial adhesins is much more complicated than for multimeric proteins, such as for example the tetrameric plant lectin concanavalin A (for a review see [82]). In multimeric proteins, a distance between the binding sites is determined by the relatively rigid quaternary organization. In contrast, the length and number of type 1 fimbriae on bacteria is highly variable, raising uncertainty in choosing optimal distances for targeting FimH lectin domains and making both the design process and the evaluation of results difficult.

Regardless of this uncertainty, soluble glycoconjugated multivalent scaffolds most often cluster several bacteria via their fimbriae, rather than roping the fimbriae of a single bacterium together. In clusters, the distance between the sugar receptors on a surface to be bridged by the fimbrial adhesins does not really have a strict upper limit; nevertheless a minimal distance between two mannose epitopes is needed to

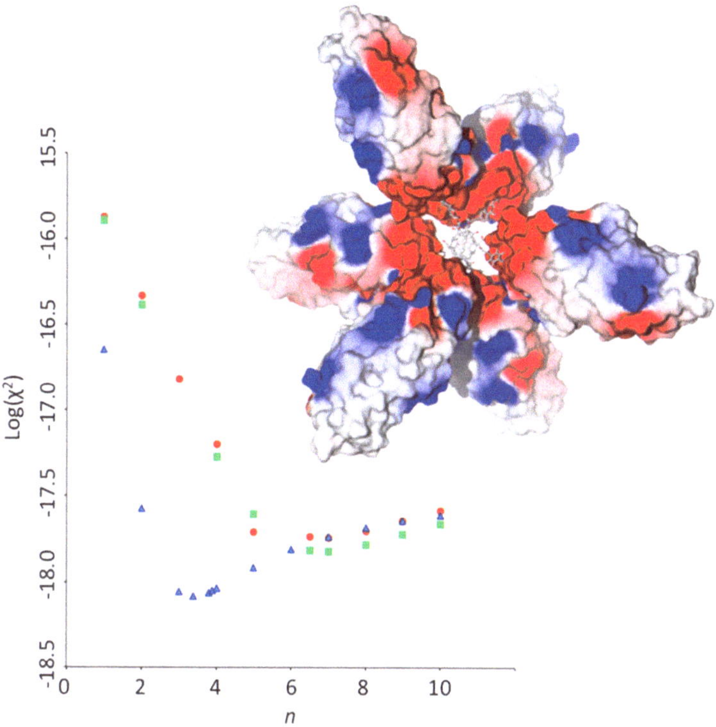

Fig. 6 Results for the short mannosidic ligand are shown by the blue triangles ($n = 3.4$), mannosides substituted with longer aglycon moieties are associated with the data points in red spheres ($n = 7$) and in green squares ($n = 7$). The number of FimH lectin domains, n, bound per fullerene was determined from the best fit, as evaluated by χ^2, of the equilibrium binding constants K_a and K_d and of the uninhibited concentration of FimH, by sampling the molar ratio n. *Inset*: Model of the hypothetical arrangement of six FimH molecules (a seventh one is not shown for clarity) around a C60 fullurene that was 12 times bearing a benzylic aromatic ring at the anomeric position of mannose and an amide group that might provide favorable contacts with the lectin binding cavity. *Red colors* represent negative charges, *blue colors* represent positive charges [36]

avoid steric hindrance of FimH lectin domains. For this reason, the first part of the aglycon in a polyvalent mannosidic compound should not be considered as a linker, because it superimposes on the hexaoses of oligomannoside-3 as a functional part of the antagonist [27]. The FimH–oligomannoside-3 complex can be used to estimate the ligand size, although the minimal distance between two FimH lectin epitopes can equally be understood by considering the binding of alkyl mannosides to the mannose-binding site of FimH. The length of an alkyl tail *O*-linked to α-D-mannose that is optimal for interaction with FimH is seven carbon atoms [33], suggesting that a linker between antagonists meant for interaction with FimH should be more than double this size to avoid steric clashes of the two approaching binding surfaces of FimH.

Biophysical methods have been explored to analyze the geometry and dynamics complexes between multivalent ligands with aglycons grafted onto α-D-mannosides and conjugated by linkers to a scaffold and the FimH lectin domain [36, 83, 84]. The molar ratio (*n*) of FimH binding to the multivalent ligand can be determined not only using ITC [83, 84] but also using SPR [36]. The stoichiometries of the FimH lectin with mannose-derivatized fullerenes revealed that a minimal size of the ligand is indeed required to optimize occupation of sterically available mannosides on a multivalent scaffold [36]. The functional valency *n*, or stoichiometry, of the FimH inhibition in solution by fullerenes was measured in an SPR competition assay, using molar concentrations of the fullurenes. This number of FimH lectin domains, *n*, bound per fullurene was determined from the best fit, as evaluated by χ^2, of the equilibrium binding constants K_a and K_d and of the uninhibited concentration of FimH, by sampling the molar ratio *n*. A fullerene derivatized with 12 α-D-mannosides binds either 3.4 or 7 FimH lectin domains, depending on the size of the mannosidic ligands (Fig. 6).

4 Chemistry and Design of FimH Antagonists

4.1 *Structural Basis for Design of FimH Inhibitors*

Mannose-binding type 1 pili are important virulence factors for the establishment of *E. coli* urinary tract infections. These infections are initiated by adhesion of uropathogenic *E. coli* to uroplakin receptors in the uroepithelium via the FimH adhesin located at the tips of type 1 pili. An alternative to adhesin-based vaccines as a means to block bacterial adhesion is the use of small compounds that interact tightly with the *E. coli* adhesins [85, 86]. Indeed, blocking of bacterial adhesion not only allows to prevent the infection but also to reduce bacterial counts and disease symptoms [37, 87, 88]. This makes FimH antagonists accessible for the development of therapeutics. The very first demonstration of the potency of mannose derivatives as FimH antagonists in vitro and in vivo goes back to 1979, only 1 year after the mannose-binding properties of *E. coli* had been discovered [21]. The blocking agent was methyl α-D-mannopyranoside, which with a dose of 18 mg per female mouse visibly reduced the number of bacteria attached to the mouse bladder mucosa [87].

Almost every *E. coli* can express type 1 fimbriae with the FimH adhesin, which displays high specificity for mannose using a binding pocket (Fig. 7) that is conserved among *E. coli* strains [9, 33]. This makes the application of mannose-based anti-adhesives applicable to therapy of a large range of inflammations and infections caused by pathogenic *E. coli*. The FimH adhesin almost completely envelops mannose in a deep pocket and involves it in an intense hydrogen bonding network (Fig. 7). The specificity is strongly geared toward mannose ($K_d = 2.3$ μM), as no other monosaccharide can fit in the FimH binding pocket and deoxymannosides pay a large penalty in affinity [33]. The only exception is the

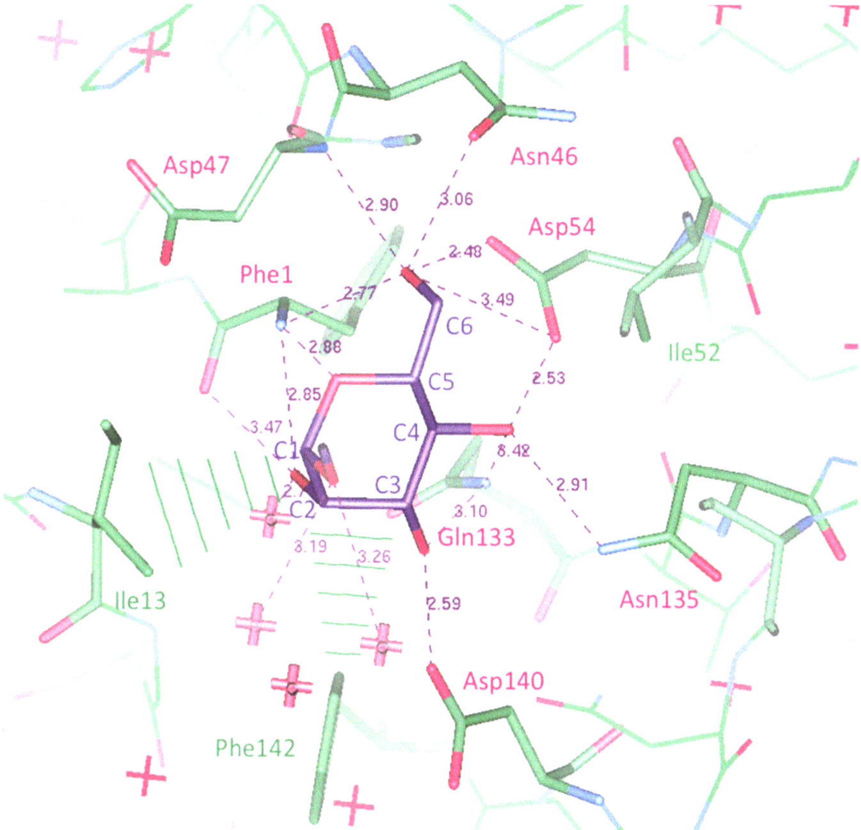

Fig. 7 Interactions of mannosidic anti-adhesives targeting FimH. The polar mannose-binding pocket is highlighted, with butyl α-D-mannose (PDB entry 1uwf) making polar (potential hydrogen bonds and their length, in Å, in *purple*, ligating residues labels in *red*) and apolar (*green cones* and labeled residues) interactions

pyranose form of L-fructose, but not fructofuranose, which can substitute for mannose with a 15-fold reduction in affinity [33, 89]. With mannose being the common recognition point for the FimH lectin, generalized anti-*E. coli* anti-adhesives have been based on the common core, mannose. A limited number of attempts have been undertaken to modify the mannose monosaccharide as the basic unit for the design of anti-adhesives. The most apparent is the change of the oxygen on the anomeric C1 position of α-D-mannose and involved in the glycosidic link with the remainder of the FimH ligand, by a nitrogen [90, 91], sulfur [35], or carbon [91, 92] atom. This elementary change of the glycosidic linker atom may lead to a minor loss in affinity. Loss in affinity has been found due to the mannose chair taking on the $^{1}C_{4}$ conformation for a series of *C*-glycosidically linked α-D-mannosides [91]; however, such chair conversion is not a general rule for *C*-glycosides [92].

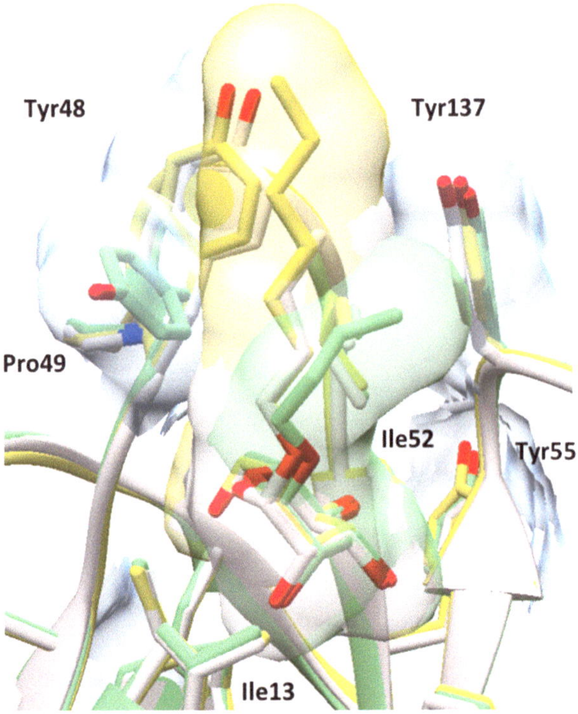

Fig. 8 First generation of mannosidic anti-adhesives targeting FimH. The small and narrow binding pocket of FimH is a negatively charged pocket that attracts polar ligands through a greasy slide: the so-called tyrosine gate, formed by Tyr48 and Tyr137 and backed up by Ile52. The two different alkyl conformations of butyl α-D-mannosides (BM, PDB entries 1tr7 (*green*) and 1uwf (*white*)) and heptyl α-D-mannose (HM, PDB entry 4buq (*yellow*)) are overlayed

Exogenous butyl α-D-mannose exhibits a significantly better affinity for FimH (BM, $K_d = 0.15$ μM) than mannose. Exploration of the binding affinities of α-D-mannosides with longer alkyl tails revealed affinities up to 5 nM for heptyl α-D-mannose (HM) [33]. Natural receptors, being high-mannose glycans, have to pass through the hydrophobic tyrosine gate (tyrosines 48 and 137, Fig. 8) to reach the polar mannose-binding pocket. The tyrosine gate of FimH is the reason why alkyl- and aryl-substituted mannosides bind so much better than just mannose. The tyrosine gate was first named as such when it was discovered in the high-resolution structure of the lectin domain of FimH that the butyl hydrophobic substituent on mannosides makes favorable interactions inside this gate [33]. Also the high affinities of 4-methylumbelliferyl α-D-mannose and *p*-nitrophenyl α-D-mannose could be explained by the docked structures of these aglycons into the tyrosine gate of FimH. The FimH binding site maintains an open gate conformation when not involved in crystal packing contacts (Fig. 4 & 5c, d). In many co-crystal structures however, the Tyr48 side chain is displaced by the Val155 side chain of a neighboring molecule, thus closing the tyrosine gate [35]. Ligands then bind over, instead of in between, the two tyrosines, a binding mode that was later called the out-docking mode [93] when recognized in the co-crystal structure of FimH in complex with a biphenyl mannoside [94].

4.2 Considerations for Adaptability to Conditions In Vivo

The potential of ligand-based design of FimH antagonists is not limited to blocking of bacterial adhesion and invasion, but is also in the blocking of intracellular bacterial growth by the inhibition of biofilm formation and consequently incapacitating *E. coli* to flux back out of the epithelial cells and cause recurrent and persistent urinary tract infections [33]. These observations signaled the start of a second generation series of modified mannosides. The aim remains to conceive pharmacophores for FimH antagonists, with pharmacophores being defined as the ensemble of steric and electronic features that is necessary to ensure the optimal supramolecular interactions with a specific biological target and to trigger (or block) its biological response [95]. FimH antagonists simulate the action of oligomannose-3, with the central mannose and the *N*-acetylglucosamines present in the chitobiose core linking to the asparagine of glycoproteins. The mannose-binding pocket and the tyrosine gate form the center of interactions for the second-generation FimH antagonists, that are in the broad sense based on the highly specific α-D-mannoside monosaccharide, substituted with aromatic aglycons optimized to improve the stacking within the tyrosine gate, especially with tyrosine 48.

Pharmacological parameters are being taken into consideration and become built in into the compounds to allow their therapeutic use in vivo. The decisions on the structures of FimH antagonists are paralleled by the required pharmacokinetics and distribution of the drug molecules to permit a therapeutic dose during a sufficient retention time in the diseased organ. An optimal combination of these parameters will allow to make full use of the activity of the therapeutic molecules and will tend to avoid undesired side effects such as cell and organ toxicity. Successful examples have recently been illustrated in the literature for the application of FimH antagonists in the mouse bladder to eradicate urinary tract infections, upon oral administration [38, 96] or upon a single intravenous injection [84].

Solubility, stability, lack of toxicity, partitioning based on lipophilicity or binding to plasma serum proteins, permeability through membranes that have to be crossed by the drug molecules to reach to infection niche have to be analyzed and optimized in relation to the route of administration and the body niche infected by *E. coli*. Also they have to be placed in the metabolic context of the patient suffering from, for example, diabetes mellitus or in pregnancy [20], Crohn's diseases [90], or cancer. The placement in a specific metabolic context is especially important to avoid unwanted side effects of the mannose-based antagonists. This is because expression of glycoprotein receptors for FimH is different, often augmented, under metabolic imbalance [40], and the activated state of the immune systems with increased involvement of macrophages that are prone to interfere with FimH antagonism. Characteristics that are commonly shared between the drug molecule candidates are the ability to clear them from the organism and to avoid retention in kidneys and liver, as evaluated using logP and logD7.4 [37]. Orally available FimH antagonists have to resist enzymatic and chemical degradation, and have to be soluble as well as contain lipophylic properties to absorb over the intestinal membrane (evaluated in, for example, a caco-2 assay) or to bind plasma proteins once in the blood circulation. The group of Beat Ernst established an ingenious prodrug approach to unite these apparently opposed characteristics in an orally administrated biphenyl α-D-mannose (bpMan) [97]. An ester group was brought onto the bpMan for the molecule to undergo intestinal absorption. Esterases present in enterocytes and in the liver cleave the ester to a carboxylate that helps its clearance by the liver and renal elimination.

In history, two varieties of synthetic FimH antagonists were particularly studied, that is, the monovalent mannosides with specific anomeric aglycons and the multivalent glycoclusters where several copies of a given mannoside are grafted to a common scaffold. From a conceptual point of view, mono- and multivalent compounds operate by specific and different binding mechanisms. Efforts in developing monovalent mannosides were nearly exclusively focused on shaping the anomeric aglycon moiety, the mannose core being too deeply buried in the lectin domain for modifications. In a first generation of mannose-based anti-adhesives, aliphatic and aromatic anomeric aglycon created hydrophobic stacking and van der Waals interactions with the side chains of Tyr48 and Tyr137, located at the entrance of the FimH binding site. Aside from this classical "key and lock" approach, several groups have developed multivalent mannosides in an attempt to increase the apparent affinity of FimH by a "glycoside cluster effect" [98]. Indeed, it is well known that multivalent ligands for lectins can be much more effective than expected from their structural ligand valency. Here, we will illustrate these two approaches with specific examples of highly potent and promising mono- and multivalent FimH inhibitors. In view of the large number of structures developed since the first proof-of-concept, the few examples reported here forms a small part of the FimH antagonists reported in literature. For a more complete overview, we invite the reader to refer to specific reviews [99, 100].

4.3 Monovalent Anti-adhesives

Pioneering work by Sharon and coworkers showed that mannosides with an aromatic aglycon were much better FimH inhibitors than methylmannoside reference (MeMan, Scheme 1) [34]. *p*-Nitro-*o*-chlorophenyl α-D-mannoside (pNPoClαMan) and 4-methylumbelliferyl α-D-mannoside (MeUmbαMan) were identified as the most potent FimH antagonists of the series. The compounds are 470- and 1,015-fold more potent than MeMan in inhibiting *E. coli* attachment to guinea pig ileal cells, with affinities relative to methyl α-D-mannose that correlate well with the relative concentrations required for inhibition of yeast and intestinal cell adhesion of type-1 piliated *E. coli* [34]. This finding highly stimulated the development of new synthetic inhibitors based on phenyl mannoside.

Starting from the potent antagonist **2**, Lindhorst and coworkers have developed a simple procedure to introduce a squarate moiety to further interact with the lectin domain exterior of FimH (Scheme 2) [101]. The nitro group of **2** was reduced with

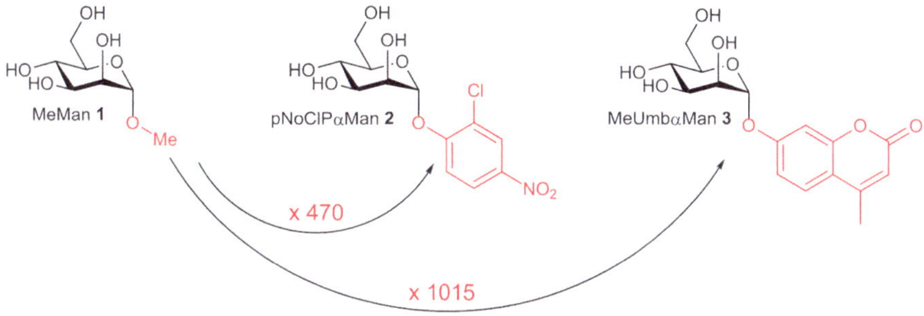

Scheme 1 Relative inhibitory values of the aromatic mannosides pNoClPαMan and MeUmbαMan compared to MeMan in removing adherent *E. coli* strain O128 from ileal epithelial cells

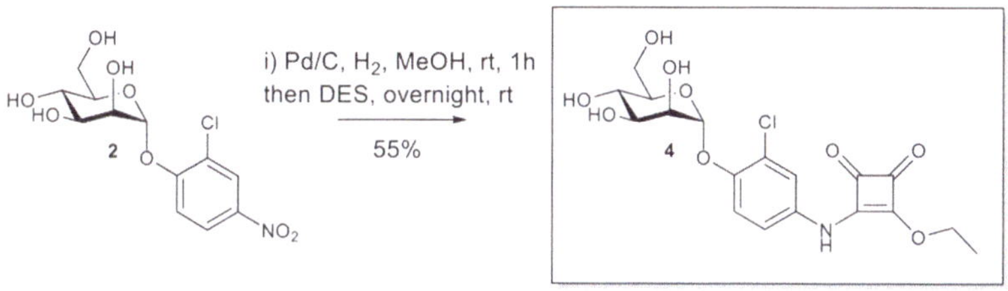

Scheme 2 Lindhorst and coworkers grafted a DES group to further enhance the binding affinity at the entrance of the FimH lectin domain [101]

H_2/Pd and the amine was subsequently reacted with diethylsquarate (DES) leading to **4** with 55% overall yield. The compound proved to be a potent *E. coli* anti-adhesive in an ELISA assay. The relative inhibitory of **4** was shown to be 6,900-fold more potent than MeMan **1** and 35-fold than **2**, showing the benefit of extending interactions beyond the phenyl ring. It should be noted that the higher anti-adhesive effect of **4** compared to **2** was also evidenced by others in an aggregation assay [102].

Introduction of a second phenyl ring in the *para* position of phenylmannoside (pMan) led to a particularly potent class of FimH antagonist. The corresponding biphenyl mannosides (bpMan) were first reported by Scott Hultgren and coworkers (Scheme 3) [94]. The authors developed a large library of bpMan and the structural basis for enhanced potency was determined with the obtention of an X-ray crystal structure of analogue **5** bound to FimH. A key π-stacking interaction was formed with Tyr48 and the biphenyl moiety, and a H-bond electrostatic interaction with the Arg98/Glu50 salt bridge. If several lead compounds were identified here, Beat Ernst and coworkers showed in an independent study that closely related bpMans have the potential to reduce uropathogenic *E. coli* levels in the bladder after oral administration to mice [97]. Reduction of colony forming units by several logs was observed in the bladder and urine with FimH antagonist **7**. These results highlight the potential of this class of compounds for the effective treatment of UTIs.

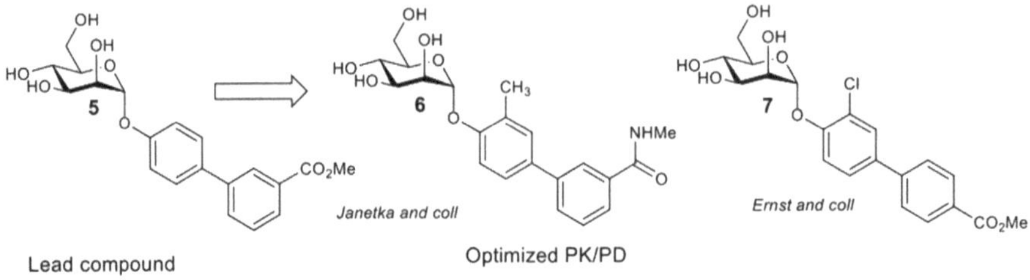

Scheme 3 Independent studies conducted by the groups of respectively Ernst and Hultgren revealed in vivo potential of bpMan for treating urinary tract infections [94, 97]

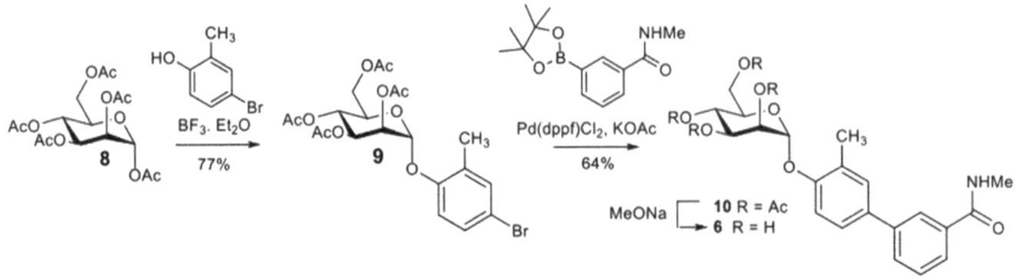

Scheme 4 Chemical design of an improved bpMan for treating urinary tract infections [38, 88]

BpMan can be obtained by a synthetic sequence involving a glycosylation step to introduce the first aromatic ring and a Pd-mediated coupling for the terminal phenyl group. This strategy was used by Hultgren and coworkers for the obtention of a second generation of optimized FimH antagonists (compound **6**, Scheme 4) [38]. Treatment of cystitis in a mouse model with **7** and related compounds showed particularly promising results. The compounds were shown to prevent urinary tract infection in a prophylactic therapy and to treat chronic *cystitis*. Diminution of bacterial levels in the bladder was also shown to be higher with bpMan than with the commonly prescribed antibiotic trimethoprim-sulfamethazole. Ernst and coworkers have also recently shown that indolinyl phenyl mannoside can reduce bladder bacterial load to a similar extent (four orders of magnitude) as a standard antibiotic treatment with ciprofloxacin [37].

Recently, a new set of FimH antagonists has been developed in the group of Sebastien Gouin, with the aim to introduce different natures of atoms at positions of interaction between the FimH adhesin and the anti-adhesive molecule (Scheme 5). A convenient addition–cyclization methodology was developed to build thiazoles, pyrimidines, and triazines on the mannose, by performing the heterocyclization reaction with differently substituted dienophiles and electrophiles [90]. Heterocycles circumvent the rather low solubility generally encountered with *O*- or *C*-mannosides bearing hydrophobic aglycons and potentially represent versatile tools for adapting partitioning (logP, logD), membrane permeability, pharmacokinetics, and organ distribution. Because the structure of the heterocyclic moiety and the R substituent can each be conveniently modulated, a large panel of potential inhibitors with good aqueous solubilities can be obtained.

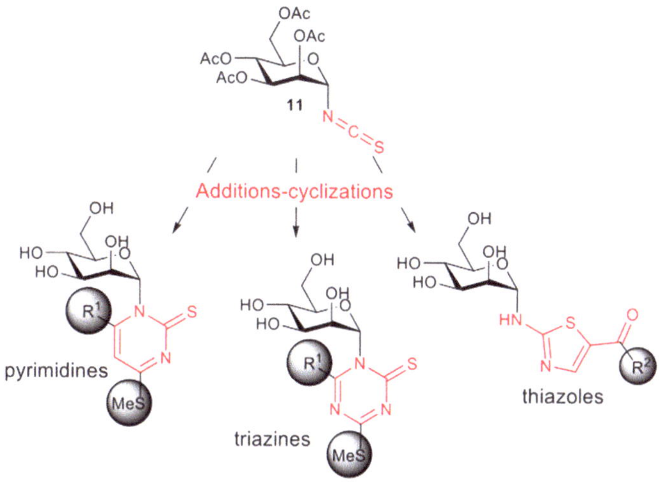

Scheme 5 Mannosides with heterocyclic aglycons for improved water solubility and affinity. Results obtained with the thiazole series (TazMans) suggested for the first time a potential anti-adhesive treatment of Crohn's disease [90]

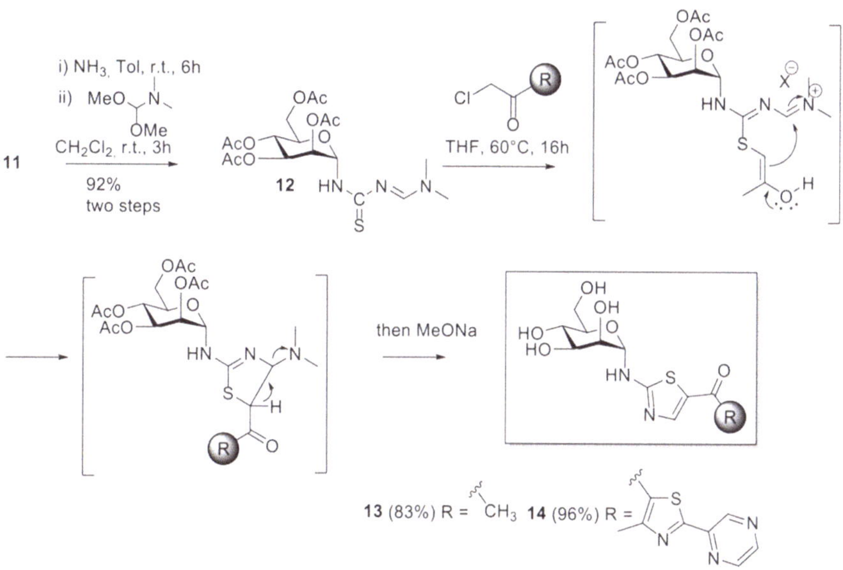

Scheme 6 The addition–cyclization strategy to design TazMans [90]

N-linked thiazolyl α-D-mannosides (TazMans) have been obtained from mannosylisothiocyanate **11** (Scheme 6). The compound was first converted into a thiazadiene **12** with NH_3 gas followed by treatment with *N,N*-dimethylformamide dimethyl acetate. The critical addition–cyclization steps with **12** and a set of chlorocetones led to the expected thiazoles with good to excellent yields (Scheme 6). TazMans were particularly potent FimH antagonists with binding affinities in the nanomolar range in an ELISA assay [90]. The high-resolution crystal structure obtained with lead compound **13** and FimH showed the thiazole moiety interacting with Tyr48 and interaction energies were calculated to be higher with the thiazolyl group than with a phenyl group (see Section 4.4). The most active TazMan **14** was shown to prevent attachment of adherent-invasive *E. coli* (AIEC) to intestinal cells at 10,000- and 100-fold lower concentrations than mannose and the potent FimH inhibitor HM, respectively. AIEC were previously shown to colonize the ileal mucosa of patients with Crohn's disease, inducing the secretion of large amount of tumor necrosis factor alpha (TNF-α) [43]. The present results suggest for the first time a potential anti-adhesive treatment against AIEC in inflammatory bowel diseases [103].

4.4 Multivalent Anti-adhesives

The serendipitous discovery of alkyl mannosides as strong binders to FimH of these "sticky" ligands led to the identification of a new class of FimH inhibitors. Being easily synthesized and highly soluble in water, they became potential blocking agents for FimH-mediated adhesion. Disappointingly, rather high concentrations (around 5 mM, see Fig. 3) were required to obtain 90% reduction of bacterial by the monovalent HM compound in a mouse model. Therefore, mannosides were also grafted in multiple copies to synthetic scaffolds to evaluate the effect of ligands clustering on *E. coli* adhesion. It is well known that so-called glycoclusters can improve affinity for lectins to a large extent, with examples reported reaching up to several logs increased affinity compared to monovalent references. A logarithmic affinity enhancement is however not a general rule, the binding mode operating and the nature of the lectin playing a critical role. The highest multivalent effects were reported with glycoclusters able to embrace simultaneously several lectin domains of a multimeric lectin. Several binding mechanisms operating independently or simultaneously were shown to stabilize glycocluster–lectin interactions [104–106].

A first one is that upon the first ligand–lectin domain interaction, subsequent binding is favored due to the spatial preorganization of the ligands during the first interaction and thus the lower entropic cost paid. This chelate scenario is not operating with FimH due to the monovalent nature of the lectin. However, *E. coli* expresses hundreds of type 1 pili at their surface. With a high density of low-affinity glycan receptors on cells, a "bind and recapture" process may prevail and diminish bacterial dissociation rates [59]. A multivalent mannosidic inhibitor can potentially cross-link multiple FimH, which either belong to fimbriae of the same bacteria and thus forms a chelate or belong to different bacteria and thus forming clusters of *E. coli*. The concentrations, required to block adherence to the cells, of multivalent inhibitors that cluster bacteria should be decreased compared to its monovalent

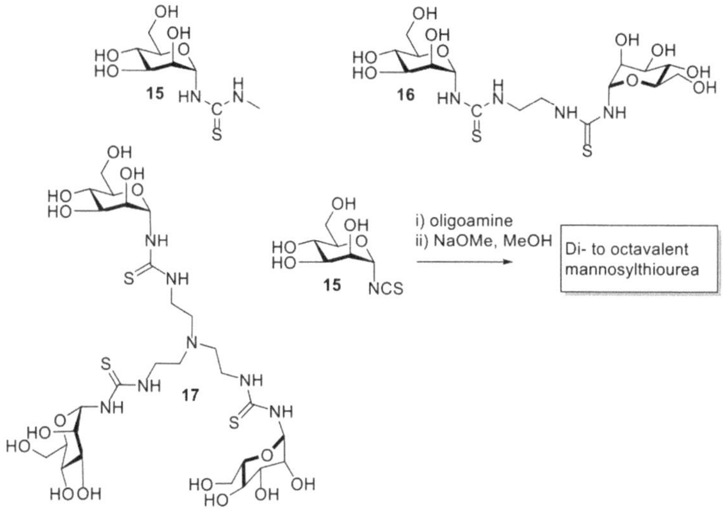

Scheme 7 Di- and trivalent mannosylthiourea probing multivalent effects on type 1 piliated *E. coli* [107]

analogues, because bacteria trapped inside the aggregates would be incapacitated to attach to a surface even if only a small part of their pili are cross-linked.

Lindhorst and coworkers first observed multivalent effects on type-1 piliated *E. coli* with synthetic glycoclusters (Scheme 7) [107]. The authors have developed a simple procedure to form di- to octavalent mannosylthiourea. Di- and trivalent compounds **16** and **17** were shown to be the most potent inhibitors of the series in a haemmaglutination assay. Relative inhibitory titers expressed in moles mannose compared to monovalent **15** reached 39 and 37 for **16** and **17**, respectively. The tetra- to octavalent analogues did not show enhanced affinity, meaning that a plateau of inhibition is readily obtained at a low valency.

Lee and Roy have studied how highly mannosylated ligands enhance affinity toward *E. coli* K-12 [108]. Mannose-containing neoglycoproteins were designed from bovine serum albumin or human serum albumin (HSA). Up to 35 mannosides copies were grafted to the amino groups of the proteins with different linking arm. L-Lysine-based mannose dendrimers developed by René Roy were also included in the study. Compounds with 2, 4, 8, or 16 phenylmannosides residues were obtained by increasing the generation of the dendrimers. Compounds were evaluated in a competitive assay with ^{125}I-labeled mannosylated HSA. The dendrimers and the neoglycoproteins with the highest mannose valency were particularly potent anti-adhesive compounds with IC_{50} values in the subnanomolar range. A significant multivalent effect was also observed with the dendrimers, DP-2, DP-4, DP-8, and DP-16 being around two-, four-, three-, and fivefold more potent inhibitors on a mannose molar basis when compared to *para*-nitrophenyl mannoside (Scheme 8). DP-16 was the most potent dendrimer evaluated in the assay with 4 orders of magnitude higher affinity than MeMan. Valency-corrected multivalent effects

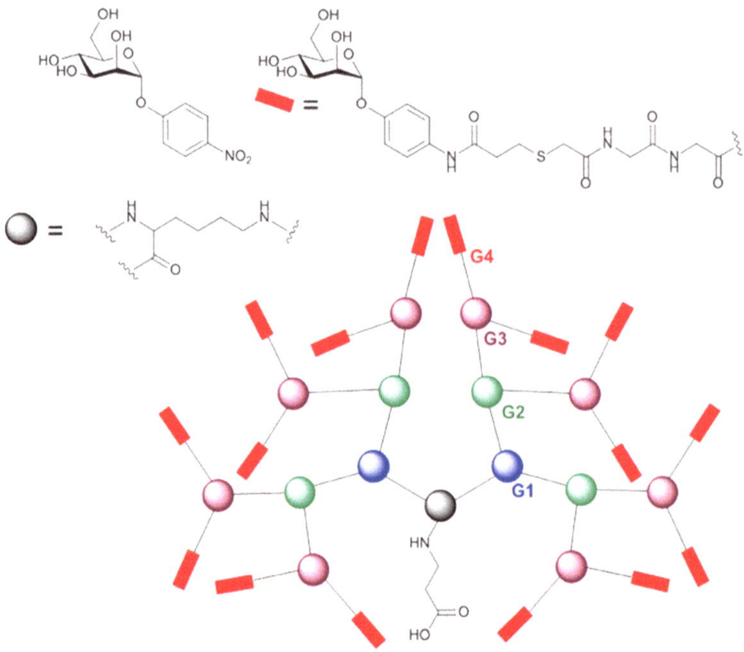

Scheme 8 Dendritic mannosides developed by Roy and coworkers [108]

reaching up to one log were also observed with mannosylated glycoproteins. Both the chemical nature and the length of the tethers were shown to be important factors on the affinity. To rationalize the affinity enhancement observed with increased arm length, the authors proposed that the compounds bind to two or three fimbriae of *E. coli*. Based on the neoglycoprotein size, they moreover suggested that the mannosyl residues that span a distance of 20 nm or longer can be sampled by more than one fimbrial tip.

Roy and coworkers also identified potent dendrimers with much lower valencies. A tetrameric glycodendrimer **18** carrying aryl mannosides had subnanomolar affinities ($K_d = 0.45$ nM) for FimH [109], compared with nanomolar affinity for the monovalent form ($K_d = 18.3$ nM) [35], in SPR experiments, but inhibited haemagglutination by type-1 fimbriated *E. coli* a factor 2 better than HM (Fig. 9).

A set of HM clusters has been designed to combine the potency of the anomeric heptyl chain, in improving intrinsic FimH affinity, with a multivalent stabilization (Scheme 9) [83, 84, 110]. Tetravalent HM **19** with ethylene glycol spacers was shown to prevent in vitro bacterial bladder cell binding at 12 nM [110]. In comparison, mannose and HM required 6,000- and 64-fold (sixteenfold in ligand molar basis) higher concentrations to achieve a similar anti-adhesive effect. In subsequent reports, the FimH binding mechanism was investigated for HM clusters **20** [84] and **21** [83]. Data obtained from ITC showed that compound **21** was able to accommodate seven FimH molecules simultaneously. The formation of larger FimH

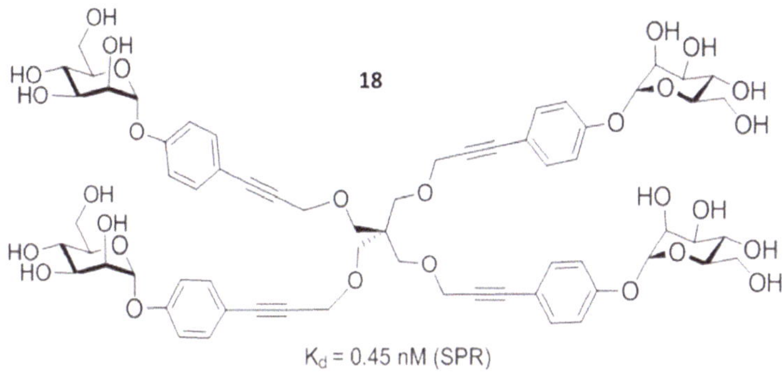

Fig. 9 A pentaerythritol scaffolds was used for the preparation of glycodendrimers bearing aryl α-D-mannopyranoside residues, the latter assembled using Sonogashira and click chemistry. Surface plasmon resonance measurements showed these two mannosylated clusters recognize FimH from *E. coli* at subnanomolar concentrations; however, they showed no obvious multivalent effect in the inhibition of haemagglutination [109]

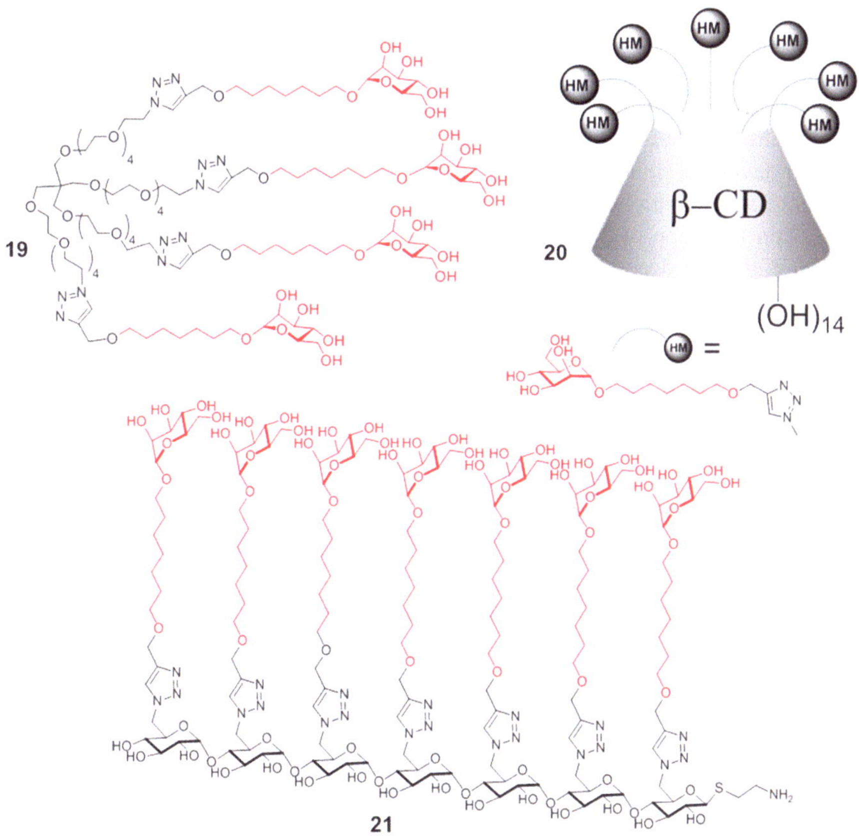

Scheme 9 Multivalent-HM with strong in vitro (**19**, **21**) and in vivo (**20**) anti-adhesive effects. **20** and **21** were shown to aggregate bacteria in solution, giving a rational basis to the multivalent effect observed [83, 84, 110]

aggregates was also shown with dodecamanosylated fullerene that can accommodate up to seven FimH molecules (Fig. 6) [36]. These results obtained with two different types of multivalent mannosides evidenced that ligands separated by only a short distance on the scaffold can be functionally binding FimH.

A first model of a glycocluster–FimH interaction, with compound **20**, was created by combining data from isothermal titration calorimetry (ITC), dynamic light scattering (DLS), and small-angle X-ray solution scattering (SAXS). Small-angle X-ray scattering, in conjunction with dynamic light scattering, is a powerful way to image shapes and solution behavior of low-resolution structures (Fig. 10). Interestingly, heterocompounds **20** and **21** were also able to promote bacterial clustering in solution by cross-linking FimH attached to different bacteria pili, giving a particularly valuable insight in the binding mode operating. This effect was not readily observed with the corresponding monovalent references; however, a head-to-tail association of FimH lectin domains was proposed from the fitting of SAXS data, when aggregation was observed to follow HM binding in solution NMR studies [111]. These results clearly showed that the capture/aggregation of living bacteria in solution can be promoted by highly dynamic glycoclusters with low valencies and that this phenomenon is not restricted to polymeric mannosides [112].

The effect of multivalency of FimH anti-adhesives, namely compound **20**, was for the first time evaluated in vivo, in the murine *cystitis* model C3H/HeN [84]. The level of infection was reduced tenfold for instillation of 10 μM of **20** in the bladder. Similar levels of inhibitions were observed with HM and a monovalent HM reference at much higher concentrations of 5 and 1 mM, respectively. A 100-fold lower concentration of **20** correspond to an injected dose by animals of less than 2 μg. With this dose, a similar anti-adhesive effect compared to monovalent analogues was obtained, an effect that was retained over more than 24 h in the bladder. The bladder targeting and retention was also assessed with a ^{99m}Tc-labeled analogue of **20**. Results showed a very rapid accumulation in the bladder (locus of infection). The rapid distribution of multivalent **20** in the bladder (20% after 5 min) avoids undesired side effects that may occur when a mannose-based FimH antagonist is applied in vivo by interference with mannose-binding lectins, such as DC-SIGN and the macrophage mannose receptor, from cells of the innate immune system and indeed suggests that the multivalent HM-derivatives are poorly retained by the human mannose-binding lectins. On the other hand, a reduction of **20** with only about 40% between 30 min and 24 h upon intravenous injection suggests a favorably long-lasting retention of the treatment that can treat during a complete pathogenic life cycle of *E. coli*, even of those strains with a slow kinetics. The pharmacokinetic and pharmacodistribution parameters (supplementary information in [84]) indicate a strong therapeutic potential of **20** by a single-shot intravenous administration.

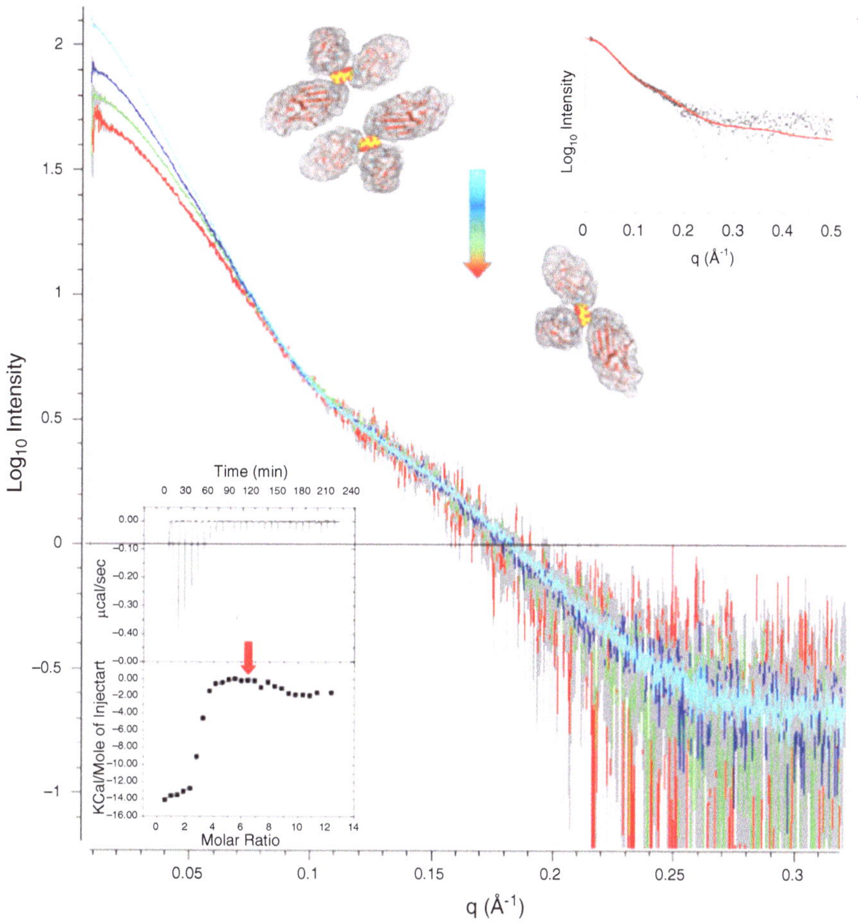

Fig. 10 Behavior of a supramolecular complex of **20** with FimH. **20** is a β-cyclodextrin that is sevenfold grafted with *n*-heptyl α-D-mannose (see Scheme 9). The SAXS data fit a model of three FimH lectin domains bound to **20** (inset *right up*). Twofold dilutions in the same buffer of 12.4 g/L (*cyan*) to 6.4 g/L (*blue*), 3.3 g/L (*green*), and 1.6 g/L (*red*) demonstrate that more of a higher-order species exists at higher FimH concentrations that readily dissociates, upon dilution, toward the stable $(FimH)_3$:cyclodextrin trimeric complex that is observed in reverse isothermal titrations ($n = 3$ in the ITC experiment in the inset *left under*). This multivalent FimH antagonist interferes at very low doses (2 μg per mouse is sufficient to reduce bacterial counts over a 24-h period) in aggregation-prone complexes with *E. coli* adhesion. The complex formation of FimH with **20** is followed by a reorganization with an enthalpic stabilization at FimH concentrations higher than six times the inhibitor concentration (decline in the ITC titration curve at $n = 6$, *red arrow* in ITC inset) into a higher-order oligomer formation based on the inter-trimer FimH–FimH interactions [84]

4.5 Optimization Using Principles from Quantum Chemistry

Ligand binding in proteins is an extremely complex subject with the binding free energy resulting from different contributions (e.g., desolvation, loss of conformational motion of the bound ligand, noncovalent interactions formed in the binding pocket upon binding). Experimental techniques can give insight in the overall binding free energy ΔG, but not in its individual enthalpic and entropic contributions, making the rational design of high affinity ligands difficult. Furthermore, experimental characterization of protein–ligand affinity is limited by several aspects as cost, efficiency, availability of biological material. Therefore, computational techniques gain more and more interest to model and elucidate protein–ligand interactions and to predict ligand affinities prior to ligand synthesis. Computational studies can help to decide the most promising candidates to be tested experimentally [113]. Docking, to generate structural information, and force-field and empirical scoring functions, to obtain insight in the underlying energetics of ligand binding (determination of ΔG and its decomposition in different components, so-called "scoring"), together with force-field-based molecular dynamics (MD) simulations enabling structure prediction, refinement and scoring, are currently the generally applied techniques to computationally study and predict protein–ligand interactions (for a review on docking, see chapter of Martin Frank in this volume).

A very promising alternative strategy to obtain a correct ranking of ligand affinities is the calculation of the total interaction energy $E_{\text{int, total}}$ (Eq. 1) using semiempirical (approximated methods that use several parameters from empirical data) or ab initio (e.g., density functional theory, DFT) quantum mechanical calculations.

$$E_{\text{int,total}} = E_{\text{complex}} - E_{\text{protein}} - E_{\text{ligand}} \tag{1}$$

The free energy G can be written as the sum of the electronic interaction energy E_{int} and the nonelectronic contributions, thermal energy, and entropy. For similar ligands, the nonelectronic contributions to the free energy (entropy, temperature corrections, solvent effects) can be considered as constant in a first approximation [114], by which a linear correlation between the calculated interaction energies and the experimentally obtained binding affinities K_d (which is related to ΔG via $\Delta G = -RT \ln K_d$) can be found. The correlation between measured affinity and interaction energy in the FimH binding pocket nicely illustrates this feature (Fig. 11).

The ligand BM (see Fig. 11 for ligand identification) is more similar to HM and ligand **25** than to ligands **23** and **24**, and thus based on the assumption that nonsimilar ligands give nonconstant nonelectronic contributions to the free energy, it is expected that BM would not fit into the correlation. However, the BM ligand still fits in the correlation, whereas HM and ligand **25** do not fit. This rather unexpected result might be due to the model system used in the calculations. The

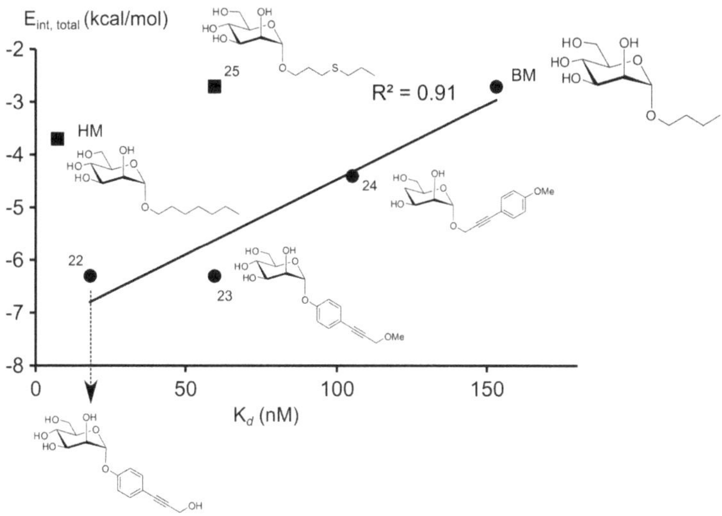

Fig. 11 Correlation between affinity K_d (nM) and the calculated interaction energy $E_{\mathrm{int,\ total}}$ (Eq. 1) calculated using FimH-ligand co-crystal structures with closed tyrosine gate [39]. The numbers refer to the ligand identification numbers used in the text

model was limited to the tyrosine gate of FimH, represented by Tyr137 and Tyr48. Errors were most likely introduced because Ile52, a residue that also contributes to the formation of the hydrophobic pocket, was not included. The model system was rather simple, which ignored explicit solvation and relaxation effects and relied only on the single binding mode found in the crystal structure. Conformational dynamics of the side chain of Tyr48, found in at least two different conformations in the crystal structures [27, 35, 115], were not taken into account.

A careful structure preparation before calculation of $E_{\mathrm{int,\ total}}$ is thus of crucial importance. DFT methods, corrected for dispersion and with an implicit continuum solvent model, are currently the most accurate to obtain a correct ranking of $E_{\mathrm{int,\ total}}$ but are restricted more by the system size than semiempirical quantum chemical methods (SQM) [116]. Advanced SQM methods, with corrections for hydrogen bond interactions, for dispersion and with an implicit continuum solvent model, demonstrate a clear improvement over the force-field-based methods and perform very similar to the previously mentioned DFT-based quantum mechanical (QM) methods to correctly rank $E_{\mathrm{int,\ total}}$ [116]. The correct ranking of calculated interaction energies $E_{\mathrm{int,\ total}}$ is already achieved for a cutoff distance of a sphere with 7 Å radius around the ligand (note that the effective distance around the ligand is about 10–12 Å, since in the set up, all residues with at least one atom within the cutoff distance were kept completely) [116]. The sizable contributions beyond a distance of 7–10 Å seem to be rather uniform and contribute similarly in all cases. As such, according to this study, quantum chemical calculations can be restricted to smaller model systems without losing predictive power.

Upon obtaining accurate $E_{\mathrm{int,\ total}}$ in a subsequent step, the strength and selectivity of the calculated $E_{\mathrm{int,\ total}}$ can be investigated and analyzed within the framework by quantitative molecular orbital (MO)-theory [117]. The MO-model and its associated energy-decomposition approach give insight in the obtained ligand–protein binding energy and form the current state of the art in explanatory chemistry [118]. Much effort has been recently undertaken to develop semiempirical QM scoring functions, in which not only $E_{\mathrm{int,\ total}}$ is present but also the desolvation free energy, the change of conformational free energy upon binding, and entropy terms. These SQM-based functions give clearly a better ranking of ΔG of, for example, HIV protease inhibitor complexes and cyclin-dependent kinase 2 (CDK2)–ligand complexes than the docking scoring functions [119]. The calculation of exact binding free energies remains problematic. To obtain exact values of ΔG in accordance with experimentally measured values, explicit solvent molecules, and not the frequently used implicit continuum solvation models, need to be included [120]. Further, desolvation energies and entropic effects need to be accurately described as correction terms upon $E_{\mathrm{int,\ total}}$.

Insight in the interactions determining $E_{\mathrm{int,\ total}}$ will help in rational drug design. Based on a well-prepared starting structure, quantum chemically obtained E_{int} using advanced calculation methods forms a promising route to at least correctly rank binding energies. Still, high-level calculations of $E_{\mathrm{int,\ total}}$ are computationally demanding and require structural information. Therefore, the development of descriptors able to predict ligand affinity based on earlier obtained insights from calculated and analyzed $E_{\mathrm{int,\ total}}$ and on the properties of the considered ligands alone is very valuable. The present authors recently validated the reactivity descriptors local hardness $\eta(\boldsymbol{r})$ [121, 122], a local version of the chemical hardness [123] and polarizability α [124] to assess affinity in the tyrosine gate of FimH, a setting of two tyrosines (Tyr48 and Tyr137) backed up by an isoleucine (Ile52) [33]. The introduction of alkyl and aryl moieties in the aglycon of synthetic α-D-mannopyranosides can mimic the stacking interactions observed of oligomannoside-3 in the tyrosine gate [27]. Using the hardness and polarizability descriptors, information on the ligand affinity might be obtained without the need of calculating the interaction energy, and thus, without the need of structural information. The dispersion and the electrostatic energy components form the major part of the noncovalent π–π interaction between aromatic rings and thus by calculating these, information on the interaction between aromatic amino acids as Tyr137 and Tyr48 of FimH and various ligands can be obtained.

From a conceptual point of view, the dispersion part of the interaction energy can be related to the polarizability α [124] and the electrostatic part of the interaction energy to the local hardness $\eta(\boldsymbol{r})$. The polarizability α gives the tendency of an electron cloud to be distorted by an electric field, caused for example by the stacking partner. A larger polarizability is favorable for the interaction. The local hardness is a measure of negative charge accumulation. A large value of $\eta(\boldsymbol{r})$ creates an unfavorable interaction energy [125, 126]. Based on the lowest local hardness and highest polarizability criteria, the interactions with ligands **24** and **22** (see Fig. 11 for ligand identification) are overall more favorable than for the

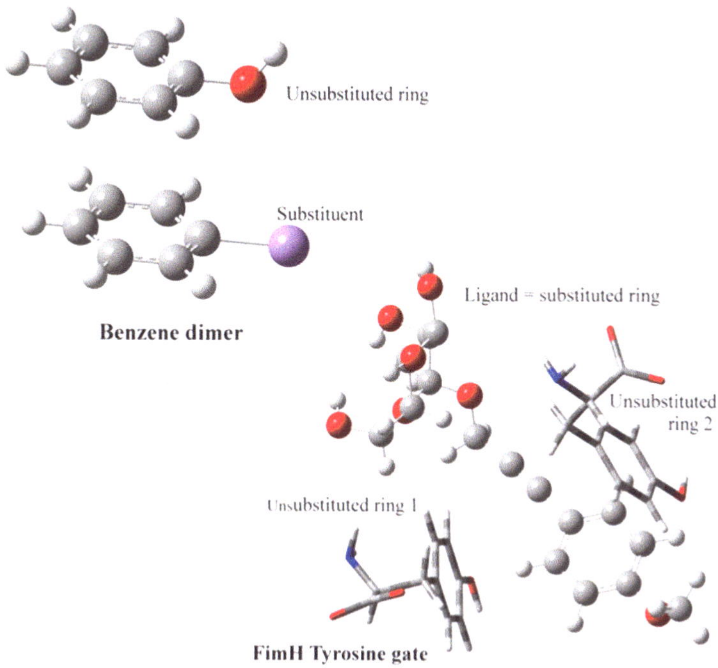

Fig. 12 Identification of substituted ring and unsubstituted rings (Tyr48 and Tyr137) in the benzene dimer and the FimH tyrosine gate, respectively

reference systems of benzene and ethyne [115]. This might be due to the delocalization of electrons (larger polarizability) between the phenyl ring and the alkyne moiety of the ligand, resulting in less negative charge around the phenyl ring and the triple bond. As such, interaction properties of large ligand systems can differ largely from the small reference systems like benzene and ethyne.

Model calculations reveal that all substituents stabilize the benzene dimer (Fig. 12) relative to the unsubstituted case, regardless of the substituent's electron-withdrawing or electron-donating character [127]. The effect of substituents on the stacking interaction does not hinge upon the changes induced in the aryl system, but are better described as arising from local, direct interactions of the substituent with the other ring. More precisely, the effect of a substituent depends on the orientation of its local dipole moment relative to the electric field of the unsubstituted ring. This is nicely illustrated for the following ligands with substituted benzenes: ligand **26** (no substitution on the phenyl), **27** (hydroxyl in *para* position), and **28** (chloride halogen in *para* position) (see Table 1 for ligand identification). The following affinity sequence was measured: **28** > **27** > **26** [115]. For these ligands, the sequence of the lowest hardness (–Cl < –OH < –H) and the highest polarizability (–Cl > –OH > –H) follows the sequence of the affinity (–Cl > –OH > –H) (Table 1). The –OH and –Cl substituents are inductive electron acceptors and mesomeric donors, with the inductive effect being the

Table 1 Local hardness and polarizability of the substituting aromatic ring [90, 115]

Ligand	SPR affinity K_d (nM)	α (Bohr^3)	η(r) P (kcal/mol)
26 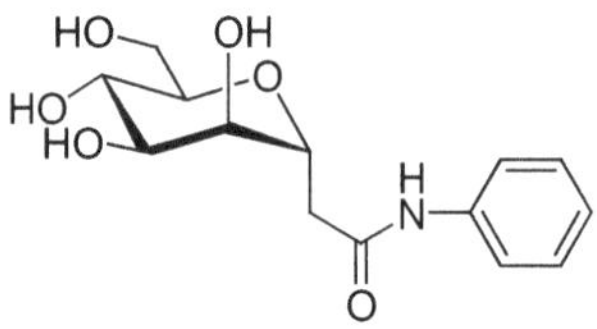	372 ± 26	66	73
27 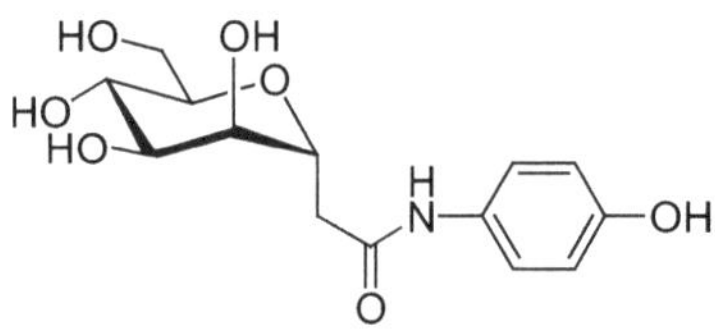	328 ± 63	67	69
28 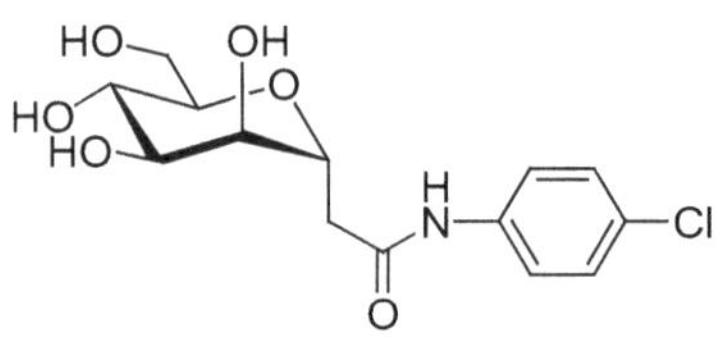	113 ± 37	70	66
29 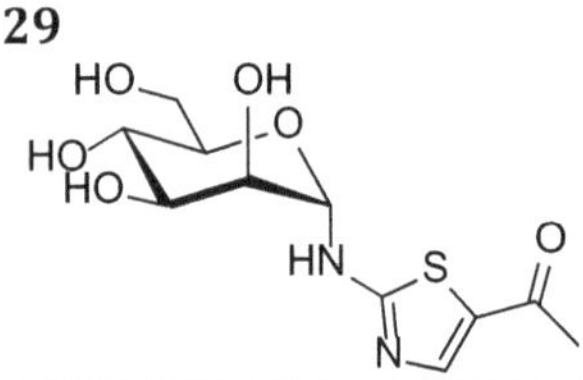	220 ± 8	99	58 (*Tyr48*) 46 (*Tyr137*)

largest for –Cl and the mesomeric effect being the largest for –OH. As such, –OH brings electrons in the ring and is a ring activator toward electrophilic substitution, whereas –Cl is a deactivator. This different behavior of the –OH and –Cl substituents toward electrophilic substitution is not reflected by the affinity, which indicates that local direct interactions between the substituent and the other ring dictate the substituent's effect [127].

In the FimH tyrosine gate, one can identify the tyrosine residues as the "unsubstituted ring" and the ligand as the substituted stacking partner (Fig. 11) [128]. The fact that the substituent depends on the orientation of its local dipole moment relative to the electric field of the unsubstituted ring might cause disagreement between affinity predictions that are based on the local hardness calculated on standard positions, without referral to structural data [115]. However, if structural information is available, predictions based on structural information of similar ligands and under the assumption that the investigated ligands bind in the same way, predictions become feasible. Keeping this in mind, the interaction between

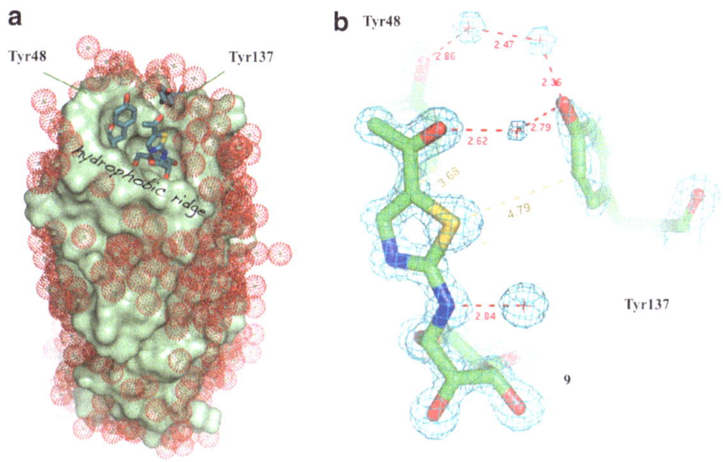

Fig. 13 Second generation mannosidic anti-adhesives targeting FimH [90]. (**a**) A thiazolylamino-mannoside (TazMan, or **13** in Scheme 6), a second generation compound, bound to the FimH lectin domain in crystal structure (PDB entry 3zl2). *Red-dotted spheres* represent water molecules that have been evacuated from the hydrophobic ridge, also because the latter is involved in crystal packing. (**b**) Aromatic stacking interactions of TazMan with Tyr48, the orientation of its heterocycle sulfur atom toward Tyr137, and the compatibility of its carboxyl group with the hydroxyl groups of the two tyrosines and the aqueous environment

ligand **29** (see Table 1 for ligand identification) and Tyr48 and Tyr137 was assessed via the local hardness on specific positions and compared with this of benzene as reference molecule located at the same position as the heterocycle of ligand **29** [90]. For the interaction with Tyr48, we found a local hardness of 58 and 104 kcal/mol for **29** and benzene, respectively. This corresponds with the 2.5 kcal/mol more favorable interaction energy calculated between **29** and Tyr48 and is in agreement with the fact that ligand **29** is at stacking distance to Tyr48 (Tyr48 at 3.68 Å, Fig. 13), but relatively far away from Tyr137 (Tyr137 at 4.79 Å, Fig. 13). For the interaction with Tyr137, we found a local hardness of 46 and 95 kcal/mol for **29** and benzene, respectively. This large difference in local hardness between benzene and **29** is not reflected in the calculated interaction energies between **29** (0.1 kcal/mol) or benzene (−0.4 kcal/mol) and Tyr137, showing no interaction is apparent at the calculated positions with Tyr137.

In conclusion, the combination of biophysical data and quantum chemical calculations forms a promising research line to develop tools for rapid ligand screening prior to synthesis [115]. Force-field-based and empirical scoring functions face a lot of complexity to rank and predict binding affinities [129]. This review article summarizes the first steps in quantum chemical research toward affinity predictions of FimH antagonists and thus further research will design improved calculation protocols and quantum-chemical-based scoring functions to successfully assess FimH-ligand affinity.

5 Perspectives

A large range of FimH inhibitors have been described during the last 25 years. Efforts focused on improving the affinity for the tyrosine gate. This hydrophobic region is quite flexible, adopting anomeric alkyl, aromatic and heterocyclic moieties that can be trapped in an open or closed gate conformation. Among the potent inhibitors developed, biphenyl α-D-mannosides are particularly studied and are promising orally available compounds for a potential treatment of UTIs. Further preclinical lead optimizations are, however, required to improve their bioavailability and limit their often insoluble nature. *E. coli* anti-adhesives that target the FimH lectin may also find new fields of applications, as recently suggested with the thiazolyl α-D-mannoside series that showed strong antagonistic effects on adherent invasive *E. coli* on intestinal cells in vitro and ex vivo [90]. Those results suggest a therapy for inflammatory bowel diseases, by hampering the *E. coli*, that maintains the inflammatory reaction in a dysbiosis [43], rather than the inflammation itself using TNF-α antagonists.

The design of multivalent mannosidic inhibitors was shown to be a relevant strategy to obtain potent anti-adhesive compounds. There are conflicting views on their *modus operandi* with the FimH lectin. Glycoclusters constructed with high-affinity ligands such as HM do not improve the intrinsic affinity for FimH, as evidenced by ITC and SPR results. These results suggest that stabilizing intramolecular chelate effects or "bind and recapture" mechanisms do not prevail for these kinds of multivalent inhibitors. The stoichiometry of a single FimH lectin domain interacting with one monovalent α-D-mannosidic derivative has been well characterized; however, significant evidence always existed that such first protein–carbohydrate interaction can trigger FimH–FimH protein aggregation into larger but easily dissociatable clusters [24, 36, 84, 111]. The spatial distribution of the ligands on the glycocluster is thereby a very important determinant for simultaneous binding of multiple FimH molecules or fimbriae on the inhibitor scaffold; it favors the co-aggregation of FimH onto the glycocluster via the, mostly hydrophobic, FimH surfaces. A first apparent "dimerization" of FimH on a glycoligand, such as HM [111], increases dynamics of FimH within the glycocluster and on its turn induces aggregation with neighboring FimH-charged glycoclusters. This aggregatory mechanism allows for bacterial clustering, with much lower concentrations of the multivalent ligand than the stoichiometric equivalent of its monovalent analogues, and explains the avidity generally observed in cell-based assays. Because of their higher water solubility and molecular weight, multivalent ligands are probably less orally available for the urinary tract but are ideally suited for therapeutic purposes in the intestines. Larger assemblies of mannosides could be tested for antagonist potencies on inflamed tissues such as the ileum in Crohn's disease patients. Nevertheless, multivalent anti-adhesive effects were recently measured in the mouse bladder, using very low doses of glycoclusters intravenously injected and with adequate pharmacokinetics and organ distributions. Mono- and multivalent FimH antagonists are therefore complementary tools to treat type-1 piliated *E. coli* infections in a context of growing antibiotic resistances.

Design based on structure, thermodynamics, and quantum chemistry of the FimH-ligand interaction is a powerful path that must be combined with optimized pharmacokinetic and pharmacodistribution parameters of the therapeutic drugs. The crystal structures of the FimH adhesin in complex with its natural receptor and butyl mannose opened the way for drug design of carbohydrate-based anti-adhesives. The soaking of the ligand-free FimH crystals with antagonists is giving a reliable picture of the interactions that allows the optimization of the calculations and the prediction of interaction modes and their energies. Calculations are being used not only for the FimH with ligands bound in the in-docking (ligand bound in the open tyrosine gate) and out-docking (ligand bound over the closed tyrosine gate) modes but in the future also for FimH in its compressed, low-affinity conformation versus its extended, high-affinity structure upon the formation of catch bonds with mannose. In conclusion, pharmacophores are being further developed to guide the drug design of FimH antagonists.

References

1. Zhou G, Mo WJ, Sebbel P, Min G, Neubert TA, Glockshuber R, Wu XR, Sun TT, Kong XP (2001) Uroplakin Ia is the urothelial receptor for uropathogenic *Escherichia coli*: evidence from in vitro FimH binding. J Cell Sci 114(Pt 22):4095–4103
2. Eto DS, Jones TA, Sundsbak JL, Mulvey MA (2007) Integrin-mediated host cell invasion by type 1-piliated uropathogenic *Escherichia coli*. PLoS Pathog 3(7):e100
3. Thumbikat P, Berry RE, Zhou G, Billips BK, Yaggie RE, Zaichuk T, Sun TT, Schaeffer AJ, Klumpp DJ (2009) Bacteria-induced uroplakin signaling mediates bladder response to infection. PLoS Pathog 5(5):e1000415
4. Hannan TJ, Mysorekar IU, Hung CS, Isaacson-Schmid ML, Hultgren SJ (2010) Early severe inflammatory responses to uropathogenic *E. coli* predispose to chronic and recurrent urinary tract infection. PLoS Pathog 6(8):e1001042
5. Anderson GG, Palermo JJ, Schilling JD, Roth R, Heuser J, Hultgren SJ (2003) Intracellular bacterial biofilm-like pods in urinary tract infections. Science 301(5629):105–107
6. Justice SS, Hung C, Theriot JA, Fletcher DA, Anderson GG, Footer MJ, Hultgren SJ (2004) Differentiation and developmental pathways of uropathogenic *Escherichia coli* in urinary tract pathogenesis. Proc Natl Acad Sci U S A 101(5):1333–1338
7. Rosen DA, Hooton TM, Stamm WE, Humphrey PA, Hultgren SJ (2007) Detection of intracellular bacterial communities in human urinary tract infection. PLoS Med 4(12):e329
8. Martinez JJ, Mulvey MA, Schilling JD, Pinkner JS, Hultgren SJ (2000) Type 1 pilus-mediated bacterial invasion of bladder epithelial cells. EMBO J 19(12):2803–2812
9. Hung CS, Bouckaert J, Hung DL, Pinkner J, Winberg C, Defusco A, Auguste CG, Strouse R, Langermann S, Waksman G, Hultgren SJ (2002) Structural basis of tropism of *Escherichia coli* to the bladder during urinary tract infection. Mol Microbiol 44:903–915
10. Mulvey MA (2002) Adhesion and entry of uropathogenic *Escherichia coli*. Cell Microbiol 4(5):257–271
11. Wright KJ, Seed PC, Hultgren SJ (2007) Development of intracellular bacterial communities of uropathogenic *Escherichia coli* depends on type 1 pili. Cell Microbiol 9(9):2230–2241
12. Chen SL, Hung CS, Pinkner JS, Walker JN, Cusumano CK, Li Z, Bouckaert J, Gordon JI, Hultgren SJ (2009) Positive selection identifies an in vivo role for FimH during urinary tract infection in addition to mannose binding. Proc Natl Acad Sci U S A 106(52):22439–22444

13. Mysorekar IU, Hultgren SJ (2006) Mechanisms of uropathogenic *Escherichia coli* persistence and eradication from the urinary tract. Proc Natl Acad Sci U S A 103(38):14170–14175
14. Mysorekar IU, Isaacson-Schmid M, Walker JN, Mills JC, Hultgren SJ (2009) Bone morphogenetic protein 4 signaling regulates epithelial renewal in the urinary tract in response to uropathogenic infection. Cell Host Microbe 5(5):463–475
15. Mysorekar IU, Mulvey MA, Hultgren SJ, Gordon JI (2002) Molecular regulation of urothelial renewal and host defenses during infection with uropathogenic *Escherichia coli*. J Biol Chem 277(9):7412–7419
16. Hannan TJ, Totsika M, Mansfield KJ, Moore KH, Schembri MA, Hultgren SJ (2012) Host-pathogen checkpoints and population bottlenecks in persistent and intracellular uropathogenic *Escherichia coli* bladder infection. FEMS Microbiol Rev 36(3):616–648
17. Hannan TJ, Mysorekar IU, Hung CS, Isaacson-Schmid ML, Hultgren SJ (2010) Early severe inflammatory responses to uropathogenic *E. coli* predispose to chronic and recurrent urinary tract infection. PLoS Pathog 6(8)
18. Geerlings SE (2008) Urinary tract infections in patients with diabetes mellitus: epidemiology, pathogenesis and treatment. Int J Antimicrob Agents 31(Suppl 1):S54–S57
19. Geerlings SE, Meiland R, van Lith EC, Brouwer EC, Gaastra W, Hoepelman AIM (2002) Adherence of type 1-fimbriated *Escherichia coli* to uroepithelial cells – more in diabetic women than in control subjects. Diabetes Care 25(8):1405–1409
20. Geerlings SE, Beerepoot MA, Prins JM (2014) Prevention of recurrent urinary tract infections in women: antimicrobial and nonantimicrobial strategies. Infect Dis Clin North Am 28(1): 135–147
21. Eshdat Y, Ofek I, Yashouv-Gan Y, Sharon N, Mirelman D (1978) Isolation of a mannose-specific lectin from *Escherichia coli* and its role in the adherence of the bacteria to epithelial cells. Biochem Biophys Res Commun 85(4):1551–1559
22. Firon N, Ofek I, Sharon N (1983) Carbohydrate specificity of the surface lectins of *Escherichia coli*, Klebsiella pneumoniae and Salmonella typhimurium. Carbohydr Res 120: 235–249
23. Firon N, Ofek I, Sharon N (1984) Carbohydrate-binding sites of the mannose-specific fimbrial lectins of enterobacteria. Infect Immun 43(3):1088–1090
24. Sharon N (1987) Bacterial Lectins, cell-cell recognition and infectious disease. FEBS Lett 217:145–157
25. Egge H, Michalski JC, Strecker G (1982) Heterogeneity of urinary oligosaccharides from mannosidosis: mass spectrometric analysis of permethylated Man9, Man8, and Man7 derivatives. Arch Biochem Biophys 213(1):318–326
26. Firon N, Ofek I, Sharon N (1982) Interaction of mannose-containing oligosaccharides with the fimbrial lectin of *Escherichia coli*. Biochem Biophys Res Commun 105(4):1426–1432
27. Wellens A, Garofalo C, Nguyen H, Van Gerven N, Slattegard R, Hernalsteens JP, Wyns L, Oscarson S, De Greve H, Hultgren S, Bouckaert J (2008) Intervening with urinary tract infections using anti-adhesives based on the crystal structure of the FimH-oligomannose-3 complex. PLoS One 3(4):e2040
28. Minion FC, Abraham SN, Beachey EH, Goguen JD (1986) The genetic determinant of adhesive function in type 1 fimbriae of *Escherichia coli* is distinct from the gene encoding the fimbrial subunit. J Bacteriol 165:1033–1036
29. Abraham SN, Sun D, Dale JB, Beachey EH (1988) Conservation of the D-mannose-adhesion protein among type 1 fimbriated members of the family enterobactericeae. Nature 336: 682–684
30. Bouckaert J, Mackenzie J, de Paz JL, Chipwaza B, Choudhury D, Zavialov A, Mannerstedt K, Anderson J, Pierard D, Wyns L, Seeberger PH, Oscarson S, De Greve H, Knight SD (2006) The affinity of the FimH fimbrial adhesin is receptor-driven and quasi-independent of *Escherichia coli* pathotypes. Mol Microbiol 61(6):1556–1568

31. Duncan MJ, Mann EL, Cohen MS, Ofek I, Sharon N, Abraham SN (2005) The distinct binding specificities exhibited by enterobacterial type 1 fimbriae are determined by their fimbrial shafts. J Biol Chem 280(45):37707–37716
32. Rosenstein IJ, Stoll MS, Mizuochi T, Childs RA, Hounsell EF, Feizi T (1988) New type of adhesive specificity revealed by oligosaccharide probes in *Escherichia coli* from patients with urinary tract infection. Lancet 2(8624):1327–1330
33. Bouckaert J, Berglund J, Schembri M, De Genst E, Cools L, Wuhrer M, Hung CS, Pinkner J, Slattegard R, Zavialov A, Choudhury D, Langermann S, Hultgren SJ, Wyns L, Klemm P, Oscarson S, Knight SD, De Greve H (2005) Receptor binding studies disclose a novel class of high-affinity inhibitors of the *Escherichia coli* FimH adhesin. Mol Microbiol 55(2):441–455
34. Firon N, Ashkenazi S, Mirelman D, Ofek I, Sharon N (1987) Aromatic alpha-glycosides of mannose are powerful inhibitors of the adherence of type 1 fimbriated *Escherichia coli* to yeast and intestinal cells. Infect Immun 55:472–476
35. Wellens A, Lahmann M, Touaibia M, Vaucher J, Oscarson S, Roy R, Remaut H, Bouckaert J (2012) The tyrosine gate as a potential entropic lever in the receptor-binding site of the bacterial adhesin FimH. Biochemistry 51(24):4790–4799
36. Durka M, Buffet K, Iehl J, Holler M, Nierengarten JF, Taganna J, Bouckaert J, Vincent SP (2011) The functional valency of dodecamannosylated fullerenes with *Escherichia coli* FimH-towards novel bacterial antiadhesives. Chem Commun 47(4):1321–1323
37. Jiang X, Abgottspon D, Kleeb S, Rabbani S, Scharenberg M, Wittwer M, Haug M, Schwardt O, Ernst B (2012) Antiadhesion therapy for urinary tract infections–a balanced PK/PD profile proved to be key for success. J Med Chem 55(10):4700–4713
38. Han Z, Pinkner JS, Ford B, Chorell E, Crowley JM, Cusumano CK, Campbell S, Henderson JP, Hultgren SJ, Janetka JW (2012) Lead optimization studies on FimH antagonists: discovery of potent and orally bioavailable ortho-substituted biphenyl mannosides. J Med Chem 55(8):3945–3959
39. Roos G, Wellens A, Touaibia M, Yamakawa N, Geerlings P, Roy R, Wyns L, Bouckaert J (2013) Validation of reactivity descriptors to assess the aromatic stacking within the tyrosine gate of FimH. ACS Med Chem Lett 4(11):1085–1090
40. Barnich N, Darfeuille-Michaud A (2010) Abnormal CEACAM6 expression in Crohn disease patients favors gut colonization and inflammation by adherent-invasive *E. coli*. Virulence 1 (4):281–282
41. Serafini-Cessi F, Monti A, Cavallone D (2005) N-Glycans carried by Tamm-Horsfall glycoprotein have a crucial role in the defense against urinary tract diseases. Glycoconj J 22 (7–9):383–394
42. Xie B, Zhou G, Chan SY, Shapiro E, Kong XP, Wu XR, Sun TT, Costello CE (2006) Distinct glycan structures of uroplakins Ia and Ib: structural basis for the selective binding of FimH adhesin to uroplakin Ia4. J Biol Chem 281(21):14644–14653
43. Darfeuille-Michaud A, Boudeau J, Bulois P, Neut C, Glasser AL, Barnich N, Bringer MA, Swidsinski A, Beaugerie L, Colombel JF (2004) High prevalence of adherent-invasive *Escherichia coli* associated with ileal mucosa in Crohn's disease. Gastroenterology 127 (2):412–421
44. Barnich N, Carvalho FA, Glasser AL, Darcha C, Jantscheff P, Allez M, Peeters H, Bommelaer G, Desreumaux P, Colombel JF, Darfeuille-Michaud A (2007) CEACAM6 acts as a receptor for adherent-invasive *E. coli*, supporting ileal mucosa colonization in Crohn disease. J Clin Invest 117(6):1566–1574
45. Hase K, Kawano K, Nochi T, Pontes GS, Fukuda S, Ebisawa M, Kadokura K, Tobe T, Fujimura Y, Kawano S, Yabashi A, Waguri S, Nakato G, Kimura S, Murakami T, Iimura M, Hamura K, Fukuoka S, Lowe AW, Itoh K, Kiyono H, Ohno H (2009) Uptake through glycoprotein 2 of FimH(+) bacteria by M cells initiates mucosal immune response. Nature 462(7270):226–230

46. Bernon C, Carre Y, Kuokkanen E, Slomianny MC, Mir AM, Krzewinski F, Cacan R, Heikinheimo P, Morelle W, Michalski JC, Foulquier F, Duvet S (2011) Overexpression of Man2C1 leads to protein underglycosylation and upregulation of endoplasmic reticulum-associated degradation pathway. Glycobiology 21(3):363–375
47. Bilyy R, Stoika R (2007) Search for novel cell surface markers of apoptotic cells. Autoimmunity 40(4):249–253
48. Bilyy RO, Shkandina T, Tomin A, Munoz LE, Franz S, Antonyuk V, Kit YY, Zirngibl M, Furnrohr BG, Janko C, Lauber K, Schiller M, Schett G, Stoika RS, Herrmann M (2012) Macrophages discriminate glycosylation patterns of apoptotic cell-derived microparticles. J Biol Chem 287(1):496–503
49. Bilyy R, Bouckaert J (2011) FimH from uropathogenic *Escherichia coli* binds to the blebs of apoptotic cells. Sepsis 4(1):96–97
50. Eto DS, Gordon HB, Dhakal BK, Jones TA, Mulvey MA (2008) Clathrin, AP-2, and the NPXY-binding subset of alternate endocytic adaptors facilitate FimH-mediated bacterial invasion of host cells. Cell Microbiol 10(12):2553–2567
51. Dhakal BK, Mulvey MA (2009) Uropathogenic *Escherichia coli* invades host cells via an HDAC6-modulated microtubule-dependent pathway. J Biol Chem 284(1):446–454
52. Thankavel K, Shah AH, Cohen MS, Ikeda T, Lorenz RG, Curtiss R, Abraham SN (1999) Molecular basis for the enterocyte tropism exhibited by Salmonella typhimurium type 1 fimbriae. J Biol Chem 274(9):5797–5809
53. Sokurenko EV, Chesnokova V, Doyle RJ, Hasty DL (1997) Diversity of the *Escherichia coli* type 1 fimbrial lectin. Differential binding to mannosides and uroepithelial cells. J Biol Chem 272(28):17880–17886
54. Lonardi E, Moonens K, Buts L, de Boer AR, Olsson JD, Weiss MWS, Fabre E, Guerardel Y, Deelder AM, Oscarson S, Wuhrer M, Bouckaert J (2013) Structural sampling of glycan interaction profiles reveals mucosal receptors for fimbrial adhesins of enterotoxigenic *Escherichia coli*. Biology 2:894–917
55. Thomas WE, Trintchina E, Forero M, Vogel V, Sokurenko EV (2002) Bacterial adhesion to target cells enhanced by shear force. Cell 109(7):913–923
56. Thomas W, Forero M, Yakovenko O, Nilsson L, Vicini P, Sokurenko E, Vogel V (2006) Catch-bond model derived from allostery explains force-activated bacterial adhesion. Biophys J 90(3):753–764
57. Pereverzev YV, Prezhdo OV, Forero M, Sokurenko EV, Thomas WE (2005) The two-pathway model for the catch-slip transition in biological adhesion. Biophys J 89(3): 1446–1454
58. Le Trong I, Aprikian P, Kidd BA, Forero-Shelton M, Tchesnokova V, Rajagopal P, Rodriguez V, Interlandi G, Klevit R, Vogel V, Stenkamp RE, Sokurenko EV, Thomas WE (2010) Structural basis for mechanical force regulation of the adhesin FimH via finger trap-like beta sheet twisting. Cell 141(4):645–655
59. Nilsson LM, Thomas WE, Trintchina E, Vogel V, Sokurenko EV (2006) Catch bond-mediated adhesion without a shear threshold – trimannose versus monomannose interactions with the FimH adhesin of *Escherichia coli*. J Biol Chem 281(24):16656–16663
60. Nilsson LM, Thornas WE, Sokurenko EV, Vogel V (2006) Elevated shear stress protects *Escherichia coli* cells adhering to surfaces via catch bonds from detachment by soluble inhibitors. Appl Environ Microbiol 72(4):3005–3010
61. Aprikian P, Tchesnokova V, Kidd B, Yakovenko O, Yarov-Yarovoy V, Trinchina E, Vogel V, Thomas W, Sokurenko E (2007) Interdomain interaction in the FimH adhesin of *Escherichia coli* regulates the affinity to mannose. J Biol Chem 282(32):23437–23446
62. Tchesnokova V, Aprikian P, Yakovenko O, Larock C, Kidd B, Vogel V, Thomas W, Sokurenko E (2008) Integrin-like allosteric properties of the catch-bond forming FimH adhesin of *E. coli*. J Biol Chem 283(12):7823–7833

63. Le Trong I, Aprikian P, Kidd BA, Thomas WE, Sokurenko EV, Stenkamp RE (2010) Donor strand exchange and conformational changes during *E. coli* fimbrial formation. J Struct Biol 172(3):380–388
64. Gong M, Makowski L (1992) Helical structure of P pili from *Escherichia Coli*. J Mol Biol 228:735–742
65. Miller E, Garcia T, Hultgren S, Oberhauser AF (2006) The mechanical properties of *E. coli* type 1 pili measured by atomic force microscopy techniques. Biophys J 91:3848–3856
66. Forero M, Yakovenko O, Sokurenko EV, Thomas WE, Vogel V (2006) Uncoiling mechanics of *Escherichia coli* type I fimbriae are optimized for catch bonds. PLoS Biol 4(9):1509–1516
67. Bullitt E, Makowski L (1995) Structural polymorphism of bacterial adhesion pili. Nature 373:164–167
68. Snyder JA, Haugen BJ, Lockatell CV, Maroncle N, Hagan EC, Johnson DE, Welch RA, Mobley HLT (2005) Coordinate expression of fimbriae in uropathogenic *Escherichia coli*. Infect Immun 73(11):7588–7596
69. Snyder JA, Lloyd AL, Lockatell CV, Johnson DE, Mobley HLT (2006) Role of phase variation of type 1 fimbriae in a uropathogenic *Escherichia coli* cystitis isolate during urinary tract infection. Infect Immun 74(2):1387–1393
70. Lindberg S, Xia Y, Sonden B, Goransson M, Hacker J, Uhlin BE (2008) Regulatory interactions among adhesin gene systems of uropathogenic *Escherichia coli*. Infect Immun 76(2):771–780
71. Abraham JM, Freitag CS, Clements JR, Eisenstein BI (1985) An invertible element of DNA controls phase variation of type 1 fimbriae of *Escherichia coli*. Proc Natl Acad Sci U S A 82: 5724–5727
72. Muller CM, Aberg A, Straseviciene J, Emody L, Uhlin BE, Balsalobre C (2009) Type 1 fimbriae, a colonization factor of uropathogenic *Escherichia coli*, are controlled by the metabolic sensor CRP-cAMP. PLoS Pathog 5(2):e1000303
73. Schneeberger C, Kazemier BM, Geerlings SE (2014) Asymptomatic bacteriuria and urinary tract infections in special patient groups: women with diabetes mellitus and pregnant women. Curr Opin Infect Dis 27(1):108–114
74. Munera D, Hulgren SJ, Fernandez LA (2007) Recognition of the N-terminal lectin domain of FimH adhesin by the usher FimD is required for type 1 pilus biogenesis. Mol Microbiol 64: 333–346
75. Phan G, Remaut H, Wang T, Allen WJ, Pirker KF, Lebedev A, Henderson NS, Geibel S, Volkan E, Yan J, Kunze MB, Pinkner JS, Ford B, Kay CW, Li H, Hultgren SJ, Thanassi DG, Waksman G (2011) Crystal structure of the FimD usher bound to its cognate FimC-FimH substrate. Nature 474(7349):49–53
76. Munera D, Palomino C, Fernández LA (2008) Specific residues in the N-terminal domain of FimH stimulate type 1 fimbriae assembly in *Escherichia coli* following the initial binding of the adhesin to FimD usher. Mol Microbiol 69(4):911–925
77. Abraham SN, Goguen JD, Beachey EH (1988) Hyperadhesive mutant of type 1 fimbriated *Escherichia coli* associated with the formation of FimH organelles (fimbriosomes). Infect Immun 56:1023–1029
78. Schwartz DJ, Kalas V, Pinkner JS, Chen SL, Spaulding CN, Dodson KW, Hultgren SJ (2013) Positively selected FimH residues enhance virulence during urinary tract infection by altering FimH conformation. Proc Natl Acad Sci U S A 110:15530–15537
79. Dennis JW, Brewer CF (2013) Density-dependent lectin-glycan interactions as a paradigm for conditional regulation by posttranslational modifications. Mol Cell Proteomics 12(4): 913–920
80. Lau KS, Partridge EA, Grigorian A, Silvescu CI, Reinhold VN, Demetriou M, Dennis JW (2007) Complex N-glycan number and degree of branching cooperate to regulate cell proliferation and differentiation. Cell 129(1):123–134
81. Eto DS, Mulvey MA (2007) Flushing bacteria out of the bladder. Nat Med 13(5):531–532

82. Chabre YM, Roy R (2008) Recent trends in glycodendrimer syntheses and applications. Curr Top Med Chem 8(14):1237–1285
83. Almant M, Moreau V, Kovensky J, Bouckaert J, Gouin SG (2011) Clustering of *Escherichia coli* type-1 fimbrial adhesins by using multimeric heptyl alpha-D-mannoside probes with a carbohydrate core. Chem. Eur. J. 17(36):10029–10038
84. Bouckaert J, Li Z, Xavier C, Almant M, Caveliers V, Lahoutte T, Weeks SD, Kovensky J, Gouin SG (2013) Heptyl alpha-D-mannosides grafted on a beta-cyclodextrin core to interfere with *Escherichia coli* adhesion: an in vivo multivalent effect. Chem Eur J 19(24):7847–7855
85. Langermann S, Palaszynski S, Barnhart M, Auguste G, Pinkner JS, Burlein J, Barren P, Koenig S, Leath S, Jones CH, Hultgren SJ (1997) Prevention of mucosal *Escherichia coli* infection by FimH-adhesin-based systemic vaccination. Science 276(5312):607–611
86. Roberts JA, Marklund BI, Ilver D, Haslam D, Kaack MB, Baskin G, Louis M, Mollby R, Winberg J, Normark S (1994) The Gal alpha(1–4) Gal-specific tip adhesin of *Escherichia coli* P-fimbriae is needed for pyelonephritis to occur in the normal urinary tract. Proc Natl Acad Sci U S A 91:11889–11893
87. Aronson M, Medalia O, Schori L, Mirelman D, Sharon N, Ofek I (1979) Prevention of colonization of the urinary-tract by blocking bacterial adherence with methyl-alpha-D-mannopyranoside. Israel J Med Sci 15(1):88
88. Cusumano CK, Pinkner JS, Han Z, Greene SE, Ford BA, Crowley JR, Henderson JP, Janetka JW, Hultgren SJ (2011) Treatment and prevention of urinary tract infection with orally active FimH inhibitors. Sci Transl Med 3(109):109ra115
89. Knight SD, Bouckaert J (2009) Structure, function, and assembly of type 1 fimbriae. Top Curr Chem 288:67–107
90. Brument S, Sivignon A, Dumych TI, Moreau N, Roos G, Guerardel Y, Chalopin T, Deniaud D, Bilyy RO, Darfeuille-Michaud A, Bouckaert J, Gouin SG (2013) Thiazolylaminomannosides as potent anti-adhesives of type 1 piliated *Escherichia coli* isolated from Crohn's disease patients. J Med Chem 56(13):5395–5406
91. Schwardt O, Rabbani S, Hartmann M, Abgottspon D, Wittwer M, Kleeb S, Zalewski A, Smiesko M, Cutting B, Ernst B (2011) Design, synthesis and biological evaluation of mannosyl triazoles as FimH antagonists. Bioorg Med Chem 19(21):6454–6473
92. Touaibia M, Wellens A, Glinschert A, Shiao TC, Wang Q, Papadopoulos A, Bouckaert J, Roy R (2014) O - and C -linked mannopyranosides to chemically disarm uropathogenic *Escherichia coli* by targeting their type-1 fimbriae mediated adhesion. To be published
93. Pang L, Kleeb S, Lemme K, Rabbani S, Scharenberg M, Zalewski A, Schadler F, Schwardt O, Ernst B (2012) FimH antagonists: structure-activity and structure–property relationships for biphenyl alpha-D-mannopyranosides. ChemMedChem 7:1404–1422
94. Han Z, Pinkner JS, Ford B, Obermann R, Nolan W, Wildman SA, Hobbs D, Ellenberger T, Cusumano CK, Hultgren SJ, Janetka JW (2010) Structure-based drug design and optimization of mannoside bacterial FimH antagonists. J Med Chem 53(12):4779–4792
95. Wermuth CG, Ganellin CR, Lindberg P, Mitscher LA (1998) Glossary of terms used in medicinal chemistry (IUPAC Recommendations 1998). Pure Appl Chem 70(5):1129–1143
96. Jiang X, Abgottspon D, Kleeb S, Rabbani S, Scharenberg M, Wittwer M, Haug M, Schwardt O, Ernst B (2012) Antiadhesion therapy for urinary tract infections-a balanced PK/PD profile proved to be key for success. J Med Chem 55(10):4700–4713
97. Klein T, Abgottspon D, Wittwer M, Rabbani S, Herold J, Jiang X, Kleeb S, Luthi C, Scharenberg M, Bezencon J, Gubler E, Pang L, Smiesko M, Cutting B, Schwardt O, Ernst B (2010) FimH antagonists for the oral treatment of urinary tract infections: from design and synthesis to in vitro and in vivo evaluation. J Med Chem 53(24):8627–8641
98. Lee YC, Lee RT (1995) Carbohydrate-protein interactions: basis of glycobiology. Acc Chem Res 28:321–327
99. Hartmann M, Lindhorst TK (2011) The bacterial lectin FimH, a target for drug discovery - carbohydrate inhibitors of type 1 fimbriae-mediated bacterial adhesion. Eur J Org Chem 2011:3583–3609

100. Touaibia M, Roy R (2007) Glycodendrimers as anti-adhesion drugs against type 1 fimbriated *E. coli* uropathogenic infections. Mini Rev Med Chem 7(12):1270–1283
101. Sperling O, Fuchs A, Lindhorst TK (2006) Evaluation of the carbohydrate recognition domain of the bacterial adhesin FimH: design, synthesis and binding properties of mannoside ligands. Org Biomol Chem 4(21):3913–3922
102. Abgottspon D, Rölli G, Hosch L, Steinhuber A, Jiang X, Schwardt O, Cutting B, Smiesko M, Jenal U, Ernst B, Trampuz A (2010) Development of an aggregation assay to screen FimH antagonists. J Microbiol Methods 82:249–255
103. Boettner B (2013) Debugging Crohn's disease. SciBX 6(26):1–2
104. Deniaud D, Julienne K, Gouin SG (2011) Insights in the rational design of synthetic multivalent glycoconjugates as lectin ligands. Org Biomol Chem 9(4):966–979
105. Lundquist JJ, Toone EJ (2002) The cluster glycoside effect. Chem Rev 102(2):555–578
106. Chabre YM, Roy R (2010) Design and creativity in synthesis of multivalent neoglycoconjugates. Adv Carbohydr Chem Biochem 63:165–393
107. Lindhorst TK, Kieburg C, Krallmann-Wenzel U (1998) Inhibition of the type 1 fimbriae-mediated adhesion of *Escherichia coli* to erythrocytes by multiantennary alpha-mannosyl clusters: the effect of multivalency. Glycoconj J 15(6):605–613
108. Nagahori N, Lee RT, Nishimura S, PagÇ D, Roy R, Lee YC (2002) Inhibition of adhesion of type 1 fimbriated *Escherichia coli* to highly mannosylated ligands. Chembiochem 3(9): 836–844
109. Touaibia M, Wellens A, Shiao TC, Wang Q, Sirois S, Bouckaert J, Roy R (2007) Mannosylated G(0) dendrimers with nanomolar affinities to *Escherichia coli* FimH. ChemMedChem 2(8):1190–1201
110. Gouin SG, Wellens A, Bouckaert J, Kovensky J (2009) Synthetic multimeric heptyl mannosides as potent antiadhesives of uropathogenic *Escherichia coli*. ChemMedChem 4 (5):749–755
111. Vanwetswinkel S, Volkov AN, Sterckx YG, Garcia-Pino A, Buts L, Vranken WF, Bouckaert J, Roy R, Wyns L, van Nuland NA (2014) Study of the structural and dynamic effects in the FimH adhesin upon alpha-d-heptyl mannose binding. J Med Chem 57(4):1416–1427
112. Disney MD, Zheng J, Swager TM, Seeberger PH (2004) Detection of bacteria with carbohydrate-functionalized fluorescent polymers. J Am Chem Soc 126(41):13343–13346
113. Klebe G (2006) Virtual ligand screening: strategies, perspectives and limitations. Drug Discov Today 11(13–14):580–594
114. Billiet L, Geerlings P, Messens J, Roos G (2012) The thermodynamics of thiol sulfenylation. Free Radic Biol Med 52(8):1473–1485
115. Roos G, Wellens A, Touaibia M, Yamakawa N, Geerlings P, Roy R, Wyns L, Bouckaert J (2013) Validation of reactivity descriptors to assess the aromatic stacking within the tyrosine gate of FimH. Med Chem Lett 4:1085–1090
116. Yilmazer ND, Korth M (2013) Comparison of molecular mechanics, semi-empirical quantum mechanical, and density functional theory methods for scoring protein-ligand interactions. J Phys Chem B 117(27):8075–8084
117. Bickelhaupt FM, Baerends EJ (2000) Kohn-Sham density functional theory: predicting and understanding chemistry. In: Lipkowitz KB, Boyd DB (eds) Reviews in computational chemistry, vol 15. Wiley-VCH, New York, pp 1–86
118. Krapp A, Bickelhaupt FM, Frenking G (2006) Orbital overlap and chemical bonding. Chem Eur J 12(36):9196–9216
119. Lepšík M, Řezáč J, Kolář M, Pecina A, Hobza P, Fanfrlík J (2013) The semiempirical quantum mechanical scoring function for in silico drug design. Chempluschem 78:921–931
120. Genheden S, Mikulskis P, Hu L, Kongsted J, Soderhjelm P, Ryde U (2011) Accurate predictions of nonpolar solvation free energies require explicit consideration of binding-site hydration. J Am Chem Soc 133(33):13081–13092

121. Chattaraj PK, Roy DR, Geerlings P, Torrent-Succarrat M (2007) Local hardness: a critical account. Theor Chem Acc 118:923–930
122. Bercowitz M, Parr RG (1985) On the concept of local hardness in chemistry. J Am Chem Soc 107:6811–6814
123. Parr RG, Pearson RG (1983) Absolute hardness: companion parameter to absolute electronegativity. J Am Chem Soc 105:7512–7516
124. Buckingham AD (1967) Permanent and induced molecular moments and long-range intermolecular forces. Adv Chem Phys 12:107
125. Mignon P, Loverix S, Steyaert J, Geerlings P (2005) Influence of the pi-pi interaction on the hydrogen bonding capacity of stacked DNA/RNA bases. Nucleic Acids Res 33(6):1779–1789
126. Mignon P, Loverix S, Geerlings P (2004) Influence of stacking on hydrogen bonding: quantum chemical study on pyridine-benzene model complexes. J Phys Chem A 108: 6038–6044
127. Wheeler SE (2011) Local nature of substituent effects in stacking interactions. J Am Chem Soc 133(26):10262–10274
128. Raju RK, Bloom JWG, Wheeler SE (2013) Broad transferability os substituent effects in p-stacking interactions provide new insight into their origin. J Chem Theory Comput 9: 3479–3490
129. Eid S, Zalewski A, Smiesko M, Ernst B, Vedani A (2013) A molecular-modeling toolbox aimed at bridging the gap between medicinal chemistry and computational sciences. Int J Mol Sci 14(1):684–700

Top Med Chem (2014) 12: 169–186
DOI: 10.1007/7355_2014_44

Published online: 11 April 2014

Carbohydrate-Based Anti-Virulence Compounds Against Chronic *Pseudomonas aeruginosa* Infections with a Focus on Small Molecules

Alexander Titz

Abstract The Gram-negative bacterium *Pseudomonas aeruginosa* can establish life-threatening chronic infections through biofilm formation. The two bacterial lectins LecA and LecB play important roles in the formation of these biofilms and the inhibition of the lectins with carbohydrate-based ligands was shown to disrupt biofilms. These effects provide a novel therapeutic option against infections caused by *P. aeruginosa*. In addition to the urgent need for novel therapeutics against *Pseudomonas* infections, two major advantages arise from these lectins as targets for therapy: (1) the extracellular localization and site of activity of LecA and LecB circumvent the bacterial cell envelope as a particularly impermeable barrier of Gram-negative pathogens, which must be overcome by drugs against intracellular targets, and (2) the lectins are targets of the so-called anti-virulence therapy and therefore a reduced appearance of resistances towards lectin-directed drugs can be anticipated. In this review, the recent development of carbohydrate-based inhibitors against both lectins is summarized with a main focus on small molecules.

Contents

A. Titz (✉)
Helmholtz Institute for Pharmaceutical Research Saarland (HIPS), Campus C 2.3, D-66123 Saarbrücken, Germany
e-mail: alexander.titz@helmholtz-hzi.de

1 Introduction

Since the discovery of penicillin by Alexander Fleming infectious diseases have lost their frightening reputation in the society. Numerous antibiotics of different classes have been discovered at a high pace, especially during the golden ages of antibiotics, the 1960s and 1970s. Consequently, the vast majority of infections could be successfully treated. In addition to the mostly beneficial but sometimes inappropriate use, for example against viral infections, antibiotics have also found widespread use in livestock and especially poultry mass production. Unfortunately, the discovery and introduction of novel classes of antibiotics into the market has stalled during the last 10–15 years and the number of pharmaceutical companies developing antibiotics has significantly dropped from 1990 to 2011 [1]. The excessive use in combination with a lack of novel drugs has led to the appearance of resistances among pathogens against existing antibiotics [2]. In recent years, multiresistant germs have appeared, pharmaceutical companies have withdrawn from antibacterial research and therefore, chemotherapeutic treatment options are consequently on the decline [1]. The Gram-positive methicillin-resistant *Staphylococcus aureus* (MRSA) is probably the best known multiresistant organism. In recent years a number of problematic organisms have appeared, and the so-called ESKAPE pathogens (*Enterococcus faecium*, *Staphylococcus aureus*, *Klebsiella pneumoniae*, *Acinetobacter baumannii*, *Pseudomonas aeruginosa*, and *Enterobacter* species) are in the center of concern regarding hospital-acquired infections [3]. Among them, highly resistant Gram-negative strains of *Pseudomonas aeruginosa* have evolved and present currently a major threat to public health [4, 5].

P. aeruginosa is a ubiquitous organism that is found in soil, water, and numerous other habitats as a consequence of its high adaptability to different environmental conditions. The genome of *P. aeruginosa* consists of a conserved core genome and a highly variable accessory genome which allows the bacterium to respond to numerous environmental factors, e.g., nutrients, niches, and defense mechanisms [6, 7]. Besides its ability to infect plants, this Gram-negative rod-shaped bacterium is an opportunistic pathogen and a major threat to hospitalized patients [8] as well as to individuals suffering from the congenital disease, cystic fibrosis [9]. Chronic pulmonary infections of cystic fibrosis patients with *P. aeruginosa* and subsequent organ failure account for 80% of deaths of these patients [10]. Among the nosocomial infections caused by this pathogen, pulmonary infections of artificially ventilated patients as well as wound infections are the most prevalent ones, but urinary tract infections or infections of implants equally pose major problems for treatment.

Contrary to the cell wall of Gram-positive bacteria like *Staphylococcus aureus*, which consist of an inner membrane and an outer layer of peptidoglycan, Gram-negatives possess an additional outer membrane covering the peptidoglycan layer and the inner membrane. This extra barrier gives an extended protection of Gram-negatives against treatment with antibiotics. Furthermore, *P. aeruginosa* has numerous so-called efflux pumps, i.e., proteins that deplete exogenous compounds like antibiotics within the cell after they have entered the cell by passive diffusion or active transport processes. In addition to the presence of various resistance genes, for example beta-lactamases, the organism can switch its lifestyle to the biofilm mode of life. In summary, *P. aeruginosa* has a vast arsenal of resistance mechanisms against antibiotic treatment [11].

The so-called biofilms are social colonies which can be formed by unicellular organisms to provide a concerted mechanism of protection or resistance against external threats, e.g., antibiotics. Such biofilms are usually located not only on surfaces such as host tissue but also on implant or non-biogenic material, e.g., catheters or respirators. The development of biofilms is usually classified into five stages from initial adhesion of single bacterial cells, to the irreversible attachment and formation of micro-colonies, two subsequent maturation stages of the biofilm and an ultimate rupture leading to a release of unicellular organisms into the environment [12]. The initiation of such biofilms is usually regulated by quorum sensing, which is a mechanism of bacterial communication to coordinate and direct joint processes against the host [13]. Pathogens embedded in such a biofilm pose severe problems for therapy, as they are placed in a self-produced matrix which provides a barrier against the host's immune defense and antibiotic treatment [14]. The matrix of bacterial biofilms is very complex and consists of extracellular nucleic acids, various polysaccharides (pel, psl, and alginate for *P. aeruginosa*), lipids, and proteinaceous material [15]. In addition to the pure physical protection, in the center of a biofilm there are the so-called persister cells in a metabolically resting or dormant state, which display a strongly decreased susceptibility towards antibiotics. This state likely comes from a lack of nutrient availability and a lack of oxygen within the biofilm, slowing down the bacterial metabolism and, therefore, antibiotics targeting metabolism are rendered ineffective [16, 17]. Making use of this mechanism, pathogenic bacteria can survive in the host for months or years and can be revived after an antibiotic therapy has ended, leading to recurrent chronic infections.

In recent years, a novel paradigm in treating infections has evolved: the so-called anti-virulence therapy aims at targeting mechanisms of infection rather than targeting bacterial viability as traditional antibiotics do [18]. By interfering with non-essential functions for viability of the organism, a strongly reduced appearance rate of resistant mutants is expected. Numerous virulence factors are known and bacterial toxins such as proteins (e.g., AB_5 toxins, proteases, and lectins) and small molecules (e.g., pyocyanin), as well as the delivery of such toxins, e.g., by targeting type three secretion systems (TTSS) or other secretory pathways, are among the most studied targets. Furthermore, the formation of biofilms is a validated target of anti-virulence therapy. The main areas of anti-biofilm strategies against Gram-negative bacteria are interfering with the quorum sensing processes, the inhibition of bacterial adhesins as well as the disruption of adhesive organelles, e.g., pili and fimbriae [19].

Because numerous adhesins are carbohydrate-binding proteins, their blocking with carbohydrate-based drugs is a validated strategy for anti-infective research [20–22]. For uropathogenic *Escherichia coli*, mannose-based inhibitors of the fimbrial adhesin FimH have been developed independently by the laboratories of Ernst and Hultgren and were shown to be active in vivo in a murine urinary tract infection model [23–29]. The literature on FimH as target for carbohydrate-based drugs is reviewed by

Julie Bouckaert in this volume. For *P. aeruginosa*, the two soluble lectins [30] LecA (formerly called PA-IL) and LecB (formerly called PA-IIL) were shown to be both cytotoxic [31–33] and involved in biofilm [34, 35] formation and are, therefore, considered as targets against chronic infections with *P. aeruginosa*.

2 The Lectins LecA and LecB as Targets Against Chronic *P. aeruginosa* Infections

2.1 Biological Relevance of LecA and LecB

Carbohydrate-binding molecules are often involved in cell-adhesion processes or host–pathogen recognition and blocking their carbohydrate binding sites may lead to novel therapies against bacterial infections [21]. In contrast to adhesins present on extracellular organelles like pili and flagella, the *P. aeruginosa* lectins LecA and LecB are soluble proteins [30, 36, 37]. In 1983, Glick and Garber reported a cytosolic localization of both, LecA and LecB [38]. However, for an adhesion process to take place, the localization of these adhesins is likely to be in the extracellular space. Therefore, it was hypothesized that the lectins are made available to the remaining population for adhesion by sacrifying bacteria through lysis. Interestingly, the expression of both lectins is controlled by quorum sensing [39], a mechanism of bacterial communication to sense the presence of surrounding fellow bacteria and to concert pathogenicity against the host. Quorum sensing leads to the expression of various virulence factors as well as to biofilm formation of the bacteria [13, 40, 41]. The lectins LecA and LecB were both shown to be necessary for biofilm formation [34, 35]. A *lecB*$^-$ mutant of *P. aeruginosa* PAOI showed strongly reduced biofilm formation on glass slides as reported by Tielker et al. [35]. Diggle and coworkers analyzed a *lecA*$^-$ mutant of PAOI and observed a similar effect on biofilms grown on stainless steel coupons [34]. In the same study the LecA-overexpressing mutant PAO-P47 showed increased biofilm formation compared to the wild-type bacteria. Interestingly, the biofilm formation of PAOI and of PAO-P47 could be inhibited by the addition of IPTG, a galactoside ligand of LecA.

P. aeruginosa has a vast array of virulence factors [42], for example the cytotoxic blue-green pigment pyocyanin, the secreted protease elastase B, which protects against host immune defense or a TTSS which injects cytotoxic effector proteins into host cells. In addition to its role in biofilm formation, it was shown in vitro that LecA displays a dose-dependent toxicity towards primary respiratory epithelial cells and the rate of ciliary beating was reduced upon exposure to LecA [33]. A similar inhibitory effect on ciliar activity was reported for LecB and ciliary beating could be restored in a carbohydrate-dependent manner by addition of fucose, a known inhibitor of LecB [32]. Since cilia are protective organelles on lung tissue, co-administration of fucose was proposed by Adam et al. to facilitate treatment of *P. aeruginosa*-mediated pneumonia. Subsequently, a significant reduction in bacterial load was reported after inhalation of aerosols consisting of fucose and galactose on chronically infected patients [43, 44]. Animal models of chronic pulmonary infection with *P. aeruginosa* relevant to biofilm-linked long-term infections are scarce [45, 46]. In an animal study of acute toxicity, mutants deficient in either LecA or LecB were tested and compared to the corresponding wild-type strain PAOI by Chemanei et al. [31]. It was shown that both lectins are equally important for adhesion and lung colonization of *P. aeruginosa* and, in addition, that LecA displays an increased toxicity towards mice, which could be blocked by addition of either D-galactose or N-acetyl D-galactosamine. Although the presence of LecB did not show an influence on acute toxicity, its important role in bacterial adhesion to A549 cells was demonstrated in this study. This cellular adhesion data together with the reported biofilm-deficiency of *lecB*⁻ strains supports the prominent role of LecB as an anti-biofilm target in chronic infections.

2.2 *LecA: Isolation, Natural Ligands, Structure, and Synthetic Inhibitors*

2.2.1 Isolation, Natural Ligands, and Derivatives

LecA was first isolated in 1972 by Gilboa-Garber from *P. aeruginosa* from a heat-stable fraction of whole bacteria [36]. The lectin was shown to bind to D-galactose (**1**, Fig. 1) and to a lower extent to *N*-acetyl D-galactosamine (**2**), both in a calcium-dependent manner. A dissociation constant of 87.5 μM was later determined for D-galactose (**1**) with LecA [47]. Numerous galactosides were then tested and the influence of their aglycon on LecA binding was analyzed [48]. Whereas methyl α-D-galactoside (**3**) is more potent than its β-anomer **4**, the binding strength is increased for β-anomers of larger and more lipophilic aglycons compared to the corresponding α-anomers. For example, 4-nitrophenyl β-D-galactoside (**5**) binds to LecA with a K_d of 14.1 μM [47, 48]. Furthermore, the introduction of thio-glycosides was beneficial for binding to LecA and **6** (IPTG) has a K_d of 32.4 μM [47, 48]. More than 30 years after the discovery of the monosaccharide specificity of LecA by Gilboa-Garber in 1972 [36], the saccharide moiety of a globo-series glycosphingolipid was identified as the potential natural ligand of LecA present on lung epithelial cells from a glycan array screen in 2008 [49]. The glycolipid **7** was reported as natural ligand and its trisaccharide Gal-α-1,4-Gal-β-1,4-Glc (**8**) showed a K_d of 77 μM for LecA. Various galabiose disaccharide derivatives of **7** showed K_d values between 37 and 132 μM [49, 50].

2.2.2 Structure

The crystal structure of LecA in complex with its galactose ligand was solved by Imberty and co-workers [51]. The protein forms a non-covalent homotetramer of

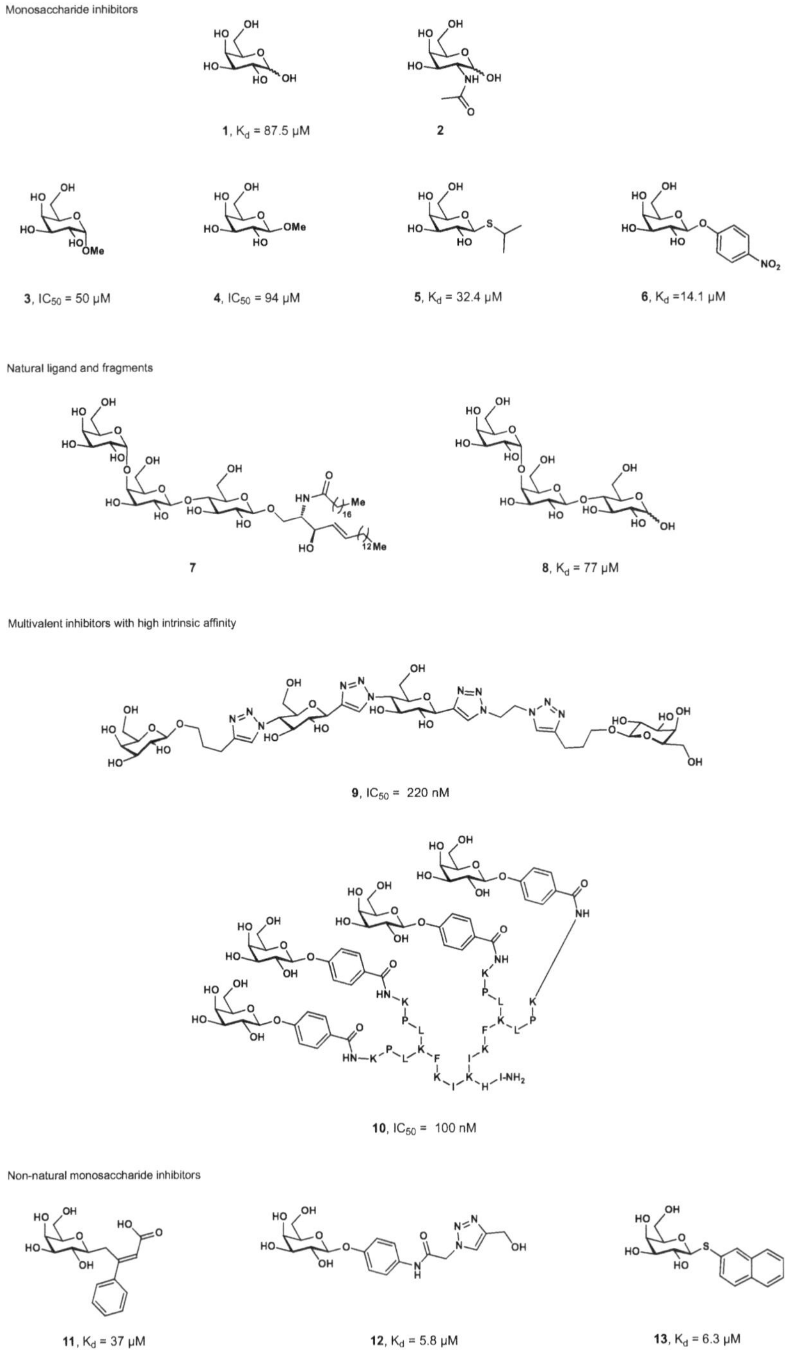

Fig. 1 Galactose-based inhibitors of LecA

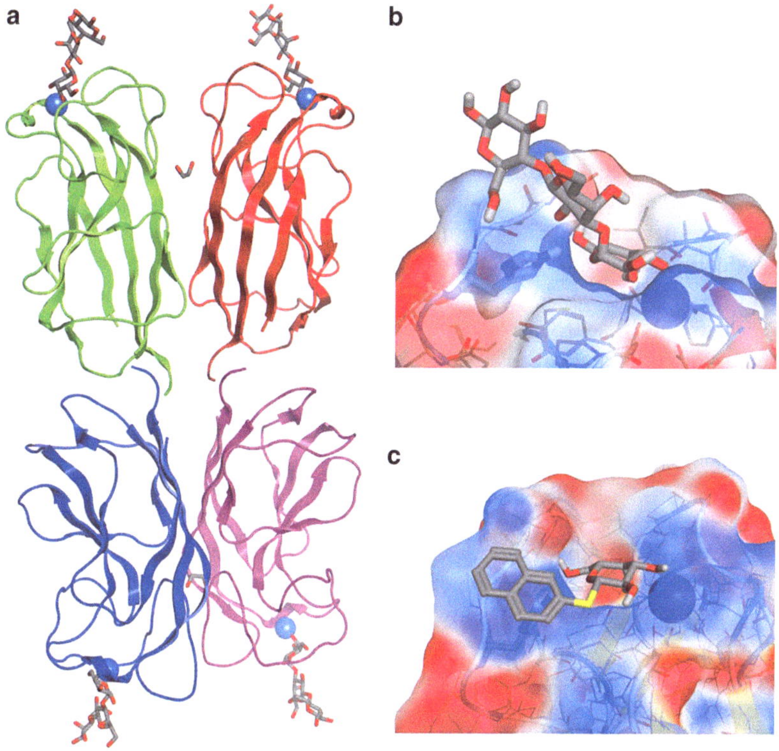

Fig. 2 (**a**) LecA forms a homotetramer with one calcium ion in the binding pocket of each monomer and coordinates D-galactose containing ligands, such as its proposed natural substrate **8** (pdb code 2VXJ). (**b**) Illustration of the coordination of the natural trisaccharide **8** in the binding site of LecA with His50 (shown in *bold sticks*) forming one hydrogen bond to the 6-OH of the reducing end galactoside (pdb code 2VXJ). (**c**) Illustration of the coordination of the nonnatural naphthyl thiogalactoside **13** in the binding site of LecA with His50 (shown in *bold sticks*) forming one hydrogen bond to the 6-OH of the galactoside as well as its T-shaped interaction through a CH contact with the protein-faced plane of the aromatic ring (pdb code 4A6S). The protein surfaces in (**b**) and (**c**) are color-coded within MOE by electrostatics: *blue* positive partial charge, *red* negative partial charge, *white*: neutral

four LecA polypeptide chains, both in solution and in the crystalline state, with the carbohydrate binding sites on opposing ends of the tetramer (Fig. 2a). The C-type lectin LecA mediates the binding to its ligands via one calcium ion in the binding site and coordinates with the cis-diol in galactose-based ligands. In addition, the galactose moiety is embedded into the binding site through extended hydrogen-bonding contacts. The crystal structure also provided an explanation for the increased activity of lipophilic β-galactosides. In α-anomers, the agylcon is positioned towards the solvent and therefore attractive interactions are minimal, whereas a patch on the surface of LecA is accessible for additional interaction with β-substituents (Fig. 2b, c).

2.2.3 Inhibitors

As a result of the low affinity of the galactose moiety alone, numerous multivalent and dendritic structures were synthesized to increase binding affinity. Especially the groups of René Roy, Sébastien Vidal, and Jean-Louis Reymond have synthesized a diverse set of core structures which provide the scaffold of galactose display. It is the focus of this review to report on the direct interaction of small molecules, especially the carbohydrate moiety or its derivatives with the *Pseudomonas* lectins. For a complete summary of the broad area of multivalent dendrimers, the interested reader is directed to some recent reviews summarizing activities in this field [52–54]. One example for a dendrimer is peptide-based **10** containing four galactose residues, which was recently reported by Kadam et al. [47] (Fig. 1). The multivalent compound is a potent inhibitor of LecA in vitro with an IC_{50} of 100 nM. Its glycopeptide residue forms contacts over a large area with the protein through recognition by the carbohydrate binding site, aromatic contact of the phenyl aglycon, and further interactions with the peptide part as determined from X-ray crystallography of the monovalent analog (Gal-β-OC_6H_4CONH-Lys-Pro-Leu-$CONH_2$) with the recently reported [55] K_d of 4.2 μM. Furthermore, the formation of biofilms by planktonic *P. aeruginosa* could be inhibited with dendrimer **10** showing the potential of such compounds. The same group recently performed a systematic chemical mutagenesis approach on the glycosylated tripeptide part of **10** and identified numerous analogs (best monovalent analogs Gal-β-OC_6H_4CONH-Lys-Arg-Leu-$CONH_2$ and Gal-β-OC_6H_4CONH-Lys-Pro-Tyr-$CONH_2$, for both $K_d = 2.7$ μM), which were active in biofilm assays when coupled to the peptide dendrimer [56]. Because the tetramer of LecA displays two of its four binding sites in a *syn* fashion ca. 32 Å apart, oligovalent inhibitors aim at a simultaneous binding of the two neighboring sites. By making use of such a chelate, binding entropy contributes significantly to the free energy of binding and optimally suited bi- or multivalent ligands display a superior affinity over their monovalent analogs. Pertici and Pieters pursued an approach to optimally structure two galactose residues for simultaneous binding to two of the four carbohydrate recognition sites of the LecA tetramer [57]. The length and nature of the linker groups between both galactosides were varied and bivalent **9** was identified as highly potent and presumably chelate-forming inhibitor of LecA with an IC_{50} or 220 nM. The authors report a 545-fold increase in relative potency of bivalent **9** compared to the corresponding monovalent analog in their assay system. Pertici et al. further extended their study on the influence of the linker motif on binding to LecA and identified one derivative of **9** with a reduced number of rotatable bonds resulting in a high affinity inhibitor with a K_d of 28 nM for LecA [58]. Recently, approaches to further modify the aglycon in β-galactosides were reported (Fig. 1). Roy and coworkers reported C-glycosides of galactose bearing an additional aromatic ring to mimic the known effect of phenyl rings for binding [59]. Compound **9** was tested by microcalorimetry and a K_d of only 37 μM was reported for this monovalent LecA inhibitor, which was further coupled to dendrimer support via its carboxylic acid to increase potency in a multivalent framework. In related studies to increase affinity of the monovalent galactoside by Vidal and coworkers, β-galactoside **12** with an amido-substituted phenyl aglycon was identified as a potent monovalent inhibitor of LecA (K_d of 5.8 μM) [60]. A screen for increased aglycons with respect to size and lipophilicity was performed by Rodrigue et al. and 2-naphthyl thio-galactoside **13** was identified as potent monovalent inhibitor of LecA (K_d of 6.3 μM) [61]. Along these lines, Reymond and coworkers analyzed further aglycons in a detailed study in combination with thermodynamic analyses and crystallography of numerous complexes with LecA [55]. Both, the Roy and Reymond groups reported a T-shape coordination of aromatic aglycons through a CH bond from a histidine side chain in close proximity to the saccharide (e.g., Fig. 2c). In summary, the interaction of LecA with galactose-based ligands has been thoroughly studied. In addition, Stoitsova et al. [62] and Boteva and coworkers [63] reported a binding of LecA to fluorescent hydrophobic ligands and the quorum sensing signaling molecules AHL, respectively. However, both compounds have not been further developed to date.

2.3 *LecB: Isolation, Natural Ligands, Structure, and Synthetic Inhibitors*

2.3.1 Isolation and Natural Ligands

In analogy to LecA, the lectin LecB has been isolated from heat-resistant fractions of *P. aeruginosa* by Gilboa-Garber and coworkers in 1977 [37]. Initially, the protein was identified as a D-mannose(**14**)-binding agglutinin (Fig. 3), which was later also shown to bind L-fucose (**15**) and other monosaccharides with the same relative orientation of their ring hydroxyl groups [30]. The dissociation constant for methyl mannoside (**16**) was reported in the intermediate micromolar range (K_d = 71 μM) [64]. L-Fucose was published as strongest monosaccharide inhibitor of LecB-mediated hemagglutination and its dissociation constant with LecB was later reported [65] in the low micromolar range (K_d = 2.9 μM) while its methyl glycoside **17** [64] displayed a sixfold stronger affinity (K_d = 430 nM). In 2005, Perret et al. analyzed the interaction of LecB with milk oligosaccharides and the trisaccharide Lewis^a (**18**) was identified as the best known monovalent inhibitor of LecB with a dissociation constant of 220 nM [65].

2.3.2 Structure

The structure of LecB in the crystalline state was first solved in complex with L-fucose by Imberty and coworkers [66] and later in complex with D-mannose and other ligands by Loris [67] and coworkers. Similar to LecA, the protein forms a

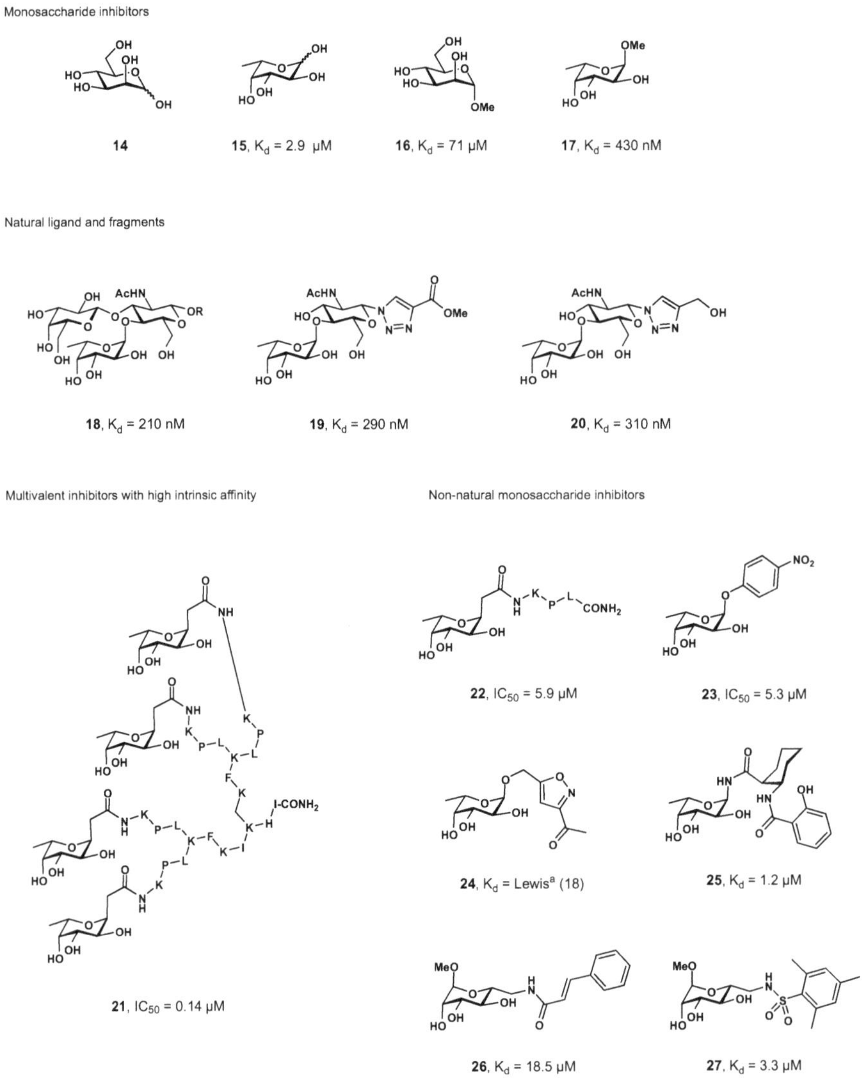

Fig. 3 Mannose- and Fucose-based inhibitors of LecB

homotetramer of its 115 amino acid- and 11.8 kDa-polypeptide chain, each monomer assembled in a β-sandwich fold with one carbohydrate binding site per subunit (Fig. 4a). Interestingly and in contrast to LecA, the binding sites are arranged in an opposing orientation where the sites are localized on the vertices of a tetrahedron and the C-terminus of one polypeptide chain contributes to carbohydrate binding of

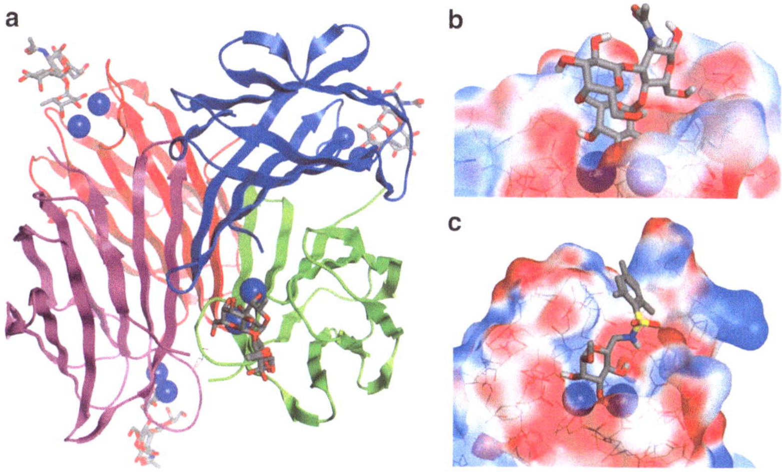

Fig. 4 (**a**) LecB forms a homotetramer with two calcium ions in the binding pocket of each monomer and coordinates L-fucose or D-mannose containing ligands, such as its proposed natural substrate Lewis[a] (pdb code 1W8H). (**b**) The interaction of Lewis[a] (**18**) is depicted in the binding site (pdb code 1W8H). (**c**) Mannose-based sulfonamide **27** closely interacts with the surface of LecB (pdb code 3ZDV). The protein surfaces in **b** and **c** are color-coded within MOE by electrostatics: *blue* positive partial charge, *red* negative partial charge, *white*: neutral

the neighboring chain. Furthermore, LecB recruits two calcium ions per binding site to coordinate its carbohydrate ligands: the 2-, 3-, and 4-hydroxy groups of fucose are coordinating simultaneously to both protein-bound calcium ions with the 3-hydroxy group acting as bridging μ^2-ligand of both calcium ions. The unusual strong interaction of LecB with L-fucose was explained by increased electrostatic interactions due to the presence of two calcium ions, as well as a lipophilic interaction of the methyl group in L-fucose with the protein surface [66–68]. The structure of D-mannose with LecB shows an additional hydrogen bond of the 6-hydroxy group of the ligand with the side chain of serine 23 of the receptor [67]. The so-called sugar-specificity loop consisting of the amino acids Ser22-Ser23-Gly24 in LecB was reported by Adam et al. to be responsible for its monosaccharide specificity for fucose and the inverted preference of the closely related *Ralstonia solanacearum* lectin RS-IIL for mannose-based ligands [69]. In 2005, Perret et al. reported the interaction of LecB with various milk oligosaccharides by microcalorimetry and X-ray crystallography [65]. In the complex of the high affinity trisaccharide ligand Lewis[a] (**18**), the terminal fucose residue of **18** is bound by LecB at its carbohydrate recognition site and the penultimate N-acetyl glucosamine moiety forms attractive interactions with the protein via its 3- and 6-hydroxy groups, whereas the galactoside moiety in **18** does not form direct contacts with LecB (Fig. 4b).

2.3.3 Inhibitors

Based on the structural data of the complex of LecB and Lewis[a] (**18**, Fig. 3), Marotte et al. subsequently truncated this trisaccharide to identify a minimal binding epitope [70]. The galactose moiety which is not in direct contact with the protein in the crystal structure of the complex was removed and an azido functionality at the reducing end of the resulting Fuc-α-1,4-GlcNAc disaccharide was introduced allowing conjugation onto multivalent scaffolds. The interaction of the two monovalent derivatives with LecB, ester **19** and alcohol **20**, was analyzed by microcalorimetry and negligible loss in binding affinity compared to the parent trisaccharide was observed (**19**: $K_d = 290$ nM, **20**: $K_d = 310$ nM). Similar to LecA-directed **10**, Reymond and coworkers developed the peptide-based dendrimer **21** carrying four C-fucosyl coupled monosaccharides [71]. Dendrimer **21** showed an IC_{50} of 140 nM in an ELLA-based in vitro assay and its monovalent C-fucosyl tripeptide **22** showed intermediate affinity ($IC_{50} = 5.9$ μM). Importantly, multivalent **21** showed successful anti-biofilm activities in vitro, both as inhibitor of biofilm formation and as disrupting agent for pre-formed biofilms of *P. aeruginosa*. The monovalent C-fucoside **22** showed weaker affinity for LecB than a comparable O-fucoside. However, a stabilization of the metabolically labile O-fucosidic linkage in previous fucose-based inhibitors can be anticipated by the introduction of such a hydrolytically stable C-glycosidic linkage. Reymond and coworkers further reported monovalent O-fucoside **23** as LecB inhibitor with intermediate affinity ($IC_{50} = 5.3$ μM). Notably, O-fucoside **23** was also shown to be active in two biofilm assays: the formation of biofilms was inhibited and disruption of existing biofilms could be achieved although at increased concentrations [71]. As a continuation of their reductionist approach for the trisaccharide Lewis[a] (**18**) leading to **19** and **20**, Imberty and Roy further simplified the LecB-bound moiety. Fucosylated oxazole **24** was reported as an equipotent ligand to LecB as Lewis[a] itself and the crystal structure of the complex was deposited at the protein database (www.pdb.org, code 2VUC) [53]. Another approach modifying the anomeric center was reported by Andreini et al. and fucosyl amides, e.g., compound **25** ($K_d = 1.2$ μM), were prepared [72]. By docking analyses, the aglycon was shown to contribute to the binding to LecB and a hydrogen bonding network between both carbonyl oxygens with the side chain of Ser23 was proposed. Recently, we reported novel mannose-based inhibitors of LecB, in which the 6-hydroxy group of mannosides was substituted as cinnamide **26** or sulfonamide **27** [73]. As a result of the unusual high affinity of fucosides with LecB, D-mannose (**14**) had not served as scaffold in the search for high affinity ligands of LecB before. We could show by X-ray crystallography and isothermal titration microcalorimetry that targeting additional patches on the surface of LecB (Fig. 4c) with substituents as in **27** could increase the binding affinity of mannosides to a K_d of 3.3 μM, which is comparable to that of L-fucose (**15**). Such terminally capped saccharides were designed with the purpose to increase specificity towards the pathogenic lectin LecB and reduce binding to the ubiquitous lectins of the host, requiring terminal mannose or fucose residues. Interestingly, both molecules offer distinct possibilities of further substitution as

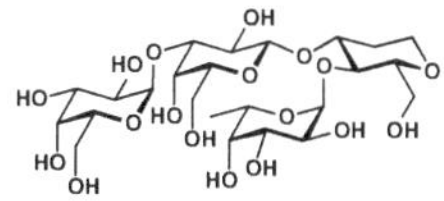

28, IC_{50} = 47 μM for LecA
IC_{50} = ~ 1 μM for LecB

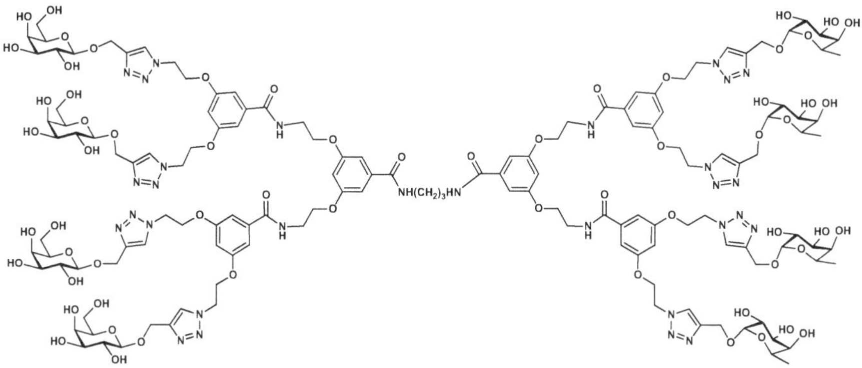

29, IC_{50} = n.a. for LecA
IC_{50} = n.a. for LecB

Fig. 5 Inhibitors designed to target LecA and LecB simultaneously

they differ significantly in the binding mode of the aromatic moieties with **27** coordinated with the surface and **26** presumably opening a cleft into the β-sandwich of LecB. Both compounds were shown to inhibit LecB-mediated adhesion of *P. aeruginosa* to glycosylated surfaces.

2.4 Attempts on Targeting Both Lectins with a Single Compound

Bifunctional inhibitors to target both lectins, LecA and LecB, simultaneously could provide an elegant way to interfere with *P. aeruginosa* biofilm formation. Single chemical entities with inhibitory groups against both lectins, i.e., broad-spectrum inhibitors targeting a single organism but two targets simultaneously, may have a regulatory advantage in the drug registration and approval process. However, experimental evidence is lacking if such compounds could block biofilm formation by blocking the individual lectins or rather act as an artificial adhesive and strengthen the biofilm architecture. In 2007, GlycoMimetics Inc. disclosed compound **28** as a mimic of Lewis[a] (**18**) with an additional α-galactoside attached [74] (Fig. 5). This compound was reported as potent monovalent inhibitor of both bacterial lectins, LecA ($IC_{50} = 47$ μM) and LecB ($IC_{50} \approx 1$ μM). Furthermore, Deguise et al. reported the heterobivalent dendrimer bearing both galactoside and fucoside residues [75]. Dendrimer **29** was reported to bind to LecA and LecB but affinities were not disclosed.

3 Conclusion

Since the discovery of the *P. aeruginosa* lectins by Gilboa-Garber and co-workers in the 1970s, their ligand binding properties and crystal structures have been solved in complex with numerous natural and synthetic compounds. The finding of their important roles in biofilm formation of this opportunistic pathogen has sparked the interest of numerous synthetic chemistry groups for the development of novel therapeutics against chronic infections with *P. aeruginosa*. Because carbohydrates usually have a low affinity to their lectin receptors, multivalent approaches to increase binding affinities are widely used. Such molecules may, however, display indiscriminate affinities for a large number of proteins in the host, such as the numerous pattern recognition receptors (PRRs) with many lectins among this class of proteins of the innate immune system [76]. Unwanted inhibition of these ubiquitous PRRs with anti-biofilm drugs can thus interfere with mechanisms of immune defense and evoke side effects. It is therefore of importance to address specificity, but attempts to analyze specificity are scarce [77] for *P. aeruginosa* directed compounds, and their interactions with human lectins have not been studied to date. To achieve specificity, a promising approach is the modification of the monovalent ligand in order to optimally exploit interactions with the lectin binding sites of LecA and LecB. Because lectins typically have shallow binding sites, high affinity inhibitors may require larger areas of interaction. Recent success on related lectin targets with shallow binding sites is encouraging [78–81].

Acknowledgements A.T. acknowledges the Helmholtz Association for financial support.

References

1. Cooper MA, Shlaes D (2011) Fix the antibiotics pipeline. Nature 472(7341):32
2. Taubes G (2008) The bacteria fight back. Science 321(5887):356–361
3. Rice LB (2008) Federal funding for the study of antimicrobial resistance in nosocomial pathogens: no ESKAPE. J Infect Dis 197(8):1079–1081
4. Chan M (2012) Antimicrobial resistance in the European Union and the world. Combating antimicrobial resistance: time for action. www.who.int/dg/speakers/2012/amr_20120314/en
5. European Centre for Disease Prevention and Control (2009) The bacterial challenge: time to react. European Centre for Disease Prevention and Control. Technical report
6. Mathee K, Narasimhan G, Valdes C, Qiu X, Matewish JM, Koehrsen M, Rokas A, Yandava CN, Engels R, Zeng E, Olavarietta R, Doud M, Smith RS, Montgomery P, White JR, Godfrey

PA, Kodira C, Birren B, Galagan JE, Lory S (2008) Dynamics of Pseudomonas aeruginosa genome evolution. Proc Natl Acad Sci U S A 105(8):3100–3105
7. Kung VL, Ozer EA, Hauser AR (2010) The accessory genome of Pseudomonas aeruginosa. Microbiol Mol Biol Rev 74(4):621–641
8. Peleg AY, Hooper DC (2010) Hospital-acquired infections due to gram-negative bacteria. New Engl J Med 362(19):1804–1813
9. Tümmler B, Kiewitz C (1999) Cystic fibrosis: an inherited susceptibility to bacterial respiratory infections. Mol Med Today 5(8):351–358
10. Cystic Fibrosis Foundation (2008) Patient Registry Report. Cystic Fibrosis Foundation. Technical report
11. Poole K (2011) Pseudomonas aeruginosa: resistance to the max. Front Microbiol 2:65
12. Monroe D (2007) Looking for chinks in the armor of bacterial biofilms. PLoS Biol 5(11):e307
13. Irie Y, Parsek MR (2008) Quorum sensing and microbial biofilms. In: Romeo T (ed) Bacterial biofilms, vol 322. Springer, Berlin/Heidelberg, p 67–84
14. Davies D (2003) Understanding biofilm resistance to antibacterial agents. Nat Rev Drug Discov 2(2):114–122
15. Flemming H, Wingender J (2010) The biofilm matrix. Nat Rev Microbiol 8(9):623–633
16. Stewart PS, Franklin MJ (2008) Physiological heterogeneity in biofilms. Nat Rev Microbiol 6 (3):199–210
17. Bjarnsholt T, Ciofu O, Molin S, Givskov M, Høiby N (2013) Applying insights from biofilm biology to drug development - can a new approach be developed? Nat Rev Drug Discov 12 (10):791–808
18. Clatworthy AE, Pierson E, Hung DT (2007) Targeting virulence: a new paradigm for antimicrobial therapy. Nat Chem Biol 3(9):541–548
19. Sommer R, Joachim I, Wagner S, Titz A (2013) New approaches to control infections: antibiofilm strategies against gram-negative bacteria. CHIMIA 67(4):286–290
20. Sharon N (1987) Bacterial lectins, cell-cell recognition and infectious disease. FEBS Lett 217 (2):145–157
21. Sharon N (2006) Carbohydrates as future anti-adhesion drugs for infectious diseases. Biochim Biophys Acta 1760(4):527–537
22. Ernst B, Magnani JL (2009) From carbohydrate leads to glycomimetic drugs. Nat Rev Drug Discov 8(8):661–677
23. Scharenberg M, Schwardt O, Rabbani S, Ernst B (2012) Target selectivity of FimH antagonists. J Med Chem 55(22):9810–9816
24. Pang L, Kleeb S, Lemme K, Rabbani S, Scharenberg M, Zalewski A, Schädler F, Schwardt O, Ernst B (2012) FimH antagonists: structure-activity and structure-property relationships for biphenyl α-D-mannopyranosides. ChemMedChem 7(8):1404–1422
25. Jiang X, Abgottspon D, Kleeb S, Rabbani S, Scharenberg M, Wittwer M, Haug M, Schwardt O, Ernst B (2012) Antiadhesion therapy for urinary tract infections–a balanced PK/PD profile proved to be key for success. J Med Chem 55(10):4700–4713
26. Han Z, Pinkner JS, Ford B, Chorell E, Crowley JM, Cusumano CK, Campbell S, Henderson JP, Hultgren SJ, Janetka JW (2012) Lead optimization studies on FimH antagonists: discovery of potent and orally bioavailable ortho-substituted biphenyl mannosides. J Med Chem 55 (8):3945–3959
27. Cusumano CK, Pinkner JS, Han Z, Greene SE, Ford BA, Crowley JR, Henderson JP, Janetka JW, Hultgren SJ (2011) Treatment and prevention of urinary tract infection with orally active FimH inhibitors. Sci Transl Med 3(109):109ra115
28. Klein T, Abgottspon D, Wittwer M, Rabbani S, Herold J, Jiang X, Kleeb S, Lüthi C, Scharenberg M, Bezençon J, Gubler E, Pang L, Smiesko M, Cutting B, Schwardt O, Ernst B (2010) FimH antagonists for the oral treatment of urinary tract infections: from design and synthesis to in vitro and in vivo evaluation. J Med Chem 53(24):8627–8641

29. Han Z, Pinkner JS, Ford B, Obermann R, Nolan W, Wildman SA, Hobbs D, Ellenberger T, Cusumano CK, Hultgren SJ, Janetka JW (2010) Structure-based drug design and optimization of mannoside bacterial FimH antagonists. J Med Chem 53(12):4779–4792
30. Gilboa-Garber N (1982) Pseudomonas aeruginosa lectins. Methods Enzymol 83:378–385
31. Chemani C, Imberty A, de Bentzmann S, Pierre M, Wimmerová M, Guery BP, Faure K (2009) Role of LecA and LecB lectins in Pseudomonas aeruginosa-induced lung injury and effect of carbohydrate ligands. Infect Immun 77(5):2065–2075
32. Adam EC, Mitchell BS, Schumacher DU, Grant G, Schumacher U (1997) Pseudomonas aeruginosa II lectin stops human ciliary beating: therapeutic implications of fucose. Am J Respir Crit Care Med 155(6):2102–2104
33. Bajolet-Laudinat O, Girod-de Bentzmann S, Tournier JM, Madoulet C, Plotkowski MC, Chippaux C, Puchelle E (1994) Cytotoxicity of Pseudomonas aeruginosa internal lectin PA-I to respiratory epithelial cells in primary culture. Infect Immun 62(10):4481–4487
34. Diggle SP, Stacey RE, Dodd C, Cámara M, Williams P, Winzer K (2006) The galactophilic lectin, LecA, contributes to biofilm development in Pseudomonas aeruginosa. Environ Microbiol 8(6):1095–1104
35. Tielker D, Hacker S, Loris R, Strathmann M, Wingender J, Wilhelm S, Rosenau F, Jaeger K (2005) Pseudomonas aeruginosa lectin LecB is located in the outer membrane and is involved in biofilm formation. Microbiology 151(Pt 5):1313–1323
36. Gilboa-Garber N (1972) Purification and properties of hemagglutinin from Pseudomonas aeruginosa and its reaction with human blood cells. Biochim Biophys Acta 273(1):165–173
37. Gilboa-Garber N, Mizrahi L, Garber N (1977) Mannose-binding hemagglutinins in extracts of Pseudomonas aeruginosa. Can J Biochem 55(9):975–981
38. Glick J, Garber N (1983) The intracellular localization of Pseudomonas aeruginosa lectins. J Gen Microbiol 129(10):3085–3090
39. Winzer K, Falconer C, Garber NC, Diggle SP, Camara M, Williams P (2000) The Pseudomonas aeruginosa lectins PA-IL and PA-IIL are controlled by quorum sensing and by RpoS. J Bacteriol 182(22):6401–6411
40. Rumbaugh KP, Griswold JA, Hamood AN (2000) The role of quorum sensing in the in vivo virulence of Pseudomonas aeruginosa. Microbes Infect 2(14):1721–1731
41. Whitehead NA, Barnard AM, Slater H, Simpson NJ, Salmond GP (2001) Quorum-sensing in Gram-negative bacteria. FEMS Microbiol Rev 25(4):365–404
42. Yahr T, Parsek M (2006) The Prokaryotes. In: Proteobacteria: Gamma Subclass, vol 6, 3rd edn. Springer
43. von Bismarck P, Schneppenheim R, Schumacher U (2001) Successful treatment of Pseudomonas aeruginosa respiratory tract infection with a sugar solution–a case report on a lectin based therapeutic principle. Klin Padiatr 213(5):285–287
44. Hauber H, Schulz M, Pforte A, Mack D, Zabel P, Schumacher U (2008) Inhalation with fucose and galactose for treatment of Pseudomonas aeruginosa in cystic fibrosis patients. Int J Med Sci 5(6):371–376
45. Kukavica-Ibrulj I, Levesque RC (2008) Animal models of chronic lung infection with Pseudomonas aeruginosa: useful tools for cystic fibrosis studies. Lab Anim 42(4):389–412
46. Egan M (2009) How useful are cystic fibrosis mouse models? Drug Discov Today Dis Models 6(2):35–41
47. Kadam RU, Bergmann M, Hurley M, Garg D, Cacciarini M, Swiderska MA, Nativi C, Sattler M, Smyth AR, Williams P, Cámara M, Stocker A, Darbre T, Reymond J (2011) A glycopeptide dendrimer inhibitor of the galactose-specific lectin LecA and of Pseudomonas aeruginosa biofilms. Angew Chem Int Ed Engl 50(45):10631–10635
48. Garber N, Guempel U, Belz A, Gilboa-Garber N, Doyle RJ (1992) On the specificity of the D-galactose-binding lectin (PA-I) of Pseudomonas aeruginosa and its strong binding to hydrophobic derivatives of D-galactose and thiogalactose. Biochim Biophys Acta 1116(3):331–333

49. Blanchard B, Nurisso A, Hollville E, Tétaud C, Wiels J, Pokorná M, Wimmerová M, Varrot A, Imberty A (2008) Structural basis of the preferential binding for globo-series glycosphingolipids displayed by Pseudomonas aeruginosa lectin I. J Mol Biol 383(4):837–853
50. Nurisso A, Blanchard B, Audfray A, Rydner L, Oscarson S, Varrot A, Imberty A (2010) Role of water molecules in structure and energetics of Pseudomonas aeruginosa PA-IL lectin interacting with disaccharides. J Biol Chem 285:20316–20327
51. Cioci G, Mitchell EP, Gautier C, Wimmerová M, Sudakevitz D, Pérez S, Gilboa-Garber N, Imberty A (2003) Structural basis of calcium and galactose recognition by the lectin PA-IL of Pseudomonas aeruginosa. FEBS Lett 555(2):297–301
52. Bernardi A, Jiménez-Barbero J, Casnati A, De Castro C, Darbre T, Fieschi F, Finne J, Funken H, Jaeger K, Lahmann M, Lindhorst TK, Marradi M, Messner P, Molinaro A, Murphy PV, Nativi C, Oscarson S, Penadés S, Peri F, Pieters RJ, Renaudet O, Reymond J, Richichi B, Rojo J, Sansone F, Schäffer C, Turnbull WB, Velasco-Torrijos T, Vidal S, Vincent S, Wennekes T, Zuilhof H, Imberty A (2012) Multivalent glycoconjugates as anti-pathogenic agents. Chem Soc Rev 42(11):4709–4727
53. Imberty A, Chabre YM, Roy R (2008) Glycomimetics and glycodendrimers as high affinity microbial anti-adhesins. Chem Eur J 14(25):7490–7499
54. Reymond J, Bergmann M, Darbre T (2013) Glycopeptide dendrimers as Pseudomonas aeruginosa biofilm inhibitors. Chem Soc Rev 42(11):4814–4822
55. Kadam RU, Garg D, Schwartz J, Visini R, Sattler M, Stocker A, Darbre T, Reymond J (2013) CH-π "T-shape" interaction with histidine explains binding of aromatic galactosides to Pseudomonas aeruginosa lectin LecA. ACS Chem Biol 8(9):1925–1930
56. Kadam RU, Bergmann M, Garg D, Gabrieli G, Stocker A, Darbre T, Reymond J (2013) Structure-based optimization of the terminal tripeptide in glycopeptide dendrimer inhibitors of pseudomonas aeruginosa biofilms targeting LecA. Chem Eur J 19(50):17054–17063
57. Pertici F, Pieters RJ (2012) Potent divalent inhibitors with rigid glucose click spacers for Pseudomonas aeruginosa lectin LecA. Chem Commun (Camb) 48(33):4008–4010
58. Pertici F, de Mol NJ, Kemmink J, Pieters RJ (2013) Optimizing divalent inhibitors of pseudomonas aeruginosa lectin LecA by using A rigid spacer. Chem Eur J 19(50):16923–16927
59. Chabre YM, Giguère D, Blanchard B, Rodrigue J, Rocheleau S, Neault M, Rauthu S, Papadopoulos A, Arnold AA, Imberty A, Roy R (2011) Combining glycomimetic and multivalent strategies toward designing potent bacterial lectin inhibitors. Chem Eur J 17(23):6545–6562
60. Cecioni S, Praly J, Matthews SE, Wimmerová M, Imberty A, Vidal S (2012) Rational design and synthesis of optimized glycoclusters for multivalent lectin-carbohydrate interactions: influence of the linker arm. Chem Eur J 18(20):6250–6263
61. Rodrigue J, Ganne G, Blanchard B, Saucier C, Giguère D, Shiao TC, Varrot A, Imberty A, Roy R (2013) Aromatic thioglycoside inhibitors against the virulence factor LecA from Pseudomonas aeruginosa. Org Biomol Chem 11(40):6906–6918
62. Stoitsova SR, Boteva RN, Doyle RJ (2003) Binding of hydrophobic ligands by Pseudomonas aeruginosa PA-I lectin. Biochim Biophys Acta 1619(2):213–219
63. Boteva RN, Bogoeva VP, Stoitsova SR (2005) PA-I lectin from Pseudomonas aeruginosa binds acyl homoserine lactones. Biochim Biophys Acta 1747(2):143–149
64. Sabin C, Mitchell EP, Pokorná M, Gautier C, Utille J, Wimmerová M, Imberty A (2006) Binding of different monosaccharides by lectin PA-IIL from Pseudomonas aeruginosa: thermodynamics data correlated with X-ray structures. FEBS Lett 580(3):982–987
65. Perret S, Sabin C, Dumon C, Pokorná M, Gautier C, Galanina O, Ilia S, Bovin N, Nicaise M, Desmadril M, Gilboa-Garber N, Wimmerová M, Mitchell EP, Imberty A (2005) Structural basis for the interaction between human milk oligosaccharides and the bacterial lectin PA-IIL of Pseudomonas aeruginosa. Biochem J 389(Pt 2):325–332

66. Mitchell E, Houles C, Sudakevitz D, Wimmerova M, Gautier C, Pérez S, Wu AM, Gilboa-Garber N, Imberty A (2002) Structural basis for oligosaccharide-mediated adhesion of Pseudomonas aeruginosa in the lungs of cystic fibrosis patients. Nat Struct Biol 9(12):918–921
67. Loris R, Tielker D, Jaeger K, Wyns L (2003) Structural basis of carbohydrate recognition by the lectin LecB from Pseudomonas aeruginosa. J Mol Biol 331(4):861–870
68. Mitchell EP, Sabin C, Snajdrová L, Pokorná M, Perret S, Gautier C, Hofr C, Gilboa-Garber N, Koca J, Wimmerová M, Imberty A (2005) High affinity fucose binding of Pseudomonas aeruginosa lectin PA-IIL: 1.0 A resolution crystal structure of the complex combined with thermodynamics and computational chemistry approaches. Proteins 58(3):735–746
69. Adam J, Pokorná M, Sabin C, Mitchell EP, Imberty A, Wimmerová M (2007) Engineering of PA-IIL lectin from Pseudomonas aeruginosa - unravelling the role of the specificity loop for sugar preference. BMC Struct Biol 7:36
70. Marotte K, Sabin C, Préville C, Moumé-Pymbock M, Wimmerová M, Mitchell EP, Imberty A, Roy R (2007) X-ray structures and thermodynamics of the interaction of PA-IIL from Pseudomonas aeruginosa with disaccharide derivatives. ChemMedChem 2(9):1328–1338
71. Johansson EMV, Crusz SA, Kolomiets E, Buts L, Kadam RU, Cacciarini M, Bartels K, Diggle SP, Cámara M, Williams P, Loris R, Nativi C, Rosenau F, Jaeger K, Darbre T, Reymond J (2008) Inhibition and dispersion of Pseudomonas aeruginosa biofilms by glycopeptide dendrimers targeting the fucose-specific lectin LecB. Chem Biol 15(12):1249–1257
72. Andreini M, Anderluh M, Audfray A, Bernardi A, Imberty A (2010) Monovalent and bivalent N-fucosyl amides as high affinity ligands for Pseudomonas aeruginosa PA-IIL lectin. Carbohydr Res 345(10):1400–1407
73. Hauck D, Joachim I, Frommeyer B, Varrot A, Philipp B, Möller HM, Imberty A, Exner TE, Titz A (2013) Discovery of two classes of potent glycomimetic inhibitors of pseudomonas aeruginosa LecB with distinct binding modes. ACS Chem Biol 8(8):1775–1784
74. Magnani JL, Patton JT, Sarkar AK (2007) patent, US2007/0037775A1
75. Deguise I, Lagnoux D, Roy R (2007) Synthesis of glycodendrimers containing both fucoside and galactoside residues and their binding properties to Pa-IL and PA-IIL lectins from Pseudomonas aeruginosa. New J Chem 31:1321–1331
76. Dam TK, Brewer CF (2010) Lectins as pattern recognition molecules: the effects of epitope density in innate immunity. Glycobiology 20(3):270–279
77. Cecioni S, Faure S, Darbost U, Bonnamour I, Parrot-Lopez H, Roy O, Taillefumier C, Wimmerová M, Praly J, Imberty A, Vidal S (2011) Selectivity among two lectins: probing the effect of topology, multivalency and flexibility of "clicked" multivalent glycoclusters. Chem Eur J 17(7):2146–2159
78. Egger J, Weckerle C, Cutting B, Schwardt O, Rabbani S, Lemme K, Ernst B (2013) Nanomolar E-selectin antagonists with prolonged half-lives by a fragment-based approach. J Am Chem Soc 135(26):9820–9828
79. Shelke SV, Cutting B, Jiang X, Koliwer-Brandl H, Strasser DS, Schwardt O, Kelm S, Ernst B (2010) A fragment-based in situ combinatorial approach to identify high-affinity ligands for unknown binding sites. Angew Chem Int Ed Engl 49(33):5721–5725
80. Rillahan CD, Schwartz E, Rademacher C, McBride R, Rangarajan J, Fokin VV, Paulson JC (2013) On-chip synthesis and screening of a sialoside library yields a high affinity ligand for siglec-7. ACS Chem Biol 8(7):1417–1422
81. Zeng Y, Rademacher C, Nycholat CM, Futakawa S, Lemme K, Ernst B, Paulson JC (2011) High affinity sialoside ligands of myelin associated glycoprotein. Bioorg Med Chem Lett 21 (17):5045–5049

Top Med Chem (2014) 12: 187–234
DOI: 10.1007/7355_2014_60

Published online: 26 August 2014

The Evolution of a Glycoconjugate Vaccine for *Candida albicans*

David R. Bundle

Abstract The monoclonal antibody C3.1 has been found to afford protection against *Candida albicans*, an opportunistic fungal pathogen. The exceptional inhibition profile of C3.1 has propelled synthetic and immunochemical studies of antibody–oligosaccharide interactions, leading to the development of candidate vaccines. The β1,2-linked mannan of the fungal-cell-wall phosphomannan complex is a protective antigen that exhibits a well-defined conformation. A β1,2-linked trisaccharide conjugated to tetanus toxoid generates protective antibodies in rabbits, and STD-NMR studies show that these antibodies approximate the binding profile of the protective monoclonal antibody. This simple trisaccharide–tetanus toxoid conjugate was a strong immunogen in rabbits, but it was poorly immunogenic in mice. However, when a carbohydrate antigen targeted at dendritic cells was incorporated in this conjugate vaccine, it improved uptake and processing of the antigen and resulted in a five-fold higher mannan-specific antibody response together with a cytokine profile appropriate for an antifungal vaccine. When the same trisaccharide was conjugated to a T-cell peptide derived from *Candida* cell-wall protein, the resulting glycopeptide–tetanus toxoid conjugate engendered a peptide- and carbohydrate-specific response that afforded protection against live challenge by *C. albicans* without requiring an adjuvant.

Keywords Adjuvant-free vaccine, Antibody reverse engineering of a protective vaccine, *Candida albicans*, Conjugate vaccine, Dendritic cell targeting of antigen, Glycopeptide vaccine, β-mannan

D.R. Bundle
Department of Chemistry, Alberta Glycomics Centre, University of Alberta, Edmonton, AB, Canada T6G 2G2
e-mail: dave.bundle@ualberta.ca

Contents

Abbreviations

Abe	Abequose (3,6-dideoxy-D-*xylo*-hexose)
Ac	Acetyl
All	Allyl
′Bn	Benzyl
Boc	*tert*-Butoxycarbonyl
BSA	Bovine serum albumin
Bu	Butyl
DMF	Dimethylformamide
DMSO	Dimethyl sulfoxide
Enol	Enolase
Fab	Fragment (of an antibody) that contains the antigen-binding region
Fba	Fructose-bisphosphate aldolase
Gal	Galactose
Gap1	Glyceraldehyde-3-phosphate dehydrogenase
Glc	Glucose
GlcNAc	*N*-acetylglucosamine
Hwp1	Hyphal wall protein-1
IC_{50}	Half-maximal inhibitory concentration
LPS	Lipopolysaccharide
mAb	Monoclonal antibody

Man	Mannose
Me	Methyl
Met6	Methyltetrahydropteroyltriglutamate
NOE	Nuclear Overhauser effect (NMR)
Pgk1	Phosphoglycerate kinase 1
Pr	Propyl
Rha	Rhamnose
STD-NMR	Saturation transfer difference nuclear magnetic resonance
t-Bu	*tert*-Butyl
Tf	Trifluoromethanesulfonyl (triflyl)
THF	Tetrahydrofuran
TMSOTf	Trimethylsilyl trifluoromethanesulfonate
Tr	Triphenylmethyl (trityl)
trNOE	Transferred nuclear Overhauser effect (NMR)
T-ROESY	Two-dimensional rotating-frame Overhauser effect spectroscopy

1 Introduction

For more than 80 years it has been known that vaccination of adults with the purified capsular polysaccharide of *Streptococcus pneumoniae* produces anti-polysaccharide antibodies [1]. These antibodies also confer protective immunity against other *S. pneumoniae* serotypes that express the same or closely related polysaccharides [2, 3]. These findings subsequently resulted in several licensed polysaccharide vaccines that afforded protection in adults and adolescents against diseases caused by encapsulated bacteria such as *S. pneumoniae* and *Neisseria meningitides*. Unfortunately, the immune responses afforded by these polysaccharides to infants under the age of 18 months and to seniors were non-existent or significantly impaired [4–7].

An equally important observation (also in the 1930s) established that capsular polysaccharides conjugated to proteins induced anti-polysaccharide antibodies that could confer active and passive immunity against the homologous strain of *S. pneumoniae* [8, 9]. It took another 50 years before this knowledge was applied to vaccine development, when the capsular antigen of *Haemophilus influenzae* type b was conjugated to a fragment of diphtheria toxoid [10]. The resulting vaccine was shown to protect infants against this agent of childhood meningitis [11]. The ensuing 30 years have seen the rapid deployment of a series of conjugate vaccines with annual sales now in the billions of dollars [12]. The use of such vaccines constitutes a highly successful and cost-effective public health measure. These successes have spurred many research initiatives into capsular polysaccharide conjugate vaccines that target a wide spectrum of infectious diseases, including Group B *Streptococcus* and *S. aureus*. Other initiatives have investigated conjugates of cell-wall polysaccharides such as lipopolysaccharides from *Shigella* species.

The bacterial diseases caused by encapsulated bacteria such as *S. pneumoniae*, *H. influenzae* type b, and *N. meningitides* have been countered by licensed vaccines.

Major producers of conjugate vaccines now face a dilemma: the cost of the research and development and marketing for new vaccines currently in the pipeline can exceed the financial capacity of even the largest manufacturers. Further, since licensed vaccines address diseases that represent the largest markets, vaccines at the research stage often address diseases that affect a smaller population than those that afford protection against *S. pneumoniae* and *N. meningitides*. In this context development of new conjugate vaccines will need to focus on diseases that are the most economically and medically critical. Many of the diseases for which conjugate vaccines have yet to be deployed represent such niche markets.

Vaccines for some microbial diseases – especially those caused by intracellular pathogens and fungal infections – will likely have to address the need for cell-mediated as well as antibody-mediated protection. Recent advances in our understanding of how the immune system processes conjugate vaccines [13] and whether carbohydrate recognition can be exploited for the purpose of cell-mediated immunity together offer intriguing possibilities for the development of chemically defined vaccines and perhaps fully synthetic vaccines.

2 Steps Toward a Vaccine for *C. albicans*

Numerous reviews have covered polysaccharide [14] and conjugate vaccines [15–22] including those developed using synthetic oligosaccharides [23–26]. This chapter describes attempts to develop a conjugate vaccine against the fungal pathogen *Candida.* The investigation evolved from studies of molecular recognition of β-mannans by two monoclonal antibodies that protected mice challenged by live *C. albicans.* In this sense, much of the initial impetus might be described as reverse engineering these antibody specificities to arrive at a potential conjugate vaccine.

2.1 Reverse Engineering Monoclonal Antibodies

The term “reverse engineering” can have a variety of subtle interpretations. In this case, it refers to using a monoclonal antibody (mAb) known to protect against a given pathogen to guide the development of an immunogen that elicits polyclonal antibodies of related specificity and – most importantly – with the capacity to confer protection following immunization with the inferred antigen [27]. Since the specificity of the mAb can be determined in some detail, molecular recognition of the native antigen can be mapped by various approaches, including saturation transfer difference (STD) NMR, crystallography, and functional group substitutions. The key recognition elements can then be incorporated into a minimally sized antigen. Depending on whether this antigenic determinant is itself immunogenic, it may be necessary to conjugate the determinant to a carrier protein to ensure a robust immune response.

The most intense attempts to reverse engineer an antibody have involved 2G12, an N-linked glycan-specific mAb [28–31]. This broadly neutralizing antibody binds the envelope protein gp120 of human immunodeficiency virus type 1 (HIV-1) [29]. It is a highly unusual antibody, the most unique feature of which is its V_H domain-exchanged structure. The multivalent binding surface has two primary glycan binding sites specific for the conserved cluster of oligomannose-type sugars on the surface of gp120. Crystal structures of the 2G12 fragment that contains the antigen-binding region (Fab) and its complexes with the disaccharide Manα(1 → 2) Man and with the oligosaccharide $Man_9GlcNAc_2$ revealed that two Fabs assemble into an interlocked V_H domain-swapped dimer [28]. This dimeric assembly was not previously observed in hundreds of other Fab structures in the protein data bank. The V_H domain swap in 2G12 results in a twist of the variable regions relative to the constant region, putting the two Fabs side-by-side with a 35 Å separation of binding sites that face in the same direction. This arrangement of immunoglobulin G (IgG) binding sites is markedly different from the norm, in which such sites are located at the tips of Y- or T-shaped antibodies. Crystal structures and models derived from them suggest that the dimeric Fab can bind up to four epitopes, two in the usual IgG sites created by the six hypervariable loops of the V_L and V_H domains and an additional two at the interface between the two V_H domains of the Fab dimer. As a result of this multivalency, 2G12 binds the saccharide epitopes with nanomolar avidity.

Heroic efforts have been expended to synthesize and conjugate $Man_9GlcNAc_2$ [30–32] and related epitopes to protein carriers. While most of the synthetic constructs have shown strong binding to 2G12, none of these immunogens have so far produced protective antibodies in immunization experiments. An epitope that binds a protective antibody will not necessarily generate an antibody that has both binding specificity and functional protective properties.

Reasons for this lack of success may include the adoption of a unique conformation by the native antigen when it is displayed at the cell surface, and synthetic conjugate antigens may lack these conformational features. Consequently, antibodies induced by immunization with such synthetic antigens may bind the native antigen but lack the essential effector functions.

An additional complication in the reverse engineering of antibodies is the existence of conformational epitopes [33–38]. Several capsular polysaccharide epitopes of this type have been identified. Typically but not exclusively, they have contained neuraminic acid. In addition, only the relatively large oligomers – often within the range of 10–20 saccharide residues – inhibit native antigen–antibody binding, whereas smaller epitopes appear to lack this ability [33]. For conformational epitopes, antigenic oligomers may need to be significantly larger to be capable of raising a protective immune response [39].

In general, one can imagine the binding site as able to capture an antigen and adapt the conformation to satisfy the primary protein–antigen interactions. Although short epitopes may fit the binding site, they may fail to adopt the conformation of the native antigen in its cell wall or membrane environment for the length of time required to induce protective antibodies.

This chapter will discuss in detail the reverse engineering of a protective carbohydrate binding antibody that binds to and protects against the pathogenic fungus *C. albicans*. The mAb C3.1 and its immunoglobulin M (IgM) counterpart B6.1 both bind the cell-wall β-mannan of *C. albicans* [40]. This substructure is a relatively small component of the much larger cell-wall phosphomannan complex.

2.2 Antibody Binding Sites

Early work by Kabat sought to determine the size of an oligosaccharide that optimally filled a binding site. These studies looked at polyclonal sera and the inhibitory power of oligomers of increasing length. In this case, the optimum size ranged from trisaccharides to hexasaccharides [41, 42]. Other work examined the size of oligosaccharide epitopes that could generate sera able to recognize the native antigen on the microbial cell wall [43–45] or that could confer protection to live microbial challenge after immunization with a conjugate vaccine prepared from the oligosaccharide [45–47].

Although these studies often used polysaccharides with grossly different characteristics (such as homo- versus heteropolysaccharides), a comparison is of interest. Although inhibition data suggested that short oligosaccharides filled the binding sites, protection data pointed to oligosaccharides that were at least twice or three times larger.

The first crystal structures of oligosaccharides bound to antibodies suggested that tri- and tetrasaccharides filled the binding sites [48, 49]. This was true for many antibodies [48–57], although exceptions were found where a single terminal residue provided nearly all of the binding specificity [58]. In another case, that of the antigen for 2,8-polysialic acid meningococcal serogroup B, the binding site required at least ten monomer residues [59]. This polysaccharide belongs to the class of carbohydrate antigens that are thought to possess a conformational epitope. These conclusions are supported by a detailed body of work [39].

Many anti-carbohydrate antibodies possess a groove-type binding site that allows entry and exit of the polysaccharide chain [50, 51, 53, 57, 59]. Given the repeating-unit character of many bacterial polysaccharides, this feature allows the antibody to bind to internal units as well as terminal epitopes. The polysaccharide can enter and exit a grooved site without impediment. An interesting example of an antibody that binds a branched lipopolysaccharide (LPS) O-antigen demonstrates the importance of this steric requirement. The *Salmonella* serogroup B O-polysaccharide has a tetrasaccharide repeating unit (Fig. 1a). The crystal structure of the antibody with bound trisaccharide shows the branching monosaccharide, an abequose residue (Abe; 3,6-dideoxy-D-*xylo*-hexopyranose) to be buried and the two residues of the main chain galactose (Gal) and mannose (Man) essentially lay across the top of the binding pocket (Fig. 1b) [48]. Modeling studies suggested that

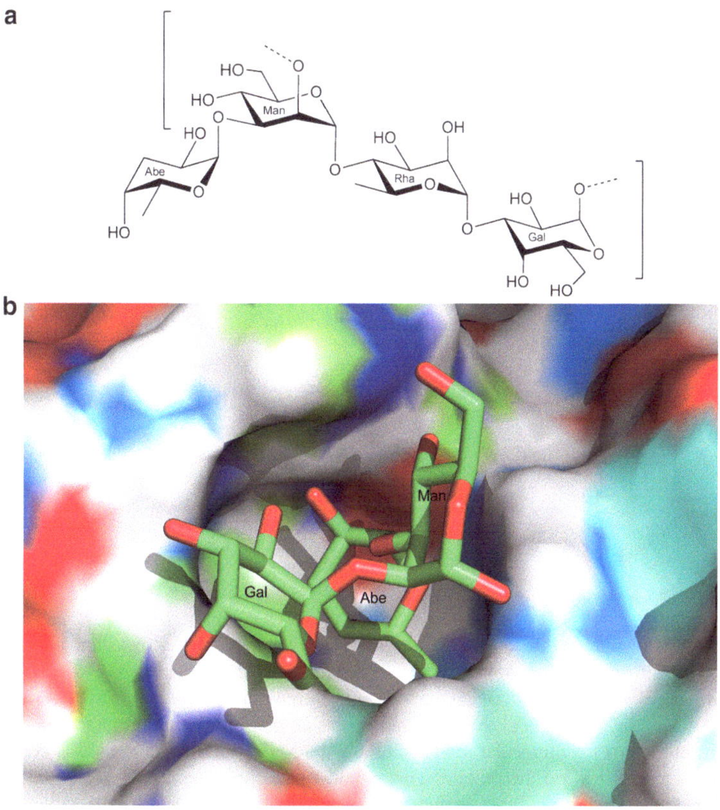

Fig. 1 (**a**) The biological repeating unit of the O-antigen of the LPS belonging to *Salmonella* serogroup B: {→2)[α-D-Abe(1 → 3)]α-D-Man(1 → 4)α-L-Rha(1 → 3)α-D-Gal(1−}. In some serogroup B antigens, an acetyl group may occur at *O*-2 of the abequose residue. (**b**) The crystal structure of the trisaccharide α-D-Gal(1 → 2)[α-D-Abe(1 → 3)]α-D-Man(1 → OCH_3 with the Fab of the monoclonal antibody Se155.4 [48]. The abequose residue is buried in the binding pocket with mannose and galactose lying across the top of the pocket

the next residue, α-L-rhamnopyranose (Rha) unit, does not contact the protein but points its anomeric bond directly toward the protein [60, 61]. This orientation would cause the next residue, a galactose, to clash with protein if the glycosidic linkage from rhamnose to galactose adopts the usual *syn* conformation (Fig. 2a, c). To avoid a steric clash, the psi angle of this Rha-Gal glycosidic bond was proposed to rotate approximately 180 degrees and adopt the *anti* glycosidic conformation. NMR experiments on the bound state confirmed this prediction (Fig. 2b, d) [60]. Further experiments using two repeating units of the polysaccharide that could each bind in only one of the frame-shifted modes also supported the requirement for an antigen conformational change on binding (Fig. 2d) [62]. Of course, the adoption of a higher energy conformation comes with an attendant decrease in intrinsic affinity [61].

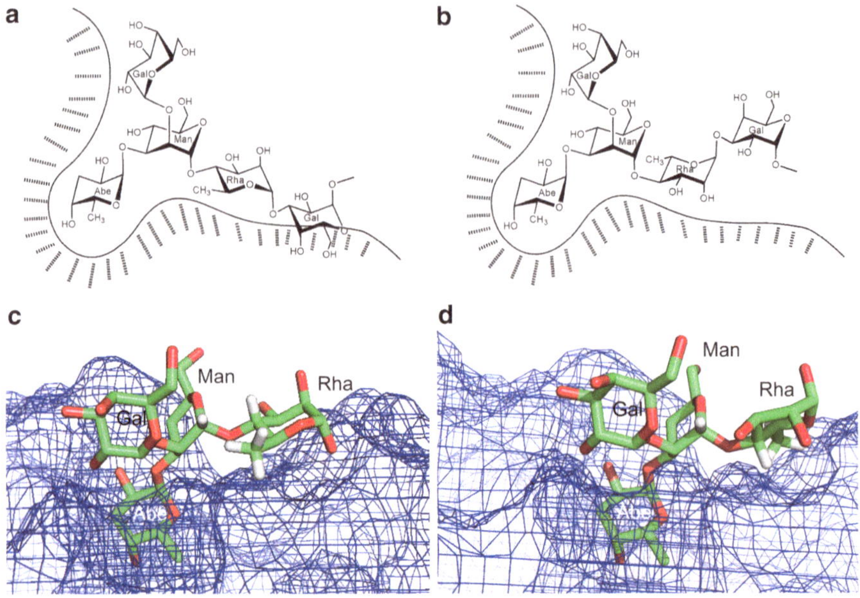

Fig. 2 (**a**) Extension of the bound trisaccharide shown in Fig. 1 shows that in the expected low-energy conformation, the Man-Rha glycosidic linkage orients the Rha O1 toward the protein. This orientation would result in the Gal residue clashing with the antibody surface. (**b**) To relieve the steric clash, the Man-Rha glycosidic bond adopts an *anti* conformation rather than the normal *syn* conformation. (**c**) A molecular model showing the Man-Rha linkage in the *syn* conformation. The Rha O1 points toward the protein. This would cause subsequent polysaccharide residues to clash with the extended antibody surface adjacent to the binding site. (**d**) A molecular model showing the Man-Rha linkage in the *anti* conformation. The Rha O1 points away from the protein, allowing the subsequent polysaccharide residues to avoid clashes with the extended antibody surface

These observations are of interest in the context of the *Candida* work to be discussed. The results with the *Salmonella* LPS antigens demonstrate that even though a significant conformation change may occur in the antigen, antigen–antibody interactions will still occur (albeit with reduced affinity) provided the energy costs in adopting this conformation are sufficiently small. The crystal structures also show that polysaccharide elements contact protein beyond the formal binding pocket.

2.3 *Candida*

Candida is a commensal organism that can shift toward pathogenesis with alterations in host immunity, physiology, or normal microflora. The acquisition of novel or hypervirulent factors is not necessary. Invasive *C. albicans* infection can be a serious clinical complication, especially for patients with an impaired immune system: it has an incidence of between 1.1 and 24 cases per 100,000 individuals and a mortality rate of more than 30% [63]. Antifungal agents have limited impact on the mortality statistics for systemic fungal infections, and new clinical approaches are needed (such as a combination of chemotherapy and immunotherapy). However, the immune response to fungal infections is complex and requires not only the production of protective antibodies but also the recruitment of cell-mediated immunity. The fungal cell wall plays an important role in directing the immune response to fungal infection [40].

Host defense against systemic candidiasis relies on the ingestion and elimination of fungal cells by cells of the innate immune system, especially neutrophils and macrophages [64]. These responses can be promoted by antibodies. Macrophages not only have direct antimicrobial activity against such organisms but also promote the presentation of antigens and the production of cytokines and chemokines. Recognition of *C. albicans* by monocytes and macrophages is mediated by systems that sense pathogen-associated molecular patterns (PAMPs) of the *C. albicans* cell wall. These systems include recognition of *N*-linked mannans by the mannose receptor (MR), O-linked mannans by Toll-like receptor 4 (TLR4), β-glucans by dectin-1/TLR2, β-mannosides by galectin-3/TLR2 complexes [65] and by the C-type lectin known as mincle [66].

2.3.1 The Cell Wall of *C. albicans*

The three major glycans of the *C. albicans* the cell wall are branched β(1 → 3) and β(1 → 6) glucans, chitin, and a phosphomannan glycoprotein [67, 68]. These polysaccharides function as a robust exoskeleton and scaffold for external protein structures. This outer layer consists of highly glycosylated mannoproteins that are modified by N-linked [69] and O-linked mannose glycans and phosphomannans [64, 69]. The phosphomannan complex has been the subject of detailed structural analyses and immunological studies.

The structural elucidation of the phosphomannan complex represents a major achievement. Unlike bacterial cell-wall polysaccharides, it lacks a regular repeating unit motif and its glycan component possesses marked microheterogeneity as well as variation across different *Candida* species [70]. An example of the substructures present in the mannan component of *C. albicans* is shown in Fig. 3. More detailed examples can be found in reviews by the Suzuki group. These structures were determined using acetolysis [71] and partial hydrolysis in conjunction with NMR and mass spectrometry [72–75]. Mannans consist largely of short α(1 → 2)- and α(1 → 3)-D-mannopyranan; less than 10% is O-linked [76]. The complex N-linked components are composed of extended α(1 → 6)-D-mannopyranan backbones containing α(1 → 2)-D-mannopyranan branches, some of which may contain α(1 → 3)-D-mannopyrannose residues. Oligomers of β(1 → 2)-mannopyranan may be attached to these side chains either glycosidically or through a phosphodiester bridge (Fig. 3). The position of attachment of the phosphodiester has yet to be

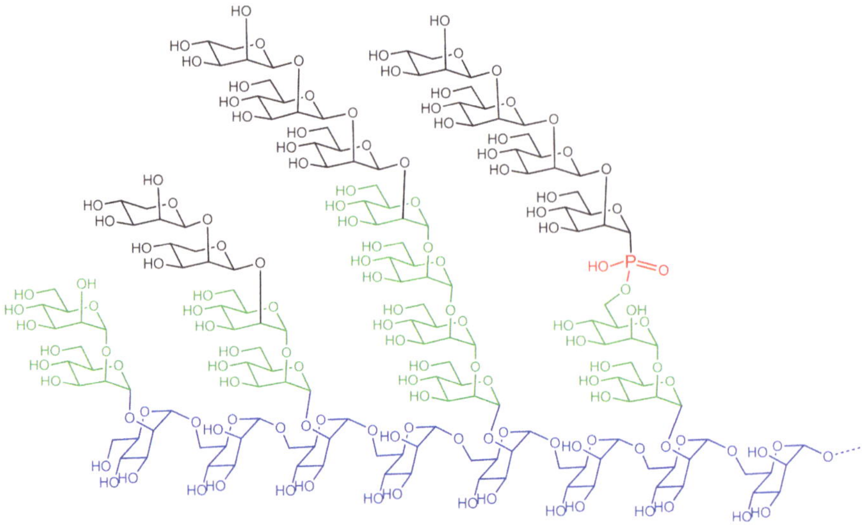

Fig. 3 The structure of the glycan chains of *C. albicans* phosphomannan glycoprotein. The composite structure varies widely with *Candida* species and with growth conditions. The $(1 \rightarrow 6)$-linked α-mannan backbone (*blue*) is attached to protein. Branching side chains consist of $(1 \rightarrow 2)$α-mannan (*green*) with occasional $(1 \rightarrow 3)$ linkages. This basic structure is shared by most *Candida* species. In *C. albicans* and several other prominent *Candida* species, $(1 \rightarrow 2)$-linked β-mannans (*black*) can be attached either via a phosphate diester (*red*), giving acid-labile β-mannan, or glycosidically linked to the α-mannan side chains

determined. Mannans are heterogeneous, with chain lengths depending on nutritional and environmental conditions [77].

The precise antigenic epitopes of β-mannan exist in three different forms. Two forms of the $\beta(1 \rightarrow 2)$-linked oligosaccharides are extensions of the $\alpha(1 \rightarrow 2)$-mannose and $\alpha(1 \rightarrow 3)$-mannose residues [55]. The third form has the $\beta(1 \rightarrow 2)$-D-mannopyranan attached via the phosphodiester to $\alpha(1 \rightarrow 2)$-D-mannopyranan branches [72, 73].

C. albicans strains are assigned to either of two major serogroups, A or B [72, 73, 75, 77]. Serotype A has β-mannose residues attached to $\alpha(1 \rightarrow 2)$ mannose side-chain residues via glycosidic bonds and via phosphodiesters. Serogroup B lacks β-mannan attached via a glycosidic bond and possesses only phosphodiester-linked β-mannan. Thus, acid-labile β-mannan (linked through a phosphodiester to the phosphomannan side chains) is thought to be a major epitope of both serotypes A and B.

C. tropicalis and *C. guilliermondii* also express the β-mannan epitope, and it may also be attached via a $(1 \rightarrow 3)$-linkage to an α-mannose residue [78–80]. Although β-mannan epitopes larger than a tetrasaccharide have been reported, it appears that $(1 \rightarrow 2)$-β-mannopyranan di-, tri-, and tetrasaccharides oligomers predominate [81–84]. Goins and Cutler found that tri- and disaccharide epitopes are most abundant in the β(1 → 2)-mannopyranan oligomers released from the phosphomannan by β elimination [85].

In their structural studies of several *Candida* species, Shibata et al. employed detailed multidimensional NMR studies and inferred from nuclear Overhauser effect (NOE) data that the β(1 → 2)-linked mannose oligosaccharides possessed a "distorted" structure [73].

2.3.2 Monoclonal Antibody Specific for the β-Mannan of *C. albicans*

Cutler and colleagues immunized mice with a liposomal preparation of the *C. albicans* cell-wall phosphomannan complex and produced a series of mAbs [86]. Those that were specific for the α-mannan component failed to protect mice against a live challenge with *C. albicans*. However, two mAbs (mAbs), C3.1 (IgG3) and B6.1 (IgM), were specific for the (1 → 2)-β-mannan epitope and protected mice against candidiasis in passive protection studies [40, 87–89]. Preliminary binding studies also suggested that these antibodies recognized a trisaccharide epitope [90]. In addition, the Cutler group reported results suggesting that relatively short β-mannan oligosaccharides (di- to tetrasaccharides) were heavily represented on the surface of growing *C. albicans* cells [85].

2.3.3 Mapping the Size of β-Mannan-Specific Antibody Binding Sites

When the mAbs C3.1 (an IgG3) and B6.1 (an IgM) were evaluated against a set of (1 → 2)-β-mannan oligosaccharide propyl glycosides ranging in size from di- to hexasaccharide, unexpected and dramatic results were observed (Fig. 4) [91]. Oligosaccharide inhibition with carbohydrate-specific antibodies normally shows affinities increasing with oligosaccharide length until a plateau of activity is reached [42, 43]. Further increases in oligosaccharide size produce only very small affinity gains. Dramatic decreases in binding are not seen. Surprisingly, however, inhibitory power for C3.1 and B6.1 mAbs reached a maximum for di- and trisaccharides. Both antibodies exhibited similar binding constants for di- and trimannosides [with half-maximal inhibitory concentration (IC_{50}) values of 44 μM and 38 μM for B6.1 and 8 μM and 16 μM for C3.1]. Larger oligosaccharides were bound progressively more weakly, with the IC_{50} of a hexamannoside being 844 μM for C3.1 and >1,000 μM for B6.1. It was speculated that the weaker binding of larger oligosaccharides was due to steric restrictions at either end of the (1 → 2)-β-mannans [91].

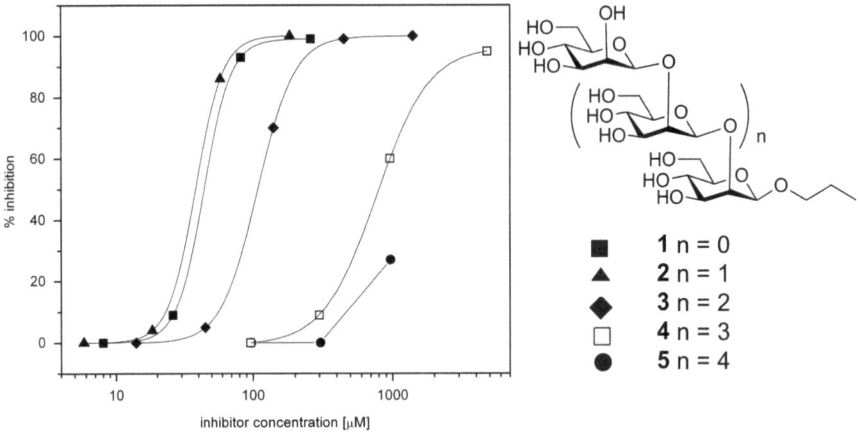

Fig. 4 The inhibition of *C. albicans* β-mannan-specific mAb (C3.1) binding to cell-wall phosphomannan by a series of β(1 → 2)-linked mannan homo-oligomer propyl glycosides disaccharide through hexasaccharide **1–5**

3 Solution Structure of β-Mannan

3.1 Conformational Studies (1 → 2)-β-Mannan Oligosaccharides

The chemical synthesis of (1 → 2)-β-mannan propyl glycosides **1–5** (Fig. 4) [92] provided a set of oligomers that were not only useful as inhibitors of antibody–antigen interactions but also permitted a thorough analysis of the solution structures of these important *C. albicans* antigens [91].

The excellent NMR signal dispersion observed for the proton resonances of the (1 → 2)-β-mannopyranan oligomers **1–5** at 800 MHz aided the assignment of resonances by gradient correlation spectroscopy (GCOSY), gradient total correlation spectroscopy (GTOCSY), and heteronuclear multiple quantum coherence (HMQC) experiments (Fig. 5). The discrete chemical shifts of each hexopyranose spin system of pentasaccharide **4** suggested that each hexose is conformationally distinct within a well-defined oligosaccharide conformation (Fig. 5a, b). The distinct ^{1}H chemical shifts observed for the H1 and H2 resonances of each mannose residue of pentasaccharide **4** are in contrast to the near identical shifts observed for H1 resonances of many homo-oligomers, (1 → 2)-α-mannopyranotetrose [93] and maltoheptose [(1 → 4)-α-glucopyranoheptose] [94]. Similarly distinct chemical shifts of anomeric resonances occur for the α- and β-1,2-linked oligomers of glucose [95, 96]. These two oligomers and (1 → 2)-β-mannopyranan were identified by early modeling studies as belonging to a class of homopolymers expected to form relatively stiff and crumpled conformations [97].

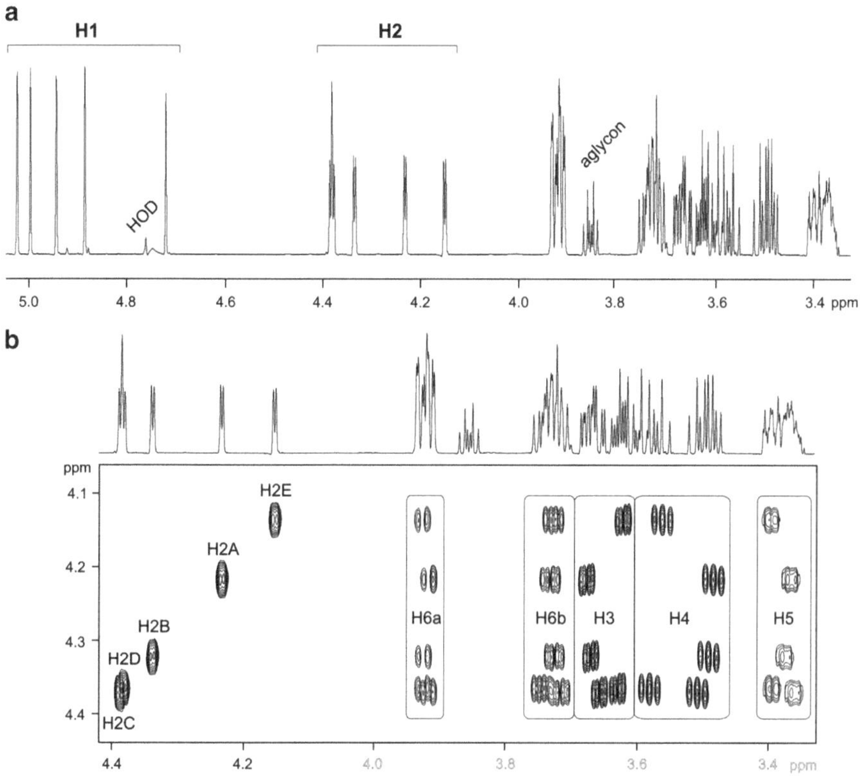

Fig. 5 (**a**) ^{1}H NMR of pentasaccharide **4**. (**b**) GTOCSY two-dimensional NMR spectrum of pentasaccharide **4** showing the spin systems H2A through H2E of the five mannose residues A through E. The mannopyranose rings are labeled A-E starting at the reducing end of the pentasaccharide

After the resonances of the ring protons had been assigned (Fig. 5b), NOE correlations between protons on either side of each glycosidic linkage could be used to establish the residue identity of the spin systems of adjacent pyranose rings and to confirm their sequence within each linear oligomer (Fig. 6) [91].

NOE correlations were then used to define solution conformations of oligosaccharides **1**–**5**. This method is especially effective when there are numerous NOE contacts between hexose residues. NOE correlations were measured from the two-dimensional rotating-frame Overhauser effect spectroscopy (T-ROESY) spectra of the oligomers (**1**–**5**). These results revealed numerous contacts between noncontiguous $n+2$ and $n+3$ residues, as illustrated for pentasaccharide **4** (Fig. 6).

NOE contacts between noncontiguous residues occur often, especially in branch structures such as the A, B and Lewis antigens in human blood [98–100]. A linear homopolysaccharide can achieve the equivalent of vicinal substitution if the hydroxyl group adjacent to the anomeric center is the site for chain extension. Many 1,2-linked homo-oligomers, such as those of the *Brucella abortus* A antigen,

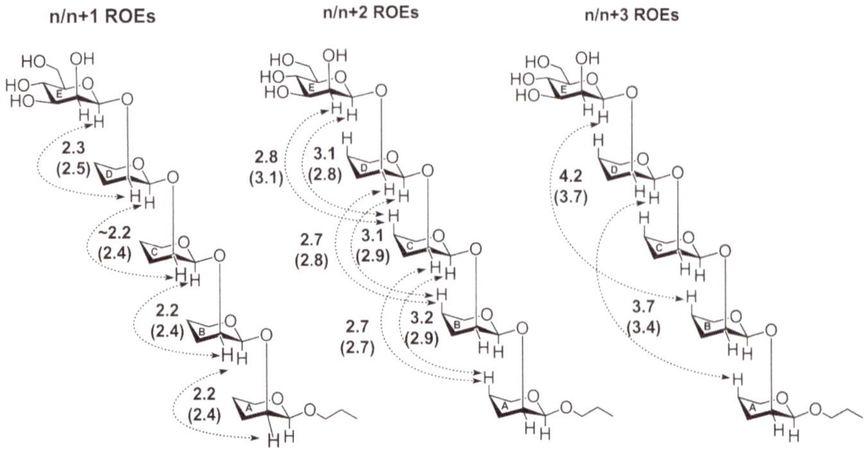

Fig. 6 NOE contacts observed for the pentasaccharide propyl glycoside **4**. Three groups of NOE are displayed: ones between adjacent residues $n/n+1$, those between residues separated by 1 pyranose ring, and those between residues separated by 2 residue. The mannose residues are labeled A through E as in Fig. 5. *Bold* numbers are the interproton distances determined by quantitation of the NOE data. Numbers in *parentheses* are the corresponding interproton distances predicted by molecular dynamics

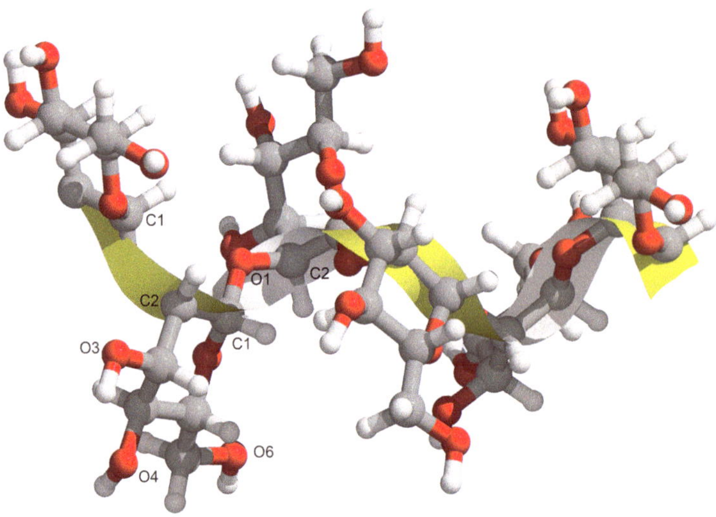

Fig. 7 Three-dimensional model of hexasaccharide **5** in solution as determined by NMR. Running from left to right atoms defining the glycosidic linkages C1-O1-C2-O2 trace out a helix shown in *yellow*. In this conformation the most prominent exposed atoms of each ring are C3 to C6 and their attached hydroxyl groups

yield few if any NOE contacts between noncontiguous residues [101]. In the conformational analysis of linear oligosaccharides observation of multiple NOE contacts between noncontiguous residues as seen for the (1 → 2)-β-mannopyranans studied here is atypical (Fig. 6). The presence of NOE interactions between residues A and D (*n* to *n* + 3 contacts) has limited precedence and is indicative of a compact structure (Fig. 7).

The T-ROESY cross-peak volumes for the tri-, tetra-, and pentasaccharide (**2**–**4**) were used to derive conformationally averaged interproton distances (Fig. 6) [91]. Similar distances observed across the glycosidic linkages of the oligomers **1**–**4** are indicative of a compact, repetitive solution structures. The distances also suggest the inferred structure is not just a result of steric interactions between distant, noncontiguous residues but represents a population of low-energy conformations with similar torsional angles about the glycosidic linkages. If steric interactions between noncontiguous residues were limiting the conformations of the oligomers, the shorter oligomers would be expected to have distances that differed from those of their counterparts with higher molecular weights, but this was not the case.

3.2 Molecular Modeling

Unrestrained molecular dynamics modeling was carried out for oligomers **2**–**4** [90]. The resulting model for pentasaccharide **4** is discussed below.

Comparisons of the ten structures of the pentasaccharide obtained from a simulated annealing protocol showed a single family of low-energy conformations. A static model of the pentasaccharide in one of these conformations shows the helical nature of this oligosaccharide (Fig. 7). The three-dimensional repeating unit approximates three mannose residues.

Theoretical interproton distances were calculated from a 5 ns molecular dynamics run at 300 K. The resulting data compared well with distances determined from experimental NOEs (Fig. 6 distances in parentheses). During the dynamics model run, the oligosaccharide stayed within the range of glycosidic torsional angles represented by the family of conformations generated by simulated annealing.

A comparison of the conformational space sampled by the oligosaccharides indicates that the tri-, tetra-, and pentasaccharides **2**–**4** explore very similar torsional angles across all their linkages [91]. The family of conformations sampled is consistent with a model that imparts helical character to this glycan chain (Fig. 7). The repeating unit is approximately three residues long, but due to the inherent flexibility about glycosidic torsional angles, the overlap of residues *n* and *n* + 3 is only approximate. In this family of conformations, hydroxyls are oriented out into the solution and a hydrophobic core forms from the α-faces of the mannose rings. Interestingly, the Crich group reported a crystal structure of a (1 → 2)-β-mannopyranan tetrasaccharide with attached organic protecting groups

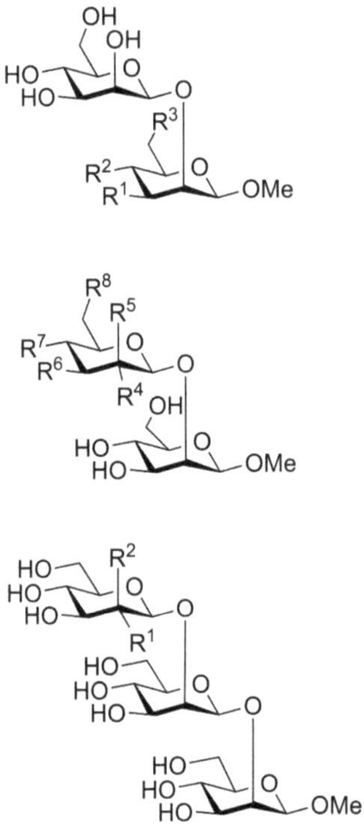

6 $R^1 = R^2 = R^3 = OH$
7 $R^1 = H, R^2 = R^3 = OH$
8 $R^1 = OCH_3, R^2 = R^3 = OH$
9 $R^2 = H, R^1 = R^3 = OH$
10 $R^2 = OCH_3, R^1 = R^3 = OH$
11 $R^3 = H, R^1 = R^2 = OH$
12 $R^3 = OCH_3, R^1 = R^2 = OH$

13 $R^4 = H, R^5 = OCH_3, R^6 = R^7 = R^8 = OH$
14 $R^4 = R^6 = H, R^5 = R^7 = R^8 = OH$
15 $R^4 = H, R^6 = OCH_3, R^5 = R^7 = R^8 = OH$
16 $R^4 = R^7 = H, R^5 = R^6 = R^8 = OH$
17 $R^4 = H, R^7 = OCH_3, R^5 = R^6 = R^8 = OH$
18 $R^4 = R^8 = H, R^5 = R^6 = R^7 = OH$
19 $R^4 = H, R^5 = R^6 = R^7 = OH, R^8 = OCH_3$
20 $R^5 = H, R^4 = R^6 = R^7 = R^8 = OH$

21 $R^1 = H, R^2 = OH$
22 $R^1 = OH, R^2 = H$

Fig. 8 Structures of monodeoxy and mono-*O*-methyl congeners **7–20** of disaccharide **6** and trisaccharides **21** and **22** used to map the binding site of mAb C3.1

[102] and the gross features of this molecule were closely related to the aqueous solution structure inferred from NMR-derived constraints.

4 Molecular Recognition of β-Mannan by Protective mAb

4.1 Mapping the C3.1 Antibody Site

Inhibition data for oligosaccharides **1–5** established that the mAb C3.1 binding site could optimally accommodate a di- or trisaccharide, but larger oligosaccharides were progressively more difficult to accommodate. To establish which hydroxyl groups were essential for binding a disaccharide epitope, the C3.1 binding site was chemically mapped. The principle of hydroxyl group replacement developed by Lemieux was adopted [103]. Synthesis of monodeoxy- and mono-*O*-methyl mannoside derivatives followed by their use in inhibition assays identified hydroxyl groups that made energetically important contacts with the mAb binding site [104, 105]. Hydroxyl groups that are involved in hydrogen bonds deep within the

Table 1 IC_{50} for inhibition by disaccharide congeners **6–20** and trisaccharides **21** and **22** of mAb C3.1 binding to *C. albicans* β-mannan

Compound	Derivative	IC_{50} (μmol/L)	Relative potency (%)	$\Delta(\Delta G)^a$ (kcal/mol)
6	Disaccharide	31	100	0
7	3-Deoxy	Inactive[b]	<1	>2.7
8	3-*O*-methyl	Inactive[c]	<1	>2.6
9	4-Deoxy	Inactive[b]	<1	>2.7
10	4-*O*-methyl	Inactive[c]	<1	>2.6
11	6-Deoxy	14	221	−0.5
12	6-*O*-methyl	81	38	0.6
13	2′-*O*-methyl	33	94	0.04
14	3′-Deoxy	119	26	0.8
15	3′-*O*-methyl	62	50	0.4
16	4′-Deoxy	Inactive[b]	<1	>2.7
17	4′-*O*-methyl	670	5	1.8
18	6′-Deoxy	426	7	1.6
19	6′-*O*-methyl	588	5	1.7
20	2′-Gluco	47	66	0.3
21	Trisaccharide	17	182	−0.4
22	2″-Gluco	52	60	0.3

[a]The molar ratio of IC_{50} values (K_{rel}) for two inhibitors is a measure of relative potencies and is used to calculate $\Delta(\Delta G) = RT\ln K_{rel}$
[b]No inhibition at 2938 μmol/L
[c]No inhibition at 2700 μmol/L

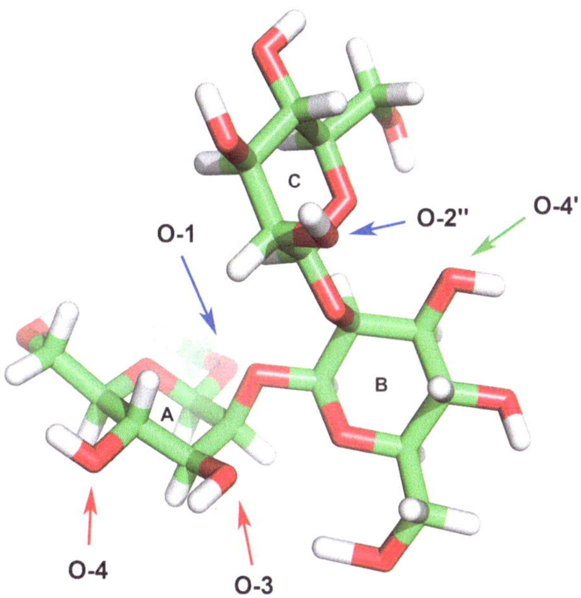

Fig. 9 The three key hydroxyl groups of disaccharide **6** (C3-OH, C4-OH and C4′-OH) that form H-bonds with the antibody. The points at which chain extension would occur in either direction for larger oligomers are labeled (*blue arrows*)

binding site cannot be deoxygenated or O-methylated without a significant loss of binding affinity. Those involved in hydrogen bonds at the periphery of the binding site may exhibit a range of changes in affinity upon deoxygenation, whereas O-methylation of such groups generally leads to a significant loss of affinity due to the challenge of accommodating the relatively bulky *O*-methyl group.

Each hydroxyl group of the disaccharide **6** (the methyl glycoside analog of the propyl glycoside **1**) was replaced to produce the disaccharide congeners **7–19** (Fig. 8) [104, 105]. Important hydrogen bonds were identified as those from the C3-, C4-, and C4′-OH groups, since the 3-deoxy (**7**), 3-*O*-methyl (**8**), 4-deoxy (**9**), and 4-*O*-methyl (**10**) congeners and the 4′-deoxy analog **16** were inactive (Table 1). We inferred that these three were buried hydroxyl groups that make essential hydrogen bonds to the antibody binding site (Fig. 9). The weak activity of the 4′-*O*-methyl disaccharide **17** is consistent with C4′ hydroxyl being relatively exposed and also a hydrogen bond acceptor. Its weak activity suggests C4′-OH is exposed at the periphery of the binding site, where it likely accepts a hydrogen bond from the protein and perhaps donates a hydrogen bond to water. If this hydroxyl group were buried in the site, the larger *O*-methyl group could not be accommodated and an inactive compound would be expected. The weak activity of the C6′-modified congeners **18** and **19**, which were also not completely inactive, suggests that the C6′-OH also forms hydrogen bonds near the periphery of the site. The *gluco* configuration of the terminal residue is well tolerated (compound **20**) suggesting little antibody–oligosaccharide interaction in this region of the epitope.

Also of interest is the only slightly higher activity [$\Delta(\Delta G) = -0.4$ kcal/mol] of trisaccharide **21** (Table 1 and Fig. 9) relative to disaccharide **6**, which showed that a third mannose residue contributes minimal additional binding affinity. This conclusion is supported by the activity of trisaccharide **22** with a β-glucose residue at the nonreducing end. These data suggest that the minimal epitope that might induce antibodies specific to β-mannan could be a disaccharide or trisaccharide.

4.2 Antigen Frameshift Revealed by Chemical Mapping

Studies with modified trisaccharide derivatives were carried out to investigate the possibility of frameshifting (Fig. 10) [106], which is the exchange of epitope binding between chemically identical binding sites on a homopolymer (assuming that these sites are also conformationally identical). Since the minimal oligosaccharide epitope appeared to be small and since β-mannan is a homopolymer, any compound consisting of more than two mannose units theoretically had the possibility of frameshifting. Monodeoxygenation data described above guided the synthesis of trisaccharides **23** and **24** in which the essential hydroxyl group at C4 of the reducing mannose or C4′ of the nonreducing mannose was deoxygenated, respectively, preventing binding to either the internal (**23**) or terminal (**24**) disaccharide epitopes [106]. Trisaccharide **23** showed the most dramatic decrease in binding

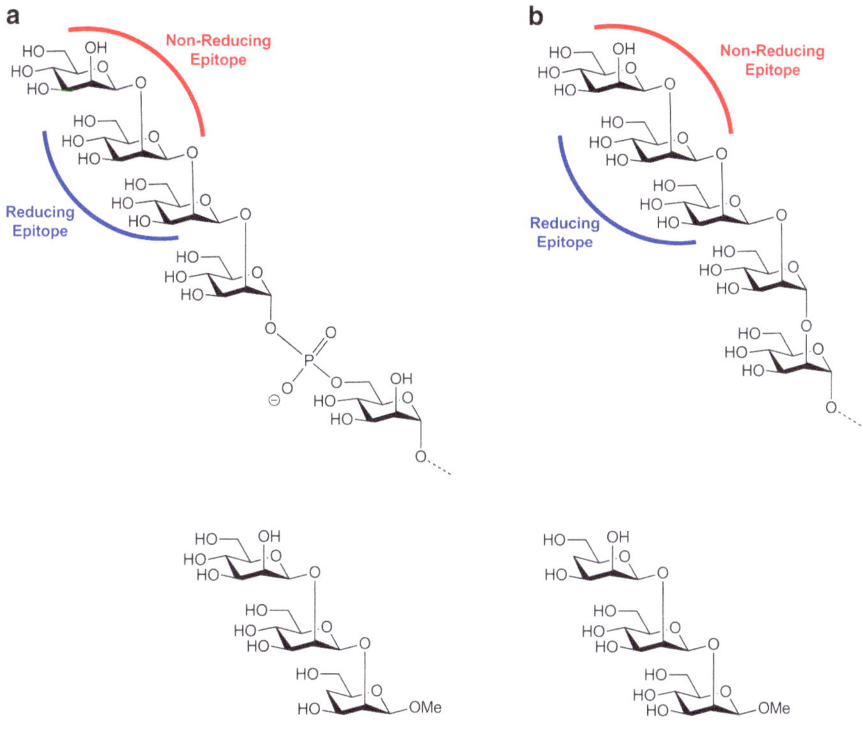

23 **24**

Fig. 10 Two β-mannan trisaccharide motifs are present in *C. albicans* phosphomannan: phosphodiester linked (**a**) and glycosidically linked (**b**) β-mannan. The possibility of frame shifting is indicated for the two modes of binding a disaccharide epitope in each of the motifs. Compounds **23** and **24** are deoxygenated at the reducing end and nonreducing end. Since deoxygenation abrogates binding of the residue in which that modification occurs **23** and **24**, respectively, provide "non-reducing" terminal and "reducing" internal disaccharide epitopes

Table 2 Comparison of the IC_{50} for inhibition by congeners **23** and **24** of mAb binding to *C. albicans* β-mannan with that of the native trisaccharide **21**

Trisaccharide	Derivative	IC_{50} (μmol/L)	$\Delta(\Delta G)$ (kcal/mol)
21	–	25	0
23	4-Deoxy	189	1.2
24	4″-Deoxy	79	0.68

affinity with $\Delta(\Delta G) = 1.2$ kcal/mol relative to the native trisaccharide **6** while **24** showed a decrease of 0.68 kcal/mol (Table 2). These results suggest that an internal disaccharide epitope is the recognition element preferred by mAb C3.1.

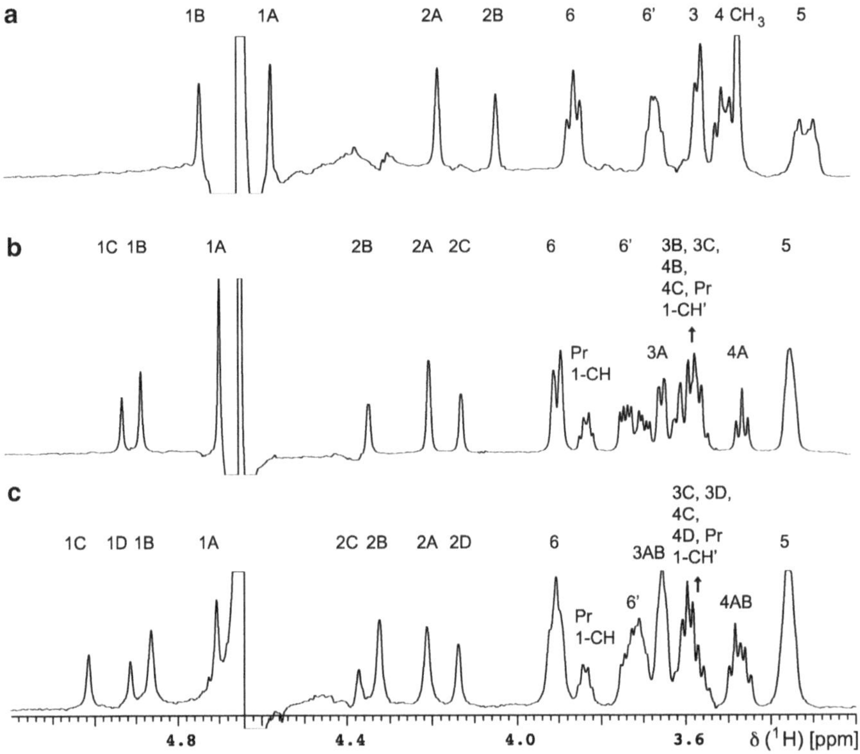

Fig. 11 STD spectra for the disaccharide methyl glycoside **6** (trace **A**), trisaccharide propyl glycoside **2** (trace **B**), and tetrasaccharide propyl glycoside **3** (trace **C**). The labels A through D refer to the saccharide residues identified as in Fig. 6. The associated digits refer to the hydrogen atoms bonded to saccharide or propyl carbon atoms numbered in the standard manner. The tetrasaccharide binds in a frame-shifted mode such that anomeric resonance H1B rather than H1A receives the strongest STD effect

4.3 *Correlation of Solution Structure with Inhibition Data*

The unusual pattern of inhibition of compounds **2–5** with C3.1 [91] was consistent with a binding site that could accommodate a trisaccharide and possibly even a fourth residue without a huge loss of binding energy. The slightly higher activity of a trisaccharide over a disaccharide suggested that a trisaccharide provided the optimum fit even though the primary polar contacts are located within a disaccharide. The well-ordered solution conformation of oligosaccharide raised the possibility that tetra-, penta-, and hexasaccharides might have to undergo a conformational change – most probably of glycosidic torsional angles – to avoid clashes with the surface of the binding site, as was seen in a *Salmonella* O-antigen-specific mAb [60, 61]. For a binding site with strict size constraints, such imposed conformational changes would be progressively more costly energetically with each additional hexose residue beyond a trisaccharide. These data also pointed to the potential of a small oligosaccharide epitope as a B-cell epitope that might be used to raise protective polyclonal sera.

4.4 NMR Studies of Bound Oligosaccharides

Observations of the oligosaccharide interfaces with the C3.1 antibody binding site were made for compounds **1–3**, **23**, and **24** using STD-NMR spectroscopy [57]. These data were consistent with and extended the conclusions of the inhibition data [91, 105]. STD-NMR experiments can provide an epitope map, with the strongest enhancements corresponding to oligosaccharide protons in close contact with protons of the antibody combining site [107]. STD-NMR patterns for compounds **2**, **3**, and **6** are shown in Fig. 11. The intensity of the ^{1}H resonances that experience the largest STD effect is seen to increase. Di- and trisaccharides **6** and **3** showed the strongest enhancements in residue A, followed by residue B. Enhancements in residue C of **6** were weaker. These results confirmed that an internal disaccharide is recognized by mAb C3.1. Further, within this disaccharide epitope, the internal monosaccharide residue (residue A) is more strongly bound than the terminal residue. Disaccharide **6** and trisaccharide **2** had similar binding modes, although some subtle differences were detectable, such as a greater enhancement of H2A over H1A in **1**.

In contrast to the results for disaccharide **6** and trisaccharide **2**, tetrasaccharide **3** (a 5-fold weaker inhibitor than trisaccharide **2**) showed maximal enhancements for residue B, followed by the other residues in the order $A > C \approx D$. These observations (Fig. 11) are interpreted as a frameshift. A frameshift results from the sequence redundancy of the homopolymer and the steric constraints of the combining site. This hypothesis is supported by the increased STD enhancement in the propyl aglycone (non-sugar portion) of **3** relative to **6**, suggesting that the increasing oligosaccharide length of **3** leads to additional contacts with the antibody at the reducing end.

Additional NMR studies based on the transferred nuclear Overhauser effect (trNOE) allowed us to determine the bound conformations of **6**, **2**, and **3**. Helical conformations with short distances between monosaccharides separated by one residue were found to be bound by the antibody. These conformations are similar to those adopted by unbound oligosaccharides in solution, suggesting that the antibody recognizes conformations similar to the solution global minimum [91, 108].

NMR data collected for the deoxygenated trisaccharide **24** (H replacing the OH at C4 on the C residue) showed similar STD enhancements to the trisaccharide glycoside **2**. Since **24** cannot bind via the B–C disaccharide, this compound provides an essentially pure representation of binding to the internal epitope (the A–B disaccharide) (Fig. 12). That the STD pattern of **24** was only minimally different from that of **2** showed that contributions from a secondary, frame-shifted binding

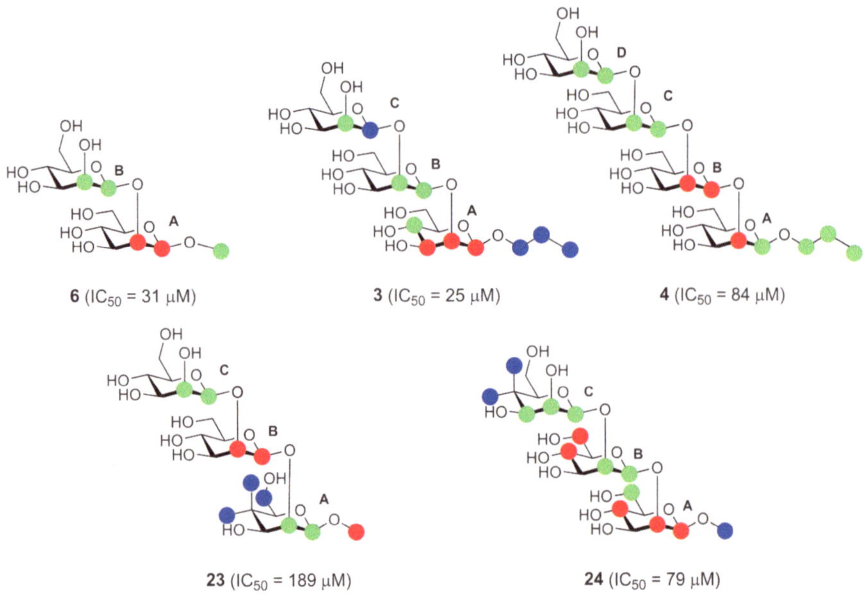

Fig. 12 Graphical representation of STD-NMR enhancements for the oligosaccharides **6**, **2**, **3**, **23**, and **24** binding to the IgG C3.1. The enhancement magnitudes of each resonance (isolated resonances only) are indicated by *colored circles* (*red* 67–100%; *green* 34–66%; *blue* 0–33%); all magnitudes are relative to the most strongly enhanced resonance in each spectrum [78]. (© American Society for Biochemistry and Molecular Biology)

mode (where B–C is recognized) are not very significant in the binding of the native trisaccharide.

In contrast, the trisaccharide **23** (deoxygenated at the 4-position of the A residue) showed a different STD enhancement pattern relative to **2**: strong enhancements of the B residue, somewhat increased enhancements of the C residue, and increased enhancements of the aglycone (Fig. 12). These observations are consistent with a frameshift similar to the pattern observed for the tetrasaccharide glycoside **3**. In total the STD data support the interpretation that molecular recognition of *C. albicans* (1 → 2)-β-mannan oligosaccharides by a protective monoclonal antibody reveals the immunodominance of internal saccharide residues.

No significant changes in the bound conformations were observed by trNOE for the deoxygenated trisaccharides, suggesting that frameshifting is not accompanied by any noteworthy conformational changes. As we discuss later, structures larger than a tetrasaccharide cannot be accommodated without significant conformational changes and resulting diminution of the binding free energy, ΔG.

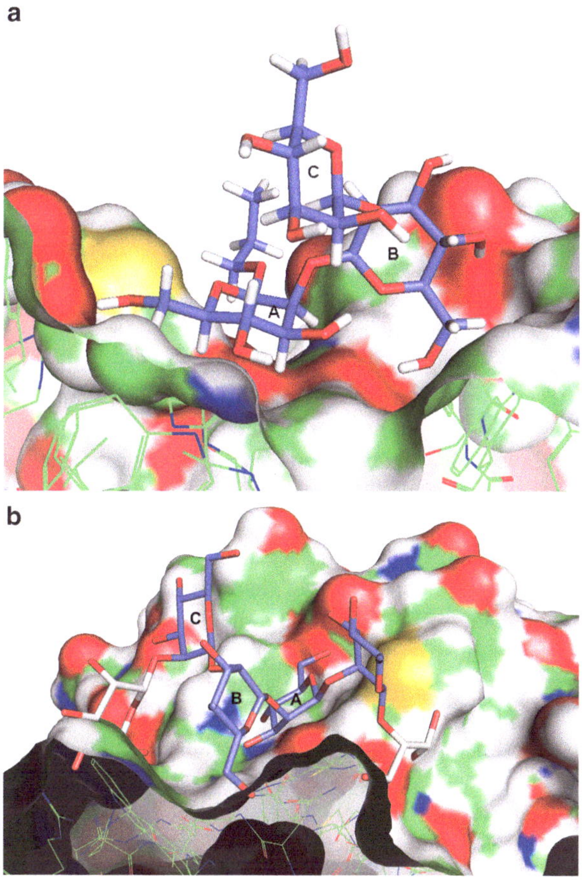

Fig. 13 (**a**) An end-on view of the C3.1 binding site with a docked trisaccharide ligand **6** and (**b**) a cutaway side view of the C3.1 binding site with a docked hexasaccharide. The residues A, B, and C displayed in Fig 13 **a** are labeled and these corresponding residues are also labeled A, B, and C in 2.13b. The cutaway side view 2.13b shows that the residue to the right of A may be accommodated without a clash with the antibody surface but not the next residue. The residue to the left of C cannot be accommodated due to a clash with protein. This view clearly shows that irrespective of frame shifting penta and hexasaccharide will experience serious obstacles to binding, thereby accounting for their weak inhibitory power

4.5 Molecular Modeling of the C3.1 Binding Site

A model of the C3.1 binding site was constructed by homology modeling and a new computational procedure was developed to dock oligosaccharides into the site [57]. This procedure used the previously developed chemical mapping and NMR results to rank computer-generated binding modes of disaccharides. Larger oligosaccharides were then superimposed in the same binding modes, and additional molecular dynamics refinement was used to generate the final model [57]. To avoid producing false-positive binding modes with the computational docking software, it was necessary to omit residues that were known to have no strong interactions with the binding site, such as pyranose C of **2** or C/D of tetramannoside **3**; the docking routine would otherwise attempt to produce interactions for these residues as well. We based our approach on disaccharide **1** and used subsequent molecular dynamics rounds to model the larger oligomers. The resulting model explained the unusual binding observations in this system, not only revealing a binding site that optimally accommodates only two mannoside residues but also explaining the preference for a disaccharide epitope (Fig. 13). In agreement with the earlier observations, residue A was found to be in closest contact with the base of the binding groove. The C3- and C4-OH hydroxyl groups of A, which are essential for binding, are near the potential hydrogen bond donors Ala H97 NH and Trp L91 NE1, and the C6-OH group forms hydrogen bonds with the Asp L50 carboxylate side chain. Residue B is partly solvent-exposed, but as predicted from the chemical mapping data, its C6-OH and C4-OH can form hydrogen bonds – the former with Trp H33 N, Asn H95 OD1, and ND1 and the latter with Asn H31 O. Residue C is highly solvent-exposed and makes only minimal contact with the antibody, consistent with the chemical mapping and NMR data.

Attempting to dock the tetramannoside **3** in the same binding mode as the trisaccharide **2** combining site did not allow room at the nonreducing end for the D residue, because of steric interactions with Trp H33 (Fig 13b). Instead, **3** binds in a frame-shifted mode where the B-C-D trisaccharide takes a position corresponding to that of the A-B-C trisaccharide in **2** (Fig. 13a). This model therefore explains the frameshifting we observed for **3** [106], which presumably must occur for any oligomer larger than a trisaccharide to bind.

The steric constraints of the binding site in the area of both the reducing and nonreducing end of the oligomer are quite strict, and when frameshifting, the ligands must avoid unfavorable interactions with Met H98, Tyr H32, and the protein backbone of CDRs H1 and H3. The shape of the binding site at the reducing end is not complementary to a longer $\beta(1 \rightarrow 2)$-linked oligosaccharide, and hence frameshifting by more than one residue without significant conformational changes is not possible. The binding of oligosaccharides larger than tetrasaccharide **3** would require several energetically unfavorable changes to glycosidic torsional angles to compensate for the lack of shape complementarity (Fig 13b), which explains the drop in binding affinity observed for penta- and hexasaccharides [91].

Taken together, these results are promising for the development of a vaccine based on an epitope of very small size, namely a di- or trisaccharide. However, we still needed to confirm the biological relevance of our hypothesis that reverse engineering the specificity of C3.1 is a valid basis for vaccine design and for establishing the structural models described here. To this end, STD-NMR experiments were performed with trisaccharide **2** and sera derived from rabbit immunization experiments with a trisaccharide-BSA (bovine serum albumin) conjugate **37**. Similar to the data for C3.1, the polyclonal sera spectra showed strongest enhancements for the A residue, with weaker enhancements for the B and C residues. The aglycone enhancement strength was also similar for both cases, as were those of H6 and H6′. Compared to the C3.1 case, however, increases were observed in the enhancement strengths of H5 and other degenerate multiplets. This observation is not surprising given that polyclonal sera contain a mixture of different antibodies. The observed STD-NMR trends thus indicate not only a detectable prevalence of C3.1-like binding but also a persistent preference for an internal epitope in experimental immunizations, which points to the biological importance of this mAb [78].

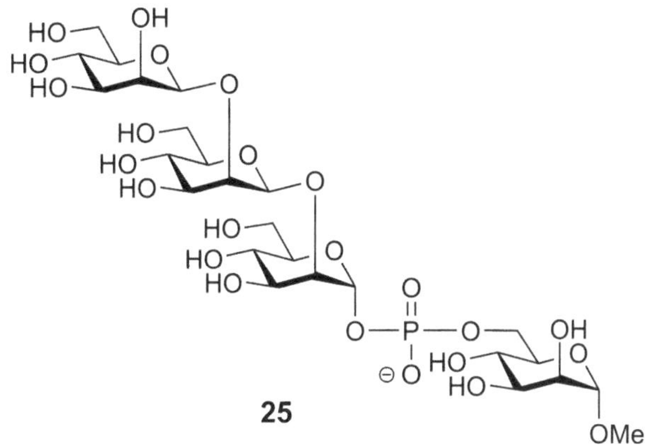

Fig. 14 The structure of the β-mannan tetrasaccharide **25**, which encompasses a phosphodiester linkage to an α-mannan side chain. This compound shows an affinity for C3.1 similar to that of the trisaccharide **21**, suggesting that the phosphate group plays no role in recognition

4.6 The β-Mannan to α-Mannan Phosphodiester Linkage

As discussed above, most structural and synthetic efforts have focused on acid-stable β-mannan epitopes. Recently, we completed synthetic and modeling studies of a *C. albicans* serotype B acid-labile β-mannan epitope that incorporates the phosphodiester: tetrasaccharide **25** (Fig. 14) [109]. This tetrasaccharide, synthesized via an anomeric H-phosphonate method, binds to C3.1 with an affinity essentially identical to that of the β-trimannoside **21**.

Modeling of **25** was carried out based on results for the β-mannosides. As described above, the C3.1 binding groove is unable to accommodate β-mannan oligosaccharides larger than a tetrasaccharide without requiring energetically unfavorable conformational adjustments to avoid steric clashes with the antibody. However, it appears that the introduction of an α-linkage at residue B in **25** allows the clashes with the narrowest part of the binding site to be avoided. The phosphodiester and the other α-linked monosaccharide, residue A, provide additional shape complementarity.

The phosphate group does not appear to engage in specific interactions; for example, no salt bridges or strong hydrogen bonds are predicted. Residues C and D of **25** can bind in a similar orientation and conformation to the residues A and B of **2**; therefore, they are predicted to participate in similar intermolecular interactions. This is consistent with the inhibition data, which show that **24** does not bind any more strongly than the β-mannotrioside **2**. Considering the greater shape complementarity of **25**, it is perhaps surprising that it is not more active than **2**. However, the introduction of charged groups at the periphery of binding sites has been shown to produce unpredictable effects, as in the case of the GS-IV lectin binding analogs of the Lewis b tetrasaccharide. In these examples, significant enthalpy and entropy changes were observed but the $\Delta(\Delta G)$ binding energy changes were close to zero [110].

We were then interested in which compound would provide a more successful conjugate vaccine [57, 111]. Our model predicts that **25** would be more complementary to the C3.1 binding site than are β-mannosides. Tetrasaccharide **25** should also contact a more diverse set of amino acid residues than the β-mannosides. This additional complementarity is provided by the arrangement of the phosphate group and residue A. These observations could mean that this compound may more efficiently stimulate the production of C3.1-like antibodies, despite having no greater binding affinity to C3.1 than propyl β-mannotrioside **2**. Because C3.1 is protective and C3.1-like binding modes are observable in experimental animal sera, C3.1-like antibodies may be desirable in a vaccine strategy. On the other hand, our candidate vaccine studies to date have been driven by inhibition studies that showed β-mannotriosides to be the most active inhibitors of C3.1 [91]. The latter approach is also favored based on synthetic efficiency as well as chemical and possibly conformational stability [91, 112]. Consequently, immunological studies have to date dealt only with β-mannosides. Syntheses and immunological studies of glycoconjugates of **25** are now in progress, however, and should provide more insight into the immunological correlations of these structural findings.

Our modeling studies with tetrasaccharide **25** also suggest that a $(1 \rightarrow 2)$-β-mannotriose joined via its α-anomer to a tether not only would provide the required specificity to β-mannans but also would allow the tether to exit the binding site unrestricted [57]. The component trisaccharide Man(β1–2)Man(β1–2) Man(α1- is common to both acid-stable and acid-labile β-mannan epitopes, and therefore might represent a common phosphomannan epitope of multiple *Candida* species and growth stages [70, 78]. This antigen would allow us to eliminate the synthetically challenging phosphate group at the same time [109]. Though some complementarity to the C3.1 binding site is predicted to be lost, this loss may be outweighed by the potential advantages of tethering the trisaccharide to other structures such as carrier proteins. The activities of vaccine constructs that incorporate some of these features have been investigated by glycopeptide antigen work (see Sect. 5.6).

5 Synthetic Conjugate Vaccines

5.1 Glycoconjugate Vaccines

Our initial vaccine design was based on the oligosaccharide inhibition data (Fig. 4) [91]. We interpreted this data as suggesting that a trisaccharide or disaccharide epitope conjugated to protein should induce antibodies able to recognize the cell-wall phosphomannan. The first test would be with phosphomannan extracted and coated on ELISA plates; the second would be against phosphomannan as it is displayed on the fungal cell wall. Provided these criteria were satisfied, the final step would be live challenge experiments.

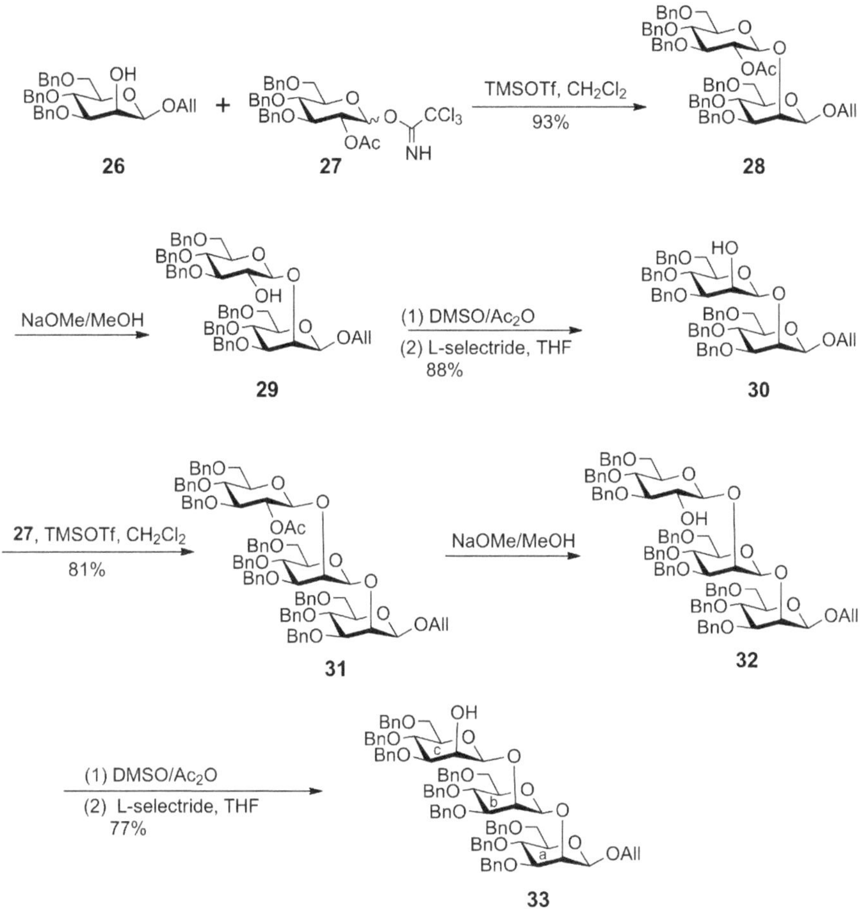

Scheme 1 (continued)

5.2 *Immunization and Challenge Experiments in Rabbits*

A trisaccharide functionalized for conjugation to a carrier protein employed an allyl group. A selectively protected allyl mannoside **26** (Scheme 1a) was glycosylated by the glucosyl imidate **27**. With persistent benzyl protection at C3, C4, and C6, the acetyl group at C2 guided efficient β-glycoside formation to yield **28**. Removal of acetate to yield **29** followed by an oxidation–reduction sequence gave the disaccharide alcohol **30**, which could be subjected to a similar sequence of glycosylation by **27** to give **31**. Then transesterification gave **32**, and the same oxidation–reduction sequence was used again to obtain the protected target trisaccharide **33**. The fully protected trisaccharide allyl glycoside **33** was reacted with cysteamine to yield an amino-terminated tether **34** after a debenzylation step (Scheme 1b) [92]. This

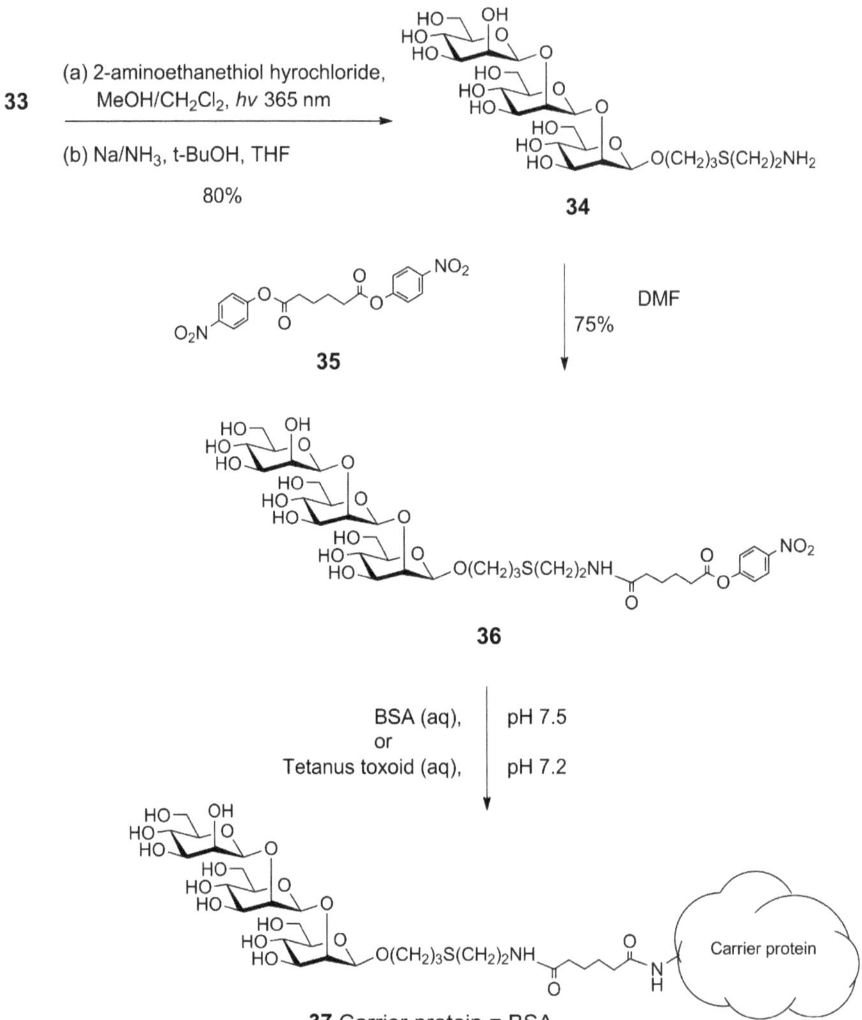

37 Carrier protein = BSA
38 Carrier protein = Tetanus toxoid

Scheme 1 (**a**) Assembly of the protected β-mannan trisaccharide **33** from monosaccharide building blocks. (**b**) Conversion of trisaccharide allyl glycoside **33** to the deprotected and amino-terminated tether trisaccharide **34**. Conjugation to BSA and tetanus toxoid to give antigens 37 and **38** employs a homobifunctional coupling agent **35**

glycoside **34** was activated for conjugation with protein by first reacting with *p*-nitrophenyl adipic acid diester **35** [112]. The half ester **36** was then reacted with protein to afford either BSA or tetanus toxoid glycoconjugates **37** and **38**. Employing a different approach, a disaccharide was similarly conjugated to chicken serum albumin [105, 113]. Both the trisaccharide and disacccharide conjugates were used to immunize rabbits.

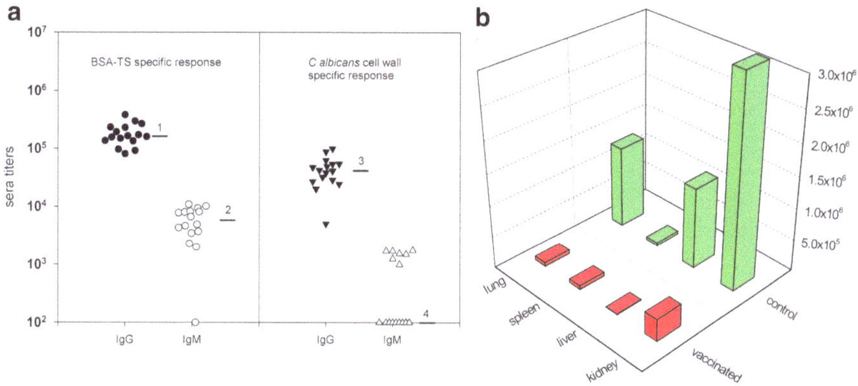

Fig. 15 (**a**) The antibody responses of rabbits to the conjugate vaccine **38**. (**b**) *C. albicans* colony count in lung, spleen, liver, and kidney of rabbits in the vaccinated and control groups. Accumulated data were collected for 16 rabbits in the vaccinated group and 13 rabbits in the control group. Colony-forming unit (CFU) counts were performed on the organs of each rabbit, and the data presented are averaged for 16 rabbits vaccinated with the conjugate vaccine and 13 rabbits vaccinated with tetanus toxoid. The statistical treatment of data with the Generalized Estimating Equations (GEE) method revealed that the reduction of Candida CFU was statistically significant in the kidney (average reduction in log counts 4.4, $p = 0.016$) and liver (average reduction in log counts 3.6, $p = 0.033$). The effect of vaccination was not observed in lungs and spleen (increase in log counts 0.002, $p = 0.99$; 0.411, $p = 0.80$, respectively) [114].

When rabbits were immunized with trisaccharide vaccine **38** (administered with alum, an adjuvant approved for use in humans), ELISA titers to the immunizing oligosaccharide conjugated to a heterologous protein exceeded 100,000 [112, 114]. Titers against the cell-wall mannan were approximately 2–3 fold lower (Fig. 15). Serum from rabbits immunized by vaccine **38** generated antibodies that stained *C. albicans* cells more intensely in fluorescent labeling studies [114] than antibodies from rabbits immunized with a disaccharide conjugate [105].

The two vaccines were also evaluated in live challenge experiments [99, 114]. Since *C. albicans* is a commensal organism, it is necessary to evaluate the vaccine in rabbits that are immunocompromised [114]. The disaccharide vaccine was studied in a smaller sample of rabbits [105], and therefore those results lack the statistical significance of the larger study performed with the trisaccharide vaccine [114]. Vaccine was administered prior to the onset of immunosuppression, allowing a persistent level of antibody directed toward the cell-wall β-mannan. Both vaccines reduced *C. albicans* counts in vital organs, but it was evident that the vaccines alone were insufficient to provide 100% protection [114].

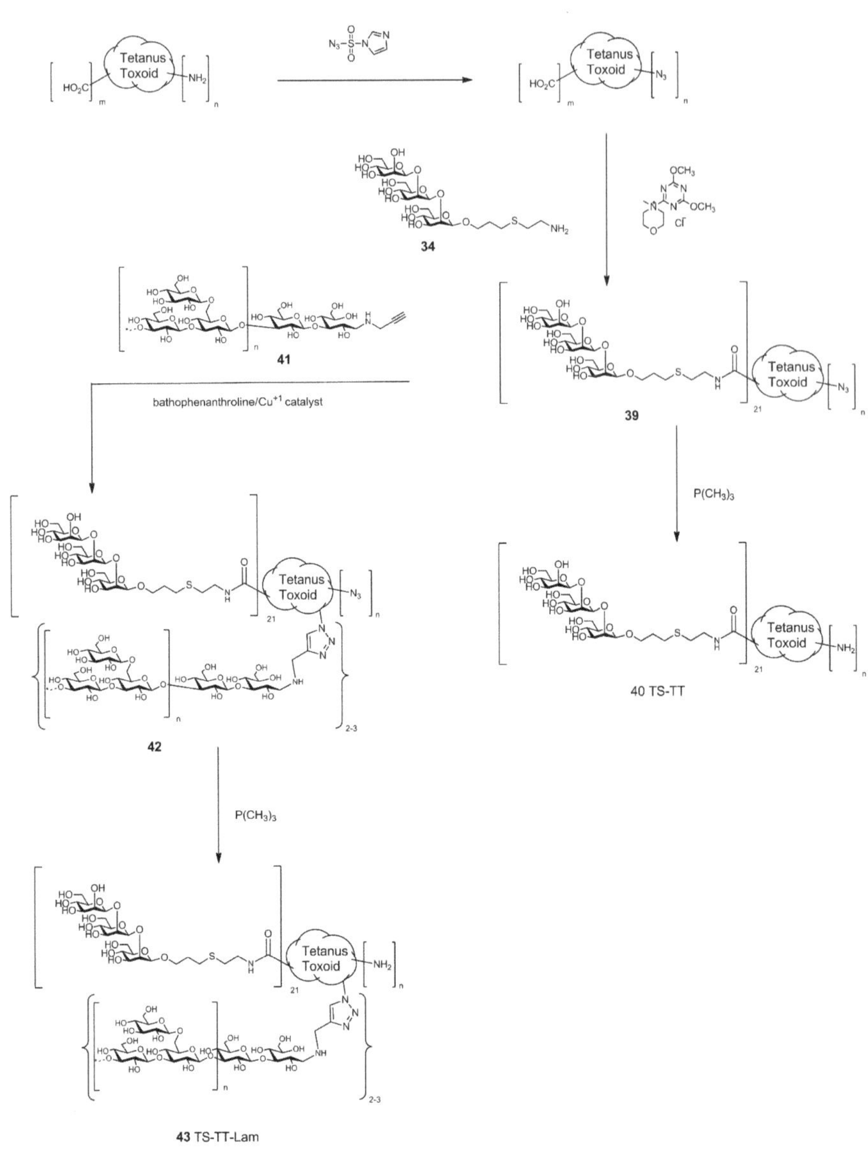

Scheme 2 The use of azido transfer reagent imidazole-1-sulfonyl azide hydrochloride converts amino groups to azides, allowing amide bond formation between **34** and the carboxylic acid residues of tetanus toxoid employing 4-(4,6-Dimethoxy-1,3,5-triazin-2-yl)-4-methylmorpholinium chloride. Trimethylphosphine is used to reduce protein azide groups back to amino groups. Circular dichroism measures suggest the tetanus toxoid retains its tertiary structure during these transformations

5.3 Immunization and Challenge Experiments in Mice

When the trisaccharide–tetanus toxoid conjugate vaccine **38** was evaluated in mice, we consistently observed only modest ELISA titers recorded on plates coated with trisaccharide-BSA conjugate [115–117]. Only in 10–20% of immunized mice did titers exceed 1:10,000, which itself was a level of antibody that we have observed to confer almost no protection to live challenge with *C. albicans*. Several different vaccine constructs were synthesized and tested [115–117]. Some approaches attempted to cluster epitopes [115], while others employed novel carrier constructs [117]. In all of these cases, the immune response as determined by antibody titer was judged to be insufficient to proceed to challenge experiments.

In work that will be described in Sect. 5.4, collaboration with the Dr. Jim Cutler and Dr. Hong Xin established that dendritic cell priming could improve the immune response to low-molecular-weight glycopeptides [118]. These data – together with accumulating literature evidence for the importance of dendritic cells in processing and presenting antigens [119–122] – suggested that targeting dendritic cells with a vaccine construct might improve the immune response to our conjugate vaccine [118].

Scientists at Novartis have shown that conjugate vaccines consisting of different β-glucans (either oligo- or polysaccharides) conjugated to CRM197, a variant of the diphtheria toxin were highly effective against fungal diseases [123, 124]. Antibodies to the β-glucan were responsible for the protection, but it is of interest to note that 1,3-β-glucans are ligands for dectin-1 [125], a C-type lectin present on the surface of dendritic cells [125–130]. In the expectation that β-glucan would deliver vaccine to dendritic cells we elected to attach β-glucan to our β-mannotriose-tetanus toxoid vaccine to examine the effect on antibody levels.

The preparation of the three-component vaccine involved unique chemistry (Scheme 2) [131]. In the first step, tetanus toxoid was reacted with a diazo transfer reagent, imidazole-1-sulfonyl azide hydrochloride, to convert amino groups into azide groups [132]. Removing amino groups in this way avoids the tendency of proteins to form intramolecular or intermolecular lactams, thus permitting efficient use of carbodiimide- or amide-forming coupling reagents. Following activation by 4-(4,6-Dimethoxy-1,3,5-triazin-2-yl)-4-methylmorpholinium chloride of the aspartic acid and glutamine side-chain carboxylic acids of the azido tetanus toxoid, coupling with the amino-terminated trisaccharide hapten **34** yields **39**. The resulting conjugate was divided into two portions. One half was reduced by trimethyl phosphine to convert the azide groups back to amines, affording a control vaccine **40** that was designated TS-TT (trisaccharide–tetanus toxoid). The second half of **39** was conjugated to propargylated laminarin **41** using click chemistry [133, 134]. The residual azide groups of the resulting conjugate **42** were reduced back to the original amino groups by trimethyl phosphine to give vaccine **43**, designated TS-TT-Lam [131].

Immunization of mice with the TS-TT-Lam vaccine resulted in an improved immune response, as manifested by high titers of antibody recognizing *C. albicans* β-mannan antigen when compared to mice immunized with TS-TT vaccine [131]. TS-TT-Lam vaccine also affected the distribution of IgG subclasses, showing that targeting the dectin-1 receptor can augment and immunomodulate the immune response. In more detailed immunological investigation of the two vaccine constructs, a macrophage cell line expressing dectin-1 was employed to reveal the binding and activation of the dectin-1 signal transduction pathway by the TS-TT-Lam vaccine. Antigen binding to dectin-1 resulted in (1) activation of Src-family-kinases and SYK, as revealed by their recruitment and phosphorylation in the vicinity of the bound conjugate, and (2) translocation of NF-κB to the nucleus. Treatment of immature bone-marrow-derived dendritic cells (BMDCs) with tricomponent or control vaccine confirmed that the vaccine containing β-glucan exerted its enhanced activity by virtue of dendritic cell targeting and uptake. Immature primary cells stimulated by the tricomponent vaccine showed activation of BMDCs, but those stimulated by the β-mannan tetanus toxoid vaccine did not. Moreover, treated BMDCs secreted increased levels of several cytokines, including TGF-β and IL 6; these cytokines activate Th17 helper T-cells, a cell type known to be important in antifungal responses.

The vaccination experiments described to this point suggest that (1) a simple trisaccharide conjugated to a potent protein carrier can reduce *C. albicans* counts in live challenge experiments, (2) targeting dendritic cells significantly augments the β-mannan-specific antibody levels, and (3) targeting dendritic cells also elevates cytokine responses, which in turn activate T-cell subsets that are important to antifungal immune responses.

5.4 Synthesis and Evaluation of Glycopeptide Vaccines

Protection against infections by *C. albicans* and other *Candida* species requires the involvement of cellular immunity, including cells of the innate immune system [135]. However, it was only recently appreciated that humoral immunity could also play a role in protection against *C. albicans* [136–138]. Consequently, fungal cell-wall peptides were sought that would serve as carriers and to provide an additional protective epitope [118]. The latter attribute would induce protection against fungal strains not producing the carbohydrate epitope, especially because that moiety is not essential for *C. albicans* growth [82, 83] and only 80% of isolates of *C. glabrata* (another important agent of candidiasis) produce β-linked mannosides [70]. To satisfy these features, efficient conjugation chemistry was developed (Scheme 3). The oligosaccharide B-cell epitope was provided in the form of trisaccharide **34**, which was acylated by the NHS ester of the acryloyl polyethylene glycol (PEG) derivative **44** to produce **45**. This allowed the facile conjugation of unprotected β-$(Man)_3$ with six candidate peptide carriers derived from the cell wall of *C. albicans* [118]. The conjugation strategy involved Michael addition of **45** to synthetic peptides that had been modified at their *N*-terminus by 3-thiopropionic acid. These structures allowed the efficient coupling of the two deprotected

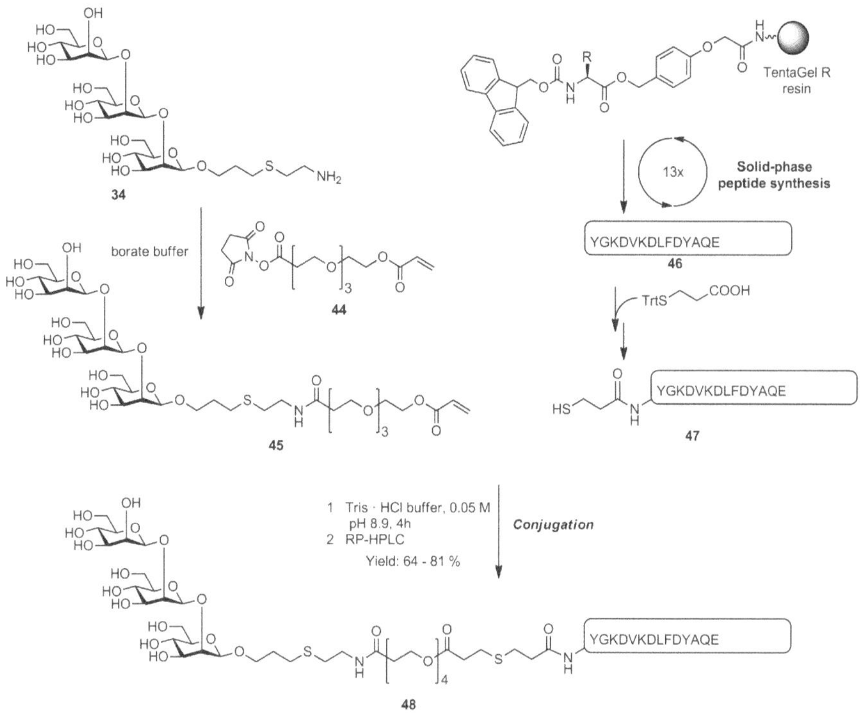

Scheme 3 Conjugation of trisaccharide **34** to the T-cell peptide Fba (YGKDVKDLFDYAQE) to give glycopeptide **48**

components – the antigenic β-(Man)$_3$ and the specific T_H peptide epitopes – as the last step of the synthesis [139]. This is shown for the Fba peptide **46**, which is first converted to **47** and then conjugated in fairly good yield to give the glycopeptide **48** (Scheme 3). This strategy permits the screening of a range of peptide carriers and facilitates the identification of the most suitable peptides. The heterobifunctional coupling reagent **44** is based on a water-soluble triethylene glycol core functionalized at one end with an *N*-hydroxysuccinimide ester and a sulfhydryl-reactive acrylate moiety at the other [139]. This linker permits efficient coupling of biomolecules under aqueous conditions while still affording, as far as possible, an immunologically silent tether.

C. albicans carrier proteins were selected from cell-wall proteins expressed during pathogenesis of human-disseminated candidiasis [140, 141]. The six candidate carriers were fructose-bisphosphate aldolase (Fba), methyltetrahydropteroyl-triglutamate (Met6), hyphal wall protein-1 (Hwp1), enolase (Enol), glyceraldehyde-3-phosphate dehydrogenase (Gap1), and phosphoglycerate kinase (Pgk1). An epitope-finding algorithm (GenScript, Antigen Design Tool, OptimumAntigen) was used to select putative epitopes composed of 14 amino acids each. Locations near the *N*-terminus were chosen, as it was presumed that this location was more likely to be accessible to the host immune system. Each peptide shows 100% homology only with *C. albicans* and none with mammalian proteins. Peptides Fba, Met6, and Pgk1 contain obligatory P1 anchor residues and are predicted to bind the human leukocyte MHC (major histocompatibility complex) class II cell surface receptor and multiple MHC class I binding sites are predicted for all six peptides [142, 143].

Antigen-pulsed dendritic-cell-based vaccination was used in all immunizations for development of antibodies in mice. Immune sera were tested for antibodies specific for the whole conjugates, synthetic β-$(Man)_3$, synthetic peptide carriers and fungal cell surface phosphomannoprotein (PMP) extract from *C. albicans* cell walls [118]. All six conjugates were immunogenic, as shown by high titers of antibodies specific to each of the test antigens. Each antiserum – but not the negative control sera – reacted directly with yeast and hyphal forms of the fungus, as evidenced by indirect immunofluorescence microscopy. The antibody response to the glycoconjugates was several-fold greater than that of sera from control groups injected with dendritic-cell adjuvant without antigen. The presence of β-$(Man)_3$ antibodies in the sera of vaccinated animals was confirmed by dose–response ELISA inhibition [118].

The conjugates were assessed for their potential in a vaccination strategy for disseminated candidiasis. Fourteen days after the second immunization, mice were challenged intravenously with a lethal dose (5×10^5) of viable yeast forms of *C. albicans* that eventually killed all control mice. Conjugate immunized animals, except for those that received the β-$(Man)_3$-Pgk1, survived longer than the control groups. The immunized groups that received the β-$(Man)_3$-Hwp1, β-$(Man)_3$-Fba, or β-$(Man)_3$-Met6 conjugates showed 80–100% survival throughout the 120-day post-challenge observation period. Mice that received either β-$(Man)_3$-Enol or β-$(Man)_3$-Gap1 showed 40–80% survival. However, control groups and mice given the β-$(Man)_3$-Pgk1 conjugate had median survival times of 5, 6, 14, 16, and 11 days, respectively [118]. Importantly, compared with animals that succumbed, the survivors had greatly reduced or even nondetectable viable fungal colony-forming units (CFUs) in kidneys, a target organ in disseminated candidiasis. The antibody dependency of protection was evidenced by the protection transferred to naive mice by immune serum, but not by serum pre-absorbed with *C. albicans* [118].

An advantage of this approach is that the peptide carrier is derived from the infectious agent of interest, providing the possibility of a dual-protective immune response against both the glycan epitope and the peptide carrier. For the trisaccharide-Fba-glycopeptide, both β-mannan and peptide antibodies contributed to protection. These experiments showed the potential for chemical synthesis of vaccines of relatively low molecular weight that induce protection against more than one unrelated epitope expressed by the infectious disease agent of interest. However, ex vivo dendritic cell stimulation is not a generally useful approach: a modified approach would be required for the practical incorporation of glycopeptides into a viable vaccine construct.

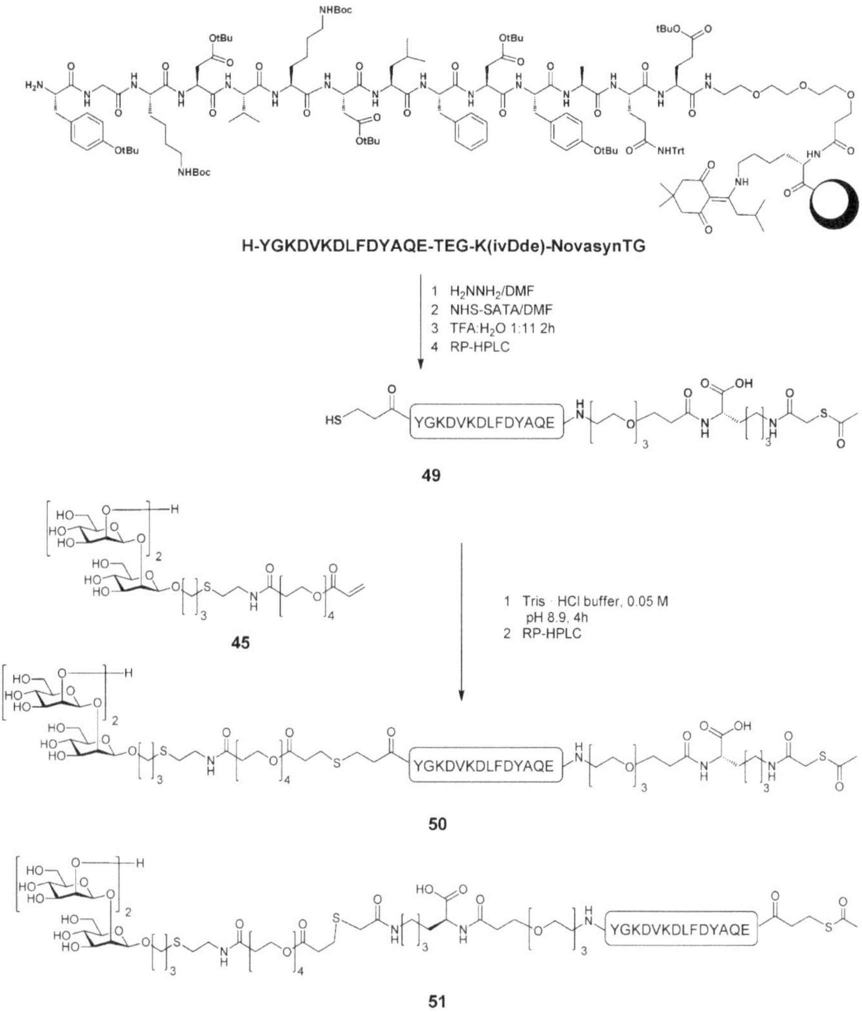

Scheme 4 (continued)

5.5 *Self-Adjuvanting Glycopeptide Conjugate Vaccine*

To improve immunogenicity and allow the incorporation of an adjuvant suitable for human use, we modified the β-(Man)$_3$-Fba conjugate by conjugating it to tetanus toxoid [144]. This necessitated a modified conjugation strategy designed to conjugate glycopeptide to the carrier protein via the *C*-terminus of the glycopeptide moiety. To this end, the peptide was synthesized with an additional lysine residue at the *C*-terminus of the peptide (Scheme 4a), and this was linked to the Fba peptide via a triethylene glycol spacer to give the peptide **49**. The target glycoconjugate

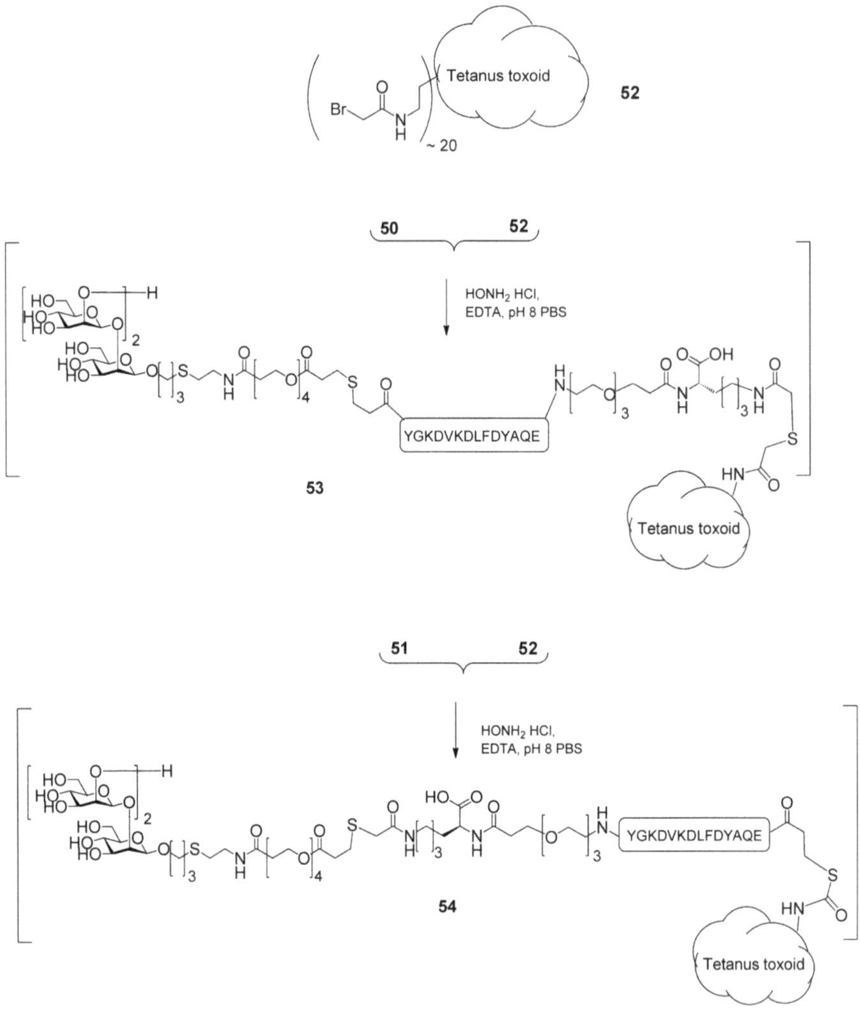

Scheme 4 (**a**) Modification of the solid-phase peptide synthesis approach shown in Scheme 3 installs a *C*-terminal lysine (K) to allow conjugation to tetanus toxoid. The crucial peptide intermediate possesses two thiol groups. The thiolglycolic acid residue at the C-terminus is protected as an acetate **49**. The basic conditions used to conjugate **45** with **49** results in acetate transfer and the formation of the two glycopeptides **50** and **51**. (**b**) Tetanus toxoid containing approximately 20 bromoacetyl groups reacts with the latent thiol groups of **50** and **51** (released under basic conjugation conditions) to form **53** and **54**, respectively

vaccine **50** was obtained by reaction of **45** and **49**. Bromoacetate groups were introduced on approximately 20 of the lysine residues present in tetanus toxoid to give **52** (Scheme 4b) and reaction of **50** with **52** was expected to give the target vaccine **53**. However, under the basic conditions used to react **49** with **45**, it was found that the acetyl group on the *C*-terminal thioglycolic acid residue used to acylate the ε-amino group of lysine can undergo an intermolecular migration to the

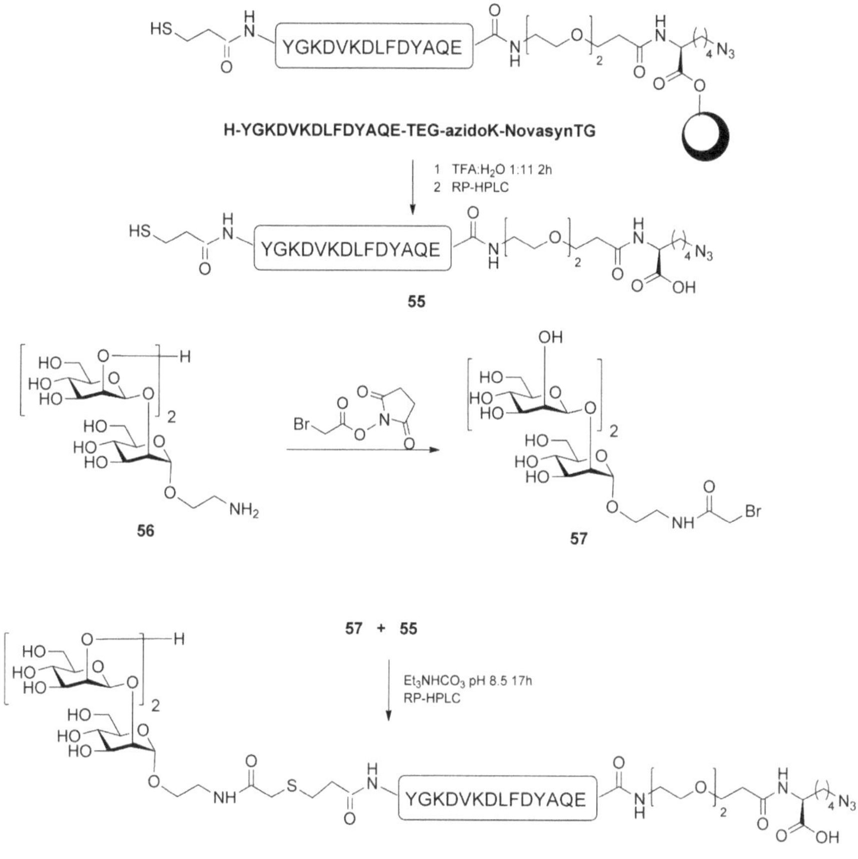

Scheme 5 (continued)

3-thiopropionic acid residue. This set up the conditions whereby two glycopeptides **50** and **51** were formed in the ratio 1:2 rather than a single product **50**. Subsequently, **51** also reacts with **52** to give a second vaccine construct **54** [145]. Thus, the study of a self-adjuvanting vaccine [144] unintentionally used a 1:2 mixture of the two constructs **53** and **54** and not the cited single vaccine **53**.

This mixture of the new glycopeptide vaccine conjugate **53**, β-$(Man)_3$-Fba-TT together with the *C*-terminal conjugated trisaccharide **54** proved to be highly immunogenic, as it induced robust antibody responses when administered with either alum or monophosphoryl lipid A as adjuvants [144]. Moreover, prior to the second booster dose, an isotype switch occurred from IgM to IgG antibodies for both the β-$(Man)_3$ and Fba peptide epitopes. This result indicated a possible memory cell response and, perhaps, a vaccine that induces long-term immunity. Most importantly, the (β-Man)$_3$-Fba-TT conjugates **53** and **54** administered with

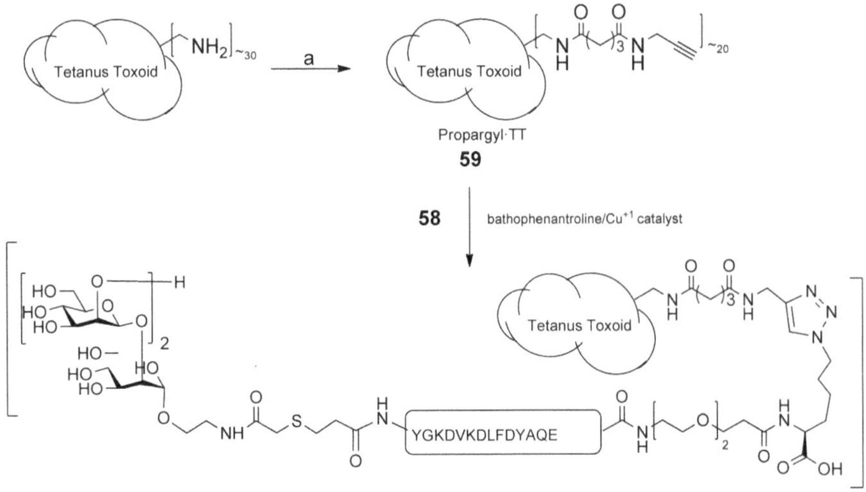

60

Scheme 5 (**a**) An alternative conjugation strategy for unambiguous attachment of glycopeptides to protein. An azido lysine residue is introduced at the *C*-terminus of the Fba peptide **55**. Conjugation of trisaccharide **56** to the peptide follows the same approach described in Schemes 3 and 4a. (**b**) Propargyl groups are introduced into tetanus toxoid to give **59**, and click chemistry is used to conjugate glycopeptides **58** with the modified tetanus toxoid **59** to produce **60**

either alum or MPL induced protection against disseminated candidiasis on a par with the high level of protection observed with the original dendritic-cell/complete Freunds adjuvant immunization approach [118]. In previous work involving the latter approach, we found that the Fba peptide itself was immunogenic, inducing not only a robust antibody response but also a protective response against disseminated candidiasis [118]. We were surprised, therefore, to find that the Fba-TT conjugate in adjuvant did not perform well in mice [144]. We were even more surprised to discover that mice that received the β-(Man)$_3$-Fba-TT without any adjuvant performed as well or better than mice that received the vaccine with adjuvant. All three groups vaccinated with glycoconjugate with or without either of the two adjuvants, alum or MPL, showed a high level of protection against a lethal challenge with *C. albicans*: survival times were significantly increased and kidney fungal burdens were reduced or nondetectable at the time of sacrifice compared to control groups that received only adjuvants or buffer prior to challenge [144]. Furthermore, sera from mice immunized with the conjugates **53** and **54** transferred protection against disseminated candidiasis to naive mice, whereas *C. albicans* pre-absorbed immune sera did not; this confirmed that the induced antibodies were protective. These results demonstrated that the addition of tetanus toxoid to the vaccine conjugate provided sufficient self-adjuvanting activity without the need for additional adjuvant. As far as we are aware, this is the first report of a self-adjuvanting glycopeptide vaccine against any infectious disease [144].

5.6 Variants of Glycopeptide Vaccines

Our modeling studies of various ligands in the binding site of mAb C3.1 suggested that if the terminal reducing residue of a trisaccharide ligand adopted the α configuration. This orientation could be favorable for allowing the extended phosphomannan to avoid contact with the antibody surface. Thus, a trisaccharide attached to a tether via an α-mannopyranosyl linkage might be as active as the β-linked analog **53**. However, it would be a synthetically preferred construct for a vaccine, since forming an α-mannopyranosyl linkage involves fewer steps. We set out to synthesize a glycopeptide ligand of this type to investigate this hypothesis and to simplify the conjugation chemistry as summarized in Scheme 5.

As in the synthesis of peptide derivative **49**, peptide **55** was synthesized on solid support. This time, however, azidolysine was attached to the resin instead of lysine and the Fba peptide was elaborated with a diethylene glycol spacer to give **55**. The trisaccharide 2-aminoethyl glycoside **56** [145] was bromoacetylated to give **57**. Conjugation to the 3-thiopropionic acid residue of **55** by **57** gave glycopeptide **58**, which in turn was conjugated to propargylated tetanus toxoid (TT) derivative **59** to afford the target vaccine **60**.

Mice were immunized with **60** using the protocol described for vaccines **53** and **54**. Sera were titrated against copovidone polymer bearing the epitope **56** and the Fba peptide [146]. While strong antibody titers of ~1:100,000 for the Fba peptide were observed, most mice exhibited only weak titers of <1:1,000–5,000 against the β-mannan.

These data raise questions about the optimum length of tether employed to attach the β-mannan epitope to Fba peptide in order to secure a strong immune response. The data from this experiment combined with other unpublished work using conjugates of glycopeptides and tetanus toxoid containing the α-linked mannan also support the conclusion that the trisaccharide epitope represented by **34** or analogs also β-linked to tether are the preferred structures to elicit protective antibodies with a binding profile of the C3.1 mAb type. This idea is consistent with the profile of antibodies observed in the sera of rabbits immunized with the simple conjugate vaccine **38**.

6 Summary, Conclusions, and Outlook

The observed binding profile of mouse mAb C3.1 with homo-oligomers of the *C. albicans* β-1,2-mannan is in stark contrast with a 40-year-old paradigm of oligosaccharide-antibody binding. This surprising finding prompted a thorough investigation of β-1,2-mannan–antibody interactions and led to the discovery of protective vaccines based on a trisaccharide epitope.

A β-1,2-mannan trisaccharide conjugated to tetanus toxoid gave high titers of antibodies specific to *C. albicans* in rabbits. Immunized rabbits rendered neutropenic and challenged intravenously with live *C. albicans* showed significantly reduced fungal burden in vital organs. However, the same conjugate vaccine was poorly immunogenic in mice.

This conjugate may be converted to a much improved vaccine by conjugation with a second carbohydrate antigen, a β-glucan, which is also a ligand for dectin-1 (a dendritic cell lectin). This vaccine is directed to antigen processing cells and induces a much improved immune response, with a cytokine profile well suited to protection against fungal infection.

A third vaccine construct is the trisaccharide conjugated to a T-cell peptide, which was conjugated in turn to tetanus toxoid. This vaccine did not require the use of an adjuvant to exhibit potent immunogenicity: it conferred protection by inducing combined carbohydrate- and peptide-specific responses.

The results summarized here suggest that highly effective and completely synthetic vaccines should be feasible by incorporating ligands targeting dendritic cells in a construct that includes B- and T-cell epitopes.

References

1. Francis T Jr, Tillett WS (1930) Cutaneous reactions in pneumonia: the development of antibodies following the intradermal injection of type-specific polysaccharide. J Exp Med 52:573–585
2. MacLeod CM, Hodges RG, Heidelberger M, Bernhard WG (1945) Prevention of pneumococcal pneumonia by immunization with specific capsular polysaccharides. J Exp Med 82:445–465
3. Heidelberger M, MacLeod CM, Di Lapi MM (1948) The human antibody response to simultaneous injection of six specific polysaccharides of pneumococcus. J Exp Med 88:369–372
4. Parke JC Jr, Schneerson R, Robbins JB, Schlesselman JJ (1977) Interim report of a controlled field trial of immunization with capsular polysaccharides of *Haemophilus influenzae* type b and group C *Neisseria meningitidis* in Mecklenburg County, North Carolina (March 1974–March 1976). J Infect Dis 136(Suppl):S51–S56
5. Peltola H, Mäkelä H, Käyhty H, Jousimies H, Herva E, Hällström K, Sivonen A, Renkonen OV, Pettay O, Karanko V, Ahvonen P, Sarna S (1977) Clinical efficacy of meningococcus group A capsular polysaccharide vaccine in children three months to five years of age. N Engl J Med 297:686–691
6. Peltola H, Käyhty H, Sivonen A, Mäkelä PH (1977) *Haemophilus influenzae* type b capsular polysaccharide vaccine in children: a double-blind field study of 100,000 vaccines 3 months to 5 Years of age in Finland. Pediatrics 60:730–737
7. Eskola J, Takala A, Käyhty H, Peltola H, Mäkelä PH (1991) Experience in Finland with *Haemophilus influenzae* type b vaccines. Vaccine S14–S16; discussion S25
8. Goebel WF, Avery OT (1931) Chemo-immunological studies on conjugated carbohydrate-proteins. IV. The synthesis of the p-aminobenzyl ether of the soluble specific substance of Type III Pneumococcus and its coupling with protein. J Exp Med 54:431–436
9. Avery OT, Goebel WF (1931) Chemo-immunological studies on conjugated carbohydrate-proteins. V. The immunological specificity of an antigen prepared by combining the capsular polysaccharide of Type III Pneumococcus with foreign protein. J Exp Med 54:437–447

10. Madore DV, Johnson CL, Phipps DC, Popejoy LA, Eby R, Smith DH (1990) Safety and immunologic response to *Haemophilus Influenzae* type b oligosaccharide-CRM197 conjugate vaccine in 1- to 6-month-old infants. Pediatrics 85:331–337
11. Yogev R, Arditi M, Chadwick EG, Amer MD, Sroka PA (1990) *Haemophilus influenzae* type b conjugate vaccine (meningococcal protein conjugate): immunogenicity and safety at various doses. Pediatrics 85:690–693
12. Gambillara V (2012) The conception and production of conjugate vaccines using recombinant DNA technology. BioPharm Int 25:28–32
13. Avci FY, Li X, Tsuji M, Kasper DL (2011) A mechanism for glycoconjugate vaccine activation of the adaptive immune system and its implications for vaccine design. Nat Med 17:1602–1609
14. Jennings HJ (1983) Capsular polysaccharides as human vaccines. Adv Carbohydr Chem Biochem 41:155–208
15. Finco O, Rappuoli R (2014) Designing vaccines for the twenty-first century society. Front Immunol 5:12
16. Robbins JB, Schneerson R, Szu SC, Fattom A, Yang Y, Lagergard T, Chu C, Sørensen US (1989) Prevention of invasive bacterial diseases by immunization with polysaccharide-protein conjugates. Curr Top Microbiol Immunol 146:169–180
17. Lesinski GB, Westerink MA (2001) Vaccines against polysaccharide antigens. Curr Drug Targets Infect Disord 1:325–334
18. Paoletti LC, Kasper DL (2003) Glycoconjugate vaccines to prevent group B streptococcal infections. Expert Opin Biol Ther 3:975–984
19. Pichichero M (2013) Protein carriers of conjugate vaccines: characteristics, development, and clinical trials. Hum Vaccin Immunother 9:2505–2523
20. Goldblatt DJ (1998) Recent developments in bacterial conjugate vaccines. Med Microbiol 47:563–567
21. Beurret M, Hamidi A, Kreeftenberg H (2012) Development and technology transfer of *Haemophilus influenzae* type b conjugate vaccines for developing countries. Vaccine 30:4897–4906
22. Terra VS, Mills DC, Yates LE, Abouelhadid S, Cuccui J, Wren BW (2012) Recent developments in bacterial protein glycan coupling technology and glycoconjugate vaccine design. J Med Microbiol 61:919–926
23. Pozsgay V (2000) Oligosaccharide-protein conjugates as vaccine candidates against bacteria. Adv Carbohydr Chem Biochem 56:153–199
24. Pozsgay V (2008) Recent developments in synthetic oligosaccharide-based bacterial vaccines. Curr Top Med Chem 8:126–140
25. Lepenies B, Seeberger PH (2010) The promise of glycomics, glycan arrays and carbohydrate-based vaccines. Immunopharmacol Immunotoxicol 32:196–207
26. Ouerfelli O, Warren JD, Wilson RM, Danishefsky SJ (2005) Synthetic carbohydrate-based antitumor vaccines: challenges and opportunities. Expert Rev Vaccines 4:677–685
27. Bundle DR, Costello C, Nycholat C, Lipinski T, Rennie R (2012) Designing a *Candida albicans* conjugate vaccine by reverse engineering protective monoclonal antibodies. In: Kosma P, Muller-Loennies S (eds) Anticarbohydrate antibodies: from molecular basis to clinical application. Springer, Vienna, pp 121–146
28. Calarese DA, Scanlan CN, Zwick MB, Deechongkit S, Mimura Y, Kunert R, Zhu P, Wormald MR, Stanfield RL, Roux KR, Kelly JW, Rudd PM, Dwek RA, Katinger H, Burton DR, Wilson IA (2003) Antibody domain exchange is an immunological solution to carbohydrate cluster recognition. Science 300:2065–2071
29. Scanlan CN, Pantophlet R, Wormald MR, Ollmann Saphire E, Stanfield R, Wilson IA, Katinger H, Dwek RA, Rudd PM, Burton DR (2002) The broadly neutralizing anti-human immunodeficiency virus type 1 antibody 2G12 recognizes a cluster of alpha1 $\rightarrow$ 2 mannose residues on the outer face of gp120. J Virol 76:7306–7321

30. Astronomo RD, Lee H-K, Scanlan CN, Pantophlet R, Huang C-Y, Wilson IA, Blixt O, Dwek RA, Wong C-H, Burton DR (2008) A glycoconjugate antigen based on the recognition motif of a broadly neutralizing human immunodeficiency virus antibody, 2G12, is immunogenic but elicits antibodies unable to bind to the self glycans of gp120. J Virol 82:6359–6368
31. Wang LX (2006) Toward oligosaccharide- and glycopeptide-based HIV vaccines. Curr Opin Drug Discov Devel 9:194–206
32. Aussedat B, Vohra Y, Park PK, Fernández-Tejada A, Alam SM, Dennison SM, Jaeger FH, Anasti K, Stewart S, Blinn JH, Liao HX, Sodroski JG, Haynes BF, Danishefsky SJ (2013) Chemical synthesis of highly congested gp120 V1V2 *N*-glycopeptide antigens for potential HIV-1-directed vaccines. J Am Chem Soc 135:13113–13120
33. Michon F, Brisson JR, Jennings HJ (1987) Conformational differences between linear alpha (2 → 8)-linked homosialooligosaccharides and the epitope of the group B meningococcal polysaccharide. Biochemistry 26:8399–8405
34. Wessels MR, Kasper DL (1989) Antibody recognition of the type 14 pneumococcal capsule. Evidence for a conformational epitopein a neutral polysaccharide. J Exp Med 169:2121–2131
35. Paoletti LC, Kasper DL, Michon F, DiFabio J, Jennings HJ, Tosteson TD, Wessels MR (1992) Effects of chain length on the immunogenicity in rabbits of group B *Streptococcus* type III oligosaccharide-tetanus toxoid conjugates. J Clin Invest 89:203–209
36. Kasper DL, Paoletti LC, Wessels MR, Guttormsen HK, Carey VJ, Jennings HJ, Baker CJ (1996) Immune response to type III group B streptococcal polysaccharide-tetanus toxoid conjugate vaccine. J Clin Invest 98:2308–2314
37. Laferriere CA, Sood RK, de Muys JM, Michon F, Jennings HJ (1998) *Streptococcus pneumoniae* type 14 polysaccharide-conjugate vaccines: length stabilization of opsonophagocytic conformational polysaccharide epitopes. Infect Immun 66:2441–2446
38. Zou W, Mackenzie R, Thérien L, Hirama T, Yang Q, Gidney MA, Jennings HJ (1999) Conformational epitope of the type III group B *Streptococcus* capsular polysaccharide. J Immunol 163:820–825
39. Jennings HJ (2012) The role of sialic acid in the formation of protective conformational bacterial polysaccharide epitopes. In: Kosma P, Muller-Loennies S (eds) Anticarbohydrate antibodies: from molecular basis to clinical application. Springer, Vienna, pp 55–74
40. Han Y, Riesselman MH, Cutler JE (2000) Protection against candidiasis by an immunoglobulin G3 (IgG3) monoclonal antibody specific for the same mannotriose as an IgM protective antibody. Infect Immun 68:1649–1654
41. Kabat EA (1956) Heterogeneity in extent of the combining regions of human antidextran. J Immunol 77:377–385
42. Kabat EA (1960) Upper limit for the size of the human antidextran combining site. J Immunol 84:82–85
43. Luderitz O, Westphal O, Staub AM, Le Minor L (1960) Preparation and immunological properties of an artificial antigen with colitose (3-deoxy-1-fucose) as the determinant group. Nature 188:556–558
44. Kleinhammer G, Himmelspach K, Westphal O (1973) Synthesis and immunological properties of an artificial antigen with the repeating oligosaccharide unit of *Salmonella illinois* as haptenic group. Eur J Immunol 3:834–838
45. Lindberg AA, Wollin R, Bruse G, Ekwall E, Svenson SB (1983) Immunology and immunochemistry of synthetic and semisynthetic *Salmonella* O-antigen-specific glycoconjugates. Am Chem Soc Symp Ser 231:83–118
46. Svenson SB, Lindberg AA (1981) Artificial *Salmonella* vaccines: *Salmonella typhimurium* O-antigen-specific oligosaccharide-protein conjugates elicit protective antibodies in rabbits and mice. Infect Immun 32:490–496
47. Svenson SB, Nurminen M, Lindberg AA (1979) Artificial *Salmonella* vaccines: O-antigenic oligosaccharide-protein conjugates induce protection against infection with *Salmonella typhimurium*. Infect Immun 25:863–872

48. Cygler M, Rose DR, Bundle DR (1991) Recognition of a cell surface oligosaccharide epitope of pathogenic *Salmonella* by an antibody Fab fragment. Science 253:442–446
49. Jeffrey PD, Bajorath J, Chang CY, Yelton D, Hellström I, Hellström KE, Sheriff S (1995) The X-ray structure of an anti-tumour antibody in complex with antigen. Nat Struct Biol 2:466–471
50. Rose DR, Przybylska M, To RJ, Kayden CS, Oomen RP, Vorberg E, Young NM, Bundle DR (1993) Crystal structure to 2.45 Å resolution of a monoclonal Fab specific for the *Brucella* A cell wall polysaccharide antigen. Protein Sci 2:1106–1113
51. Vyas NK, Vyas MN, Chervenak MC, Johnson MA, Pinto BM, Bundle DR, Quiocho FA (2002) Molecular recognition of oligosaccharide epitopes by a monoclonal Fab specific for *Shigella flexneri* Y lipopolysaccharide: X-ray structures and thermodynamics. Biochemistry 41:13575–13586
52. van Roon AMM, Pannu NS, de Vrind JPM, van der Marel GA, van Boom JH, Hokke CH, Deelder AM, Abrahams JP (2004) Structure of an anti-Lewis X Fab fragment in complex with its Lewis X antigen. Structure 12:1227–1236
53. Vulliez-Le Normand B, Saul FA, Phalipon A, Bélot F, Guerreiro C, Mulard LA, Bentley GA (2008) Structures of synthetic O-antigen fragments from serotype 2a *Shigella flexneri* in complex with a protective monoclonal antibody. Proc Natl Acad Sci U S A 105:9976–9981
54. Murase T, Zheng RB, Joe M, Bai Y, Marcus SL, Lowary TL, Ng KK (2009) Structural insights into antibody recognition of mycobacterial polysaccharides. J Mol Biol 392:381–392
55. Evans DW, Müller-Loennies S, Brooks CL, Brade L, Kosma P, Brade H, Evans SV (2011) Structural insights into parallel strategies for germline antibody recognition of lipopolysaccharide from *Chlamydia*. Glycobiology 21:1049–1059
56. Blackler RJ, Müller-Loennies S, Brade L, Kosma P, Brade H, Evans SV (2012) Antibody recognition of *Chlamydia* LPS: Structural insights of inherited immune responses. In: Kosma P, Muller-Loennies S (eds) Anticarbohydrate antibodies: from molecular basis to clinical application. Springer, Vienna, pp 75–120
57. Johnson MA, Cartmell J, Weisser NE, Woods RJ, Bundle DR (2012) Molecular recognition of *Candida albicans* (1 → 2)-β-mannan oligosaccharides by a protective monoclonal antibody reveals the immunodominance of internal saccharide residue. J Biol Chem 287:18078–18090
58. Villeneuve S, Souchon H, Riottot M-M, Mazié J-C, Lei P-S, Glaudemans CPJ, Kováč P, Fournier J-M, Alzari PM (2000) Crystal structure of an anti-carbohydrate antibody directed against *Vibrio cholerae* O1 in complex with antigen: molecular basis for serotype specificity. Proc Natl Acad Sci U S A 97:8433–8438
59. Evans SV, Sigurskjold BW, Jennings HJ, Brisson J-R, To R, Altman E, Frosch M, Weisgerber C, Kratzin H, Klebert S, Vaesen M, Bitter-Suermann D, Rose DR, Young NM, Bundle DR (1995) Evidence for the extended helical nature of polysaccharide epitopes. The 2.8 Å resolution structure and thermodynamics of ligand binding of an antigen biding fragment specific for α-(2 → 8)-polysialic acid. Biochemistry 34:6737–6744
60. Milton MJ, Bundle DR (1998) Observation of the *anti*- conformation of a glycosidic linkage in an antibody-bound oligosaccharide. J Am Chem Soc 120:10547–10548
61. Bundle DR (1998) Recognition of carbohydrate antigens by antibody binding sites. In: Hecht S (ed) Carbohydrates. Oxford University Press, Oxford, pp 370–440
62. Baumann H, Altman E, Bundle DR (1993) Controlled acid hydrolysis of an O-antigen fragment yields univalent heptasaccharide haptens containing one 3,6-dideoxyhexose epitope. Carbohydr Res 247:347–354
63. Wisplinghoff H, Bischoff T, Tallent SM, Seifert H, Wenzel RP, Edmond MB (2004) Nosocomial bloodstream infections in US hospitals: analysis of 24,179 cases from a prospective nationwide surveillance study. Clin Infect Dis 39:309–317
64. Netea MG, Brown GD, Kullberg BJ, Gow NAR (2008) An integrated model of the recognition of *Candida albicans* by the innate immune system. Nat Rev Microbiol 6:67–78

65. Netea MG, Gow NAR, Munro CA, Bates S, Collins C, Ferwerda G, Hobson RP, Bertram G, Hughes HB, Jansen T, Jacobs L, Buurman ET, Gijzen K, Williams DL, Torensma R, Van der Meer JWM, McKinnon A, Odds FC, Brown AJP, Kullberg B (2006) Immune sensing of *Candida albicans* requires cooperative recognition of mannans and glucans by lectin and Toll-like receptors. J Clin Invest 6:1642–1650
66. Wells CA, Salvage-Jones JA, Li X, Hitchens K, Butcher S, Murray RZ, Beckhouse AG, Lo YL, Manzanero S, Cobbold C, Schroder K, Ma B, Orr S, Stewart L, Lebus D, Sobieszczuk P, Hume DA, Stow J, Blanchard H, Ashman RB (2008) The macrophage-inducible C-type lectin, mincle, is an essential component of the innate immune response to *Candida albicans*. J Immunol 180:7404–7413
67. Chaffin WL, López-Ribot JL, Casanova M, Gozalbo D, Martínez JP (1998) Cell wall and secreted proteins of *Candida albicans*: identification, function, and expression. Microbiol Mol Biol Rev 62:130–180
68. Bishop CT, Blank F, Gardner PE (1960) The cell wall polysaccharides of *Candida albicans*: glucan, mannan, and chitin. Can J Chem 38:869–881
69. Cutler JE (2001) *N*-Glycosylation of yeast, with emphasis on *Candida albicans*. Med Mycol 39:75–86
70. Shibata N, Kobayashi H, Suzuki S (2012) Immunochemistry of pathogenic yeast, *Candida* species, focusing on mannan. Proc Jpn Acad Ser B Phys Biol Sci 88:250–265
71. Shibata N, Kobayashi H, Takahashi S, Okawa Y, Hisamichi K, Suzuki S, Suzuki S (1991) Structural study on a phosphorylated mannotetraose obtained from the phosphomannan of *Candida albicans* NIH B-792 strain by acetolysis. Arch Biochem Biophys 290:535–542
72. Shibata N, Arai M, Haga E, Kikuchi T, Najima M, Satoh T, Kobayashi H, Suzuki S (1992) Structural identification of an epitope of antigenic factor 5 in mannans of *Candida albicans* NIH B-792 (serotype B) and J-1012 (serotype A) as beta-1,2-linked oligomannosyl residues. Infect Immun 60:4100–4110
73. Shibata N, Hisamichi K, Kikuchi T, Kobayashi H, Okawa Y, Suzuki S (1992) Sequential nuclear magnetic resonance assignment of beta-1,2-linked mannooligosaccharides isolated from the phosphomannan of the pathogenic yeast *Candida albicans* NIH B-792 strain. Biochemistry 31:5680–5686
74. Kobayashi H, Shibata N, Nakada M, Chaki S, Mizugami K, Ohkubo Y, Suzuki S (1990) Structural study of cell wall phosphomannan of *Candida albicans* NIH B-792 (serotype B) strain, with special reference to ^{1}H and ^{13}C NMR analyses of acid-labile oligomannosyl residues. Arch Biochem Biophys 278:195–204
75. Shibata N, Ikuta K, Imai T, Satoh Y, Richi S, Suzuki A, Kojima C, Kobayashi H, Hisamichi K, Suzuki S (1995) Existence of branched side chains in the cell wall mannan of pathogenic yeast *Candida albicans*. Structure-antigenicity relationship between the cell wall mannans of *Candida albicans* and *Candida parapsilosis*. J Biol Chem 270:1113–1122
76. Cassone A (1989) Cell wall of *Candida albicans*: its functions and its impact on the host. Curr Top Med Mycol 3:248–314
77. Chaffin WL, Ringler L, Larsen HS (1988) Interactions of monospecific antisera with cell surface determinants of *Candida albicans*. Infect Immun 56:3294–3296
78. Kobayashi H, Shibata N, Suzuki S (1992) Evidence for oligomannosyl residues containing both beta-1,2 and alpha-1,2 linkages as a serotype A-specific epitope(s) in mannans of *Candida albicans*. Infect Immun 60:2106–2109
79. Kobayashi H, Takahashi S, Shibata N, Miyauchi M, Ishida M, Sato J, Maeda K, Suzuki S (1994) Structural modification of cell wall mannans of *Candida albicans* serotype A strains grown in yeast extract-Sabouraud liquid medium under acidic conditions. Infect Immun 62:968–973
80. Shibata N, Akagi R, Hosoya T, Kawahara K, Suzuki A, Ikuta K, Kobayashi H, Hisamichi K, Okawa Y, Suzuki S (1996) Existence of novel branched side chains containing beta-1,2 and alpha-1,6 linkages corresponding to antigenic factor 9 in the mannan of *Candida guilliermondii*. J Biol Chem 271:9259–9266

81. Shibata N, Onozawa M, Tadano N, Hinosawa Y, Suzuki A, Ikuta K, Kobayashi H, Suzuki S, Okawa Y (1996) Structure and antigenicity of the mannans of *Candida famata* and *Candida saitoana*: comparative study with the mannan of *Candida guilliermondii*. Arch Biochem Biophys 336:49–58
82. Shibata N, Kobayashi H, Tojo M, Suzuki S (1986) Characterization of phosphomannan-protein complexes isolated from viable cells of yeast and mycelial forms of *Candida albicans* NIH B-792 strain by the action of Zymolyase-100 T. Arch Biochem Biophys 251:697–708
83. Kobayashi H, Giummelly P, Takahashi S, Ishida M, Sato J, Takaku M, Nishidate Y, Shibata N, Okawa Y, Suzuki S (1991) *Candida albicans* serotype A strains grow in yeast extract-added Sabouraud liquid medium at pH 2.0, elaborating mannans without beta-1,2 linkage and phosphate group. Biochem Biophys Res Commun 175:1003–1009
84. Faille C, Wieruszeski JM, Lepage G, Michalski JC, Poulain D, Strecker G (1991) ^{1}H-NMR spectroscopy of manno-oligosaccharides of the beta-1,2-linked series released from the phosphopeptidomannan of *Candida albicans* VW-32 (serotype A). Biochem Biophys Res Commun 181:1251–1258
85. Goins TL, Cutler JE (2000) Relative abundance of oligosaccharides in *Candida* species as determined by fluorophore-assisted carbohydrate electrophoresis. J Clin Microbiol 38:2862–2869
86. Han Y, Cutler JE (1995) Antibody response that protects against disseminated candidiasis. Infect Immun 63:2714–2719
87. Han Y, Cutler JE (1997) Assessment of a mouse model of neutropenia and the effect of an anti-candidiasis monoclonal antibody in these animals. J Infect Dis 175:1169–1175
88. Han Y, Morrison RP, Cutler JE (1998) A vaccine and monoclonal antibodies that enhance mouse resistance to *Candida albicans* vaginal infection. Infect Immun 66:5771–5776
89. Han Y, Kozel TR, Zhang MX, MacGill RS, Carroll MC, Cutler JE (2001) Complement is essential for protection by an IgM and an IgG3 monoclonal antibody against experimental, hematogenously disseminated candidiasis. J Immunol 167:1550–1557
90. Han Y, Kanbe T, Cherniak R, Cutler JE (1997) Biochemical characterization of *Candida albicans* epitopes that can elicit protective and nonprotective antibodies. Infect Immun 65:4100–4107
91. Nitz M, Ling CC, Otter A, Cutler JE, Bundle DR (2002) The unique solution structure and immunochemistry of the *Candida albicans* β-1,2-mannopyranan cell wall antigen. J Biol Chem 277:3440–3446
92. Nitz M, Bundle DR (2001) Synthesis of di- to hexasaccharide 1,2-linked beta-mannopyranan oligomers, a terminal S-linked tetrasaccharide congener and the corresponding BSA glycoconjugates. J Org Chem 66:8411–8423
93. Ogawa T, Yamamoto H (1982) Synthesis of linear D-mannotetraose and D-mannohexaose, partial structures of the cell-surface D-mannan of *Candida albicans* and *Candida utilis*. Carbohydr Res 104:271–283
94. Sugiyama H, Toyohiko N, Horii M, Motohashi K, Sakai J, Usui T, Hisamichi K, Ishiyama J (2000) The conformation of α-(1 → 4)-linked glucose oligomers from maltose to maltoheptaose and short-chain amylose in solution. Carbohydr Res 325:177–182
95. Ogawa T, Takanashi Y (1983) Synthesis of β-D-(1 → 2)-linked D-glucopentaose, a part of the structure of the exocellular β-D-glucan of *Agrobacterium tumefaciens*. Carbohydr Res 123: C16–C18
96. Pozsgay V, Robbins JB (1995) Synthesis of a pentasaccharide fragment of Polysaccharide II of *Mycobacterium tuberculosis*. Carbohydr Res 277:51–66
97. Rees DA, Scott WE (1971) Polysaccharide conformation. Part VI. Computer model-building for linear and branched pyranoglycans. Correlations with biological function. Preliminary assessment of inter-residue forces in aqueous solution. Further interpretation of optical rotation in terms of chain conformation. J Chem Soc B 469–479

98. Lemieux RU, Bock K, Delbaere LTJ, Koto S, Rao V (1980) The conformations of oligosaccharides related to the ABH and Lewis human blood group determinants. Can J Chem 58:631–653
99. Otter A, Lemieux RU, Ball RG, Venot AP, Hindsgaul O, Bundle DR (1999) Crystal state and solution conformation of the B blood group trisaccharide α-L-Fuc*p*-(1 → 2)-[α-D-Gal*p*]-(1 → 3)]-β-D-Gal*p*-OCH_3 Eur J Biochem 259:295–303
100. Zierke M, Smieško M, Rabbani S, Aeschbacher T, Cutting B, Allain FH-T, Schubert M, Ernst B (2013) Stabilization of branched oligosaccharides: Lewis^x benefits from a non-conventional C–H···O hydrogen bond. J Am Chem Soc 135:13464–13472
101. Peters T, Brisson J-R, Bundle DR (1990) Conformational analysis of key disaccharide components of *Brucella* A and M antigens. Can J Chem 68:979–988
102. Crich D, Li H, Yao Q, Wink DJ, Sommer RD, Rheingold AL (2001) Direct synthesis of β-mannans. A hexameric [→3)-β-D-Man-(1 → 4)-β-D-Man-(1]$_3$ subunit of the antigenic polysaccharides from *Leptospira biflexa* and the octameric (1 → 2)-linked β-D-mannan of the *Candida albicans* phospholipomannan. X-ray crystal structure of a protected tetramer. J Am Chem Soc 123:5826–5828
103. Nikrad PV, Beierbeck H, Lemieux RU (1992) Molecular recognition X. A novel procedure for the detection of the intermolecular hydrogen bonds present in a protein · oligosaccharide complex. Can J Chem 70:241–253
104. Nycholat CM, Bundle DR (2009) Synthesis of mono-deoxy and mono-*O*-methyl congeners of methyl β-D-mannopyranosyl-(1 → 2)-β-D-mannopyranoside for epitope mapping of anti-*Candida albicans* antibodies. Carbohydr Res 344:555–569
105. Bundle DR, Nycholat C, Costello C, Rennie R, Lipinski T (2012) Design of a *Candida albicans* disaccharide conjugate vaccine by reverse engineering a protective monoclonal antibody. ACS Chem Biol 7:1754–1763
106. Costello C, Bundle DR (2012) Synthesis of three trisaccharide congeners to investigate frame shifting of β1,2-mannan homo-oligomers in an antibody binding site. Carbohydr Res 357:7–15
107. Meyer B, Peters T (2003) NMR spectroscopy techniques for screening and identifying ligand binding to protein receptors. Angew Chem Int Ed Engl 42:864–890
108. Nitz M, Bundle DR (2002) The unique solution structure and immunochemistry of the *Candida albicans* β1,2-mannopyranan cell wall antigen. In: Jiménez-Barbero J, Peters T (eds) NMR spectroscopy of glycoconjugates. Wiley-VCH Verlag, Weinheim, pp 145–187
109. Dang A-T, Johnson MA, Bundle DR (2012) Synthesis of a *Candida albicans* tetrasaccharide spanning the β1,2-mannan phosphodiester α-mannan junction. Org Biomol Chem 10:8348–8360
110. Lemieux RU, Du M-H, Spohr U (1994) Relative effects of ionic and neutral substituents on the binding of an oligosaccharide by a protein. J Am Chem Soc 116:9803–9804
111. Johnson MA, Bundle DR (2013) Designing a new antifungal glycoconjugate vaccine. Chem Soc Rev 42:4327–4344
112. Wu X, Bundle DR (2005) Synthesis of glycoconjugate vaccines for *Candida albicans* using novel linker methodology. J Org Chem 70:7381–7388
113. Lipinski T, Luu T, Kitov PI, Szpacenko A, Bundle DR (2011) A structurally diversified linker enhances the immune response to a small carbohydrate hapten. Glycoconj J 28:149–164
114. Lipinski T, Wu X, Sadowska J, Kreiter E, Yasui Y, Cheriaparambil S, Rennie R, Bundle DR (2012) A β-mannan trisaccharide conjugate vaccine aids clearance of *Candida albicans* in immunocompromised rabbits. Vaccine 30:6263–6269
115. Wu X, Lipinski T, Carrell F, Bailey JJ, Bundle DR (2007) Synthesis and immunochemical studies on a *Candida albicans* clustered glycoconjugate vaccine. Org Biomol Chem 5:3477–3485
116. Wu X, Lipinski T, Paszkiewicz E, Bundle DR (2008) Synthesis and immunochemical characterization of *S*-linked glycoconjugate vaccines against *Candida albicans*. Chem Eur J 14:6474–6482

117. Lipinski T, Kitov P, Szpacenko A, Paszkiewicz E, Bundle DR (2011) Synthesis and immunogenicity of a glycopolymer conjugate. Bioconjug Chem 22:274–281
118. Xin H, Dziadek S, Bundle DR, Cutler J (2008) Synthetic glycopeptide vaccines combining β-mannan and peptide epitopes induce protection against candidiasis. Proc Natl Acad Sci U S A 105:13526–13531
119. Stubbs AC, Martin KS, Coeshott C, Skaates SV, Kuritzkes DR, Bellgrau D, Franzusoff A, Duke RC, Wilson CC (2001) Whole recombinant yeast vaccine activates dendritic cells and elicits protective cell-mediated immunity. Nat Med 7:625–629
120. Banchereau J, Steinman RM (1998) Dendritic cells and the control of immunity. Nature 392:245–252
121. Steinman RM, Banchereau J (2007) Taking dendritic cells into medicine. Nature 449:419–426
122. Tacken PJ, Figdor CG (2011) Targeted antigen delivery and activation of dendritic cells in vivo: steps towards cost effective vaccines. Semin Immunol 23:12–20
123. Torosantucci A, Bromuro C, Chiani P, De Bernardis F, Berti F, Galli C, Norelli F, Bellucci C, Polonelli L, Costantino P, Rappuoli R, Cassone A (2005) A novel glycoconjugate vaccine against fungal pathogens. J Exp Med 202:597–606
124. Bromuro C, Romano M, Chiani P, Berti F, Tontini M, Proietti D, Mori E, Torosantucci A, Costantino P, Rappuoli R, Cassone A (2010) Beta-glucan-CRM197 conjugates as candidates antifungal vaccines. Vaccine 28:2615–2623
125. Brown GD, Herre J, Williams DL, Willment JA, Marshall A, Gordon S (2003) Dectin-1 mediates the biological effects of beta-glucans. J Exp Med 197:1119–1124
126. Curtis MM, Way SS (2009) Interleukin-17 in host defence against bacterial, mycobacterial and fungal pathogens. Immunology 126:177–185
127. LeibundGut-Landmann S, Gross O, Robinson MJ, Osorio F, Slack EC, Tsoni SV, Schweighoffer E, Tybulewicz V, Brown GD, Ruland J, Reis e Sousa C (2007) Syk- and CARD9-dependent coupling of innate immunity to the induction of T helper cells that produce interleukin 17. Nat Immunol 8:630–638
128. Osorio F, LeibundGut-Landmann S, Lochner M, Lahl K, Sparwasser T, Eberl G, Reise e Sousa C (2008) DC activated via dectin-1 convert Treg into IL-17 producers. Eur J Immunol 38:3274–3281
129. Lin L, Ibrahim AS, Xu X, Farber JM, Avanesian V, Baquir B, Fu Y, French SW, Edwards JE Jr, Spellberg B (2009) Th1-Th17 cells mediate protective adaptive immunity against *Staphylococcus aureus* and *Candida albicans* infection in mice. PLoS Pathog 5:e1000703
130. Gaffen SL, Hernandez-Santos N, Peterson AC (2011) IL-17 signaling in host defense against *Candida albicans*. Immunol Res 50:181–187
131. Lipinski T, Fitieh A, St. Pierre J, Ostergaard HL, Bundle DR, Touret N (2013) Enhanced immunogenicity of a tricomponent mannan tetanus toxoid conjugate vaccine targeted to dendritic cells via Dectin-1 by incorporating β-glucan. J Immunol 190:4116–4128
132. Goddard-Borger ED, Stick RV (2011) An efficient, inexpensive, and shelf-stable diazotransfer reagent: imidazole-1-sulfonyl azide hydrochloride. Org Lett 13:2514–2514
133. Rostovtsev VV, Green LG, Fokin VV, Sharpless KB (2002) A stepwise Huisgen cycloaddition process: copper(I)-catalyzed regioselective "ligation" of azides and terminal alkynes. Angew Chem Int Ed Engl 41:2596–2599
134. Tornøe CW, Christensen C, Meldal M (2002) Peptidotriazoles on solid phase: [1,2,3]-triazoles by regiospecific copper(I)-catalyzed 1,3-dipolar cycloadditions of terminal alkynes to azides. J Org Chem 67:3057–3064
135. Fidel PL Jr, Sobel JD (1994) The role of cell-mediated immunity in candidiasis. Trends Microbiol 2(6):202–206
136. Casadevall A, Cassone A, Bistono F, Cutler JE, Magliani W, Murphy JW, Polonelli L, Romani L (1998) Antibody and/or cell-mediated immunity, protective mechanisms in fungal disease: an ongoing dilemma or an unnecessary dispute? Med Mycol 36(Suppl 1):95–105
137. Casadevall A (1995) Antibody immunity and invasive fungal infections. Infect Immun 63:4211–4218

138. Cutler JE, Deepe GS Jr, Klein BS (2007) Advances in combating fungal diseases: vaccines on the threshold. Nat Rev Microbiol 5:13–28
139. Dziadek S, Jacques S, Bundle DR (2008) A novel linker methodology for the synthesis of tailored conjugate vaccines composed of complex carbohydrate antigens and specific T_H-cell peptide epitopes. Chem Eur J 14:5908–5917
140. Pitarch A, Abian J, Carrascal M, Sánchez M, Nombela C, Gil C (2004) Proteomics-based identification of novel *Candida albicans* antigens for diagnosis of systemic candidiasis in patients with underlying hematological malignancies. Proteomics 4:3084–3106
141. Clancy CJ, Nguyen ML, Cheng S, Huang H, Fan G, Jaber RA, Wingard JR, Cline C, Nguyen MH (2007) Immunoglobulin G responses against a panel of *Candida albicans* antigens as accurate and early markers for the presence of systemic candidiasis. J Clin Microbiol 46:1647–1654
142. Singh H, Raghava GPS (2001) Propred: Predication of HLA-DR binding sites. Bioinformatics 17:1236–1237
143. Sing H, Raghava GPS (2003) Propred1: prediction of promiscuous MHC class-I binding sites. Bioinformatics 19:1009–1014
144. Xin H, Cartmell J, Bailey JJ, Dziadek S, Bundle DR, Cutler JE (2012) Self-adjuvanting glycopeptide conjugate vaccine against disseminated candidiasis. PLoS One 7:e35106
145. Cartmell J, Paszkiewicz E, Tam P-H, Luu T, Sarkar S, Lipinski T, Bundle DR (2014) Synthesis of antifungal vaccines by conjugation of β-1,2 trimannosides with T-cell peptides and covalent anchoring of neoglycopeptide to tetanus toxoid. Carbohydr Res. doi:10.1016/j.carres.2014.06.24
146. Bundle DR, Tam P-H, Tran H-A, Paszkiewicz E, Cartmell J, Sadowska JM, Sarkar S, Joe M, Kitov PI (2014) Oligosaccharides and peptide displayed on an amphiphilic polymer enable solid phase assay of hapten specific antibodies. Bioconjug Chem 25:685–697

Index

MIX
Papier aus verantwortungsvollen Quellen
Paper from responsible sources
FSC® C105338

If you have any concerns about our products,
you can contact us on
ProductSafety@springernature.com

In case Publisher is established outside the EU,
the EU authorized representative is:
Springer Nature Customer Service Center GmbH
Europaplatz 3, 69115 Heidelberg, Germany

Printed by Libri Plureos GmbH
in Hamburg, Germany